Analyzing Experimental Data by Regression

Analyzing Experimental Data by Regression

David M. Allen, Ph.D.
Foster B. Cady, Ph.D.

LIFETIME LEARNING PUBLICATIONS
A division of Wadsworth, Inc.
Belmont, California

London, Singapore, Sydney
Toronto, Mexico City

Jacket Designer: Ben Wong

Designer: Rick Chafian

Copy Editor: Carol Beal

Illustrator: John Foster

Composition: Science Typographers, Inc.

© 1982 by Wadsworth, Inc. All rights reserved. No part of this book may be reproduced, stored in a retrieval system, or transcribed, in any form or by any means, electronic, mechanical, photocopying, recording, or otherwise, without the prior written permission of the publisher, Lifetime Learning Publications, Belmont, California 94002, a division of Wadsworth, Inc.

Printed in the United States of America

1 2 3 4 5 6 7 8 9 10—85 84 83 82

Library of Congress Cataloging in Publication Data

Allen, David M., 1938–
   Analyzing experimental data by regression.

     Bibliography: p.
     Includes index.
     1. Regression analysis. 2. Experimental design.   I. Cady, Foster B., 1931–   II. Title.
QA278.2.A433   519.5'36   81-15576
ISBN 0-534-97963-7   AACR2

Contents

Preface   xv

Unit 1   **Strategy of Data Analysis**   2

Unit 2   **Data Link Between Research Objectives and Analysis**   8

     New York Rivers Data   10
     Thoroughbred Auction Data   10
     Dairy Hypomagnesaemia Data   13
     Summary   15

Unit 3   **Model Formulation and Inference**   16

     Response and Explanatory Variables   17
     Formulation of a Regression Model   18
     Regression Inference   20
     Notation for Regression Data   22
     Computational Considerations   25

Unit 4   **Fitting the Mean Model**   30

     Least Squares Model Fitting   31
     Arsenic Data   32
     Fitting the Mean Model   34
     Residual Plotting   36
     Summary   37

Unit 5   **Fitting a Straight-Line Model**   38

     Fitting the Mean and Slope Model   39
     Fitting the Intercept and Slope Model   42
     Summary   46

Unit 6 — Analysis of Variance — 48

Decomposition of Regression Data   49
Sums of Squares and Analysis of Variance Table   51
Fitting a Straight Line Through the Origin   53
Summary   54

Unit 7 — Computations for Multivariable Models — 56

Abbreviated Doolittle Calculations with Several x Variables   58
Model Sequencing with Several x Variables   61
Sequential and Partial Regression Coefficients   64
Weighted Regression   65
Summary   67

Unit 8 — Coping with Model Violations — 68

Vital Statistics Data   71
Curves That Are Always Up and Have One Bend   72
Curves That Are Always Down and Have One Bend   74
Scaling and Shifting   75
Transformation Selection   75
Final Selection   77
Summary   79

Unit 9 — Case Study: Dose-Dependent Blood Pressure Data — 80

Modeling   81
Diagnostic Plots   87
Conclusions   87

Unit 10 — Interpreting Regression Coefficients in Two-Variable Models — 90

Firefly Data   92
Fitting a Two-Variable Model with ABDO   93
Interpretation of Sequential Regression Coefficients   98
Interpretation of Partial Regression Coefficients   100
Summary   101

Contents

Unit 11   ***Data Reduction by Regression***   104

     Linear Combinations of the Observations   105
     Rules for Linear Combinations   106
     Numerical Illustration of Data Reduction Using ABDO   107
     Quadratic Polynomial Model   108
     Soymilk Data   111
     Mean and Variance of Linear Combinations in Regression   115
     Summary   117

Unit 12   ***Generalizing Beyond the Data***   120

     The Inference Model   122
     Linear Combinations of Regression Coefficients   123
     A General Approach to Data Analysis and Generalization   124
     Elimination (ELIM) Algorithm for Calculating Estimates and Standard Errors   125
     Interesting Linear Combinations for the Firefly Data   129
     Summary   133

Unit 13   ***Comparison of Candidate Models***   134

     Decomposition of the Firefly Data   135
     Analysis of Variance for the Firefly Data   137
     Comparing Candidate Models   138
     Electricity Load Data   141
     Summary   146

Unit 14   ***Case Study: Land Evaluation Data***   150

     Evaluation of General Model   153
     Improved Modeling   157
     Summary: An Approach to Data Analysis for Several x Variables   159

Unit 15   ***Analysis of Sample Means***   162

     History   164
     A General Means Model   164
     Residual and Experimental Errors   165

Construction of Indicator Variables   166
Estimation of Means and Standard Errors   168
Sequencing Candidate Models   169
Extension to Three Treatments   174
Summary   176

Unit 16 Case Study: Potato Leafhopper Survival Data   178

General Means Model   179
Selection of an Adequate Model   182
Estimation and Interpretation of Estimated Coefficients   184
Constructing Contrast Variables   187
Summary   189

Unit 17 Analysis of Cross-Classified Sample Means   192

Lymphocyte Data   193
Fat Digestibility Data   199
Comparison of the Lymphocyte and Fat Digestibility Data Analyses   204
Summary   211

Unit 18 Analysis of Sample Means with Unequal Numbers of Observations   214

Protein Nutrition Data   215
Swamp pH Data   221
General Data Analysis Approach   228

Unit 19 Covariance Analysis and Comparison of Regression Lines   232

Soybean Physiological Data   233
Testing the Assumption of a Common Slope   238
Comparing Regression Lines in Factorial Studies   239
Summary   245

Unit 20 Model Selection with Many Explanatory Variables   246

Need for Computer-assisted Model Selection   248
The SWEEP Operator   248
Stepwise Approaches   250
All Possible Regressions   251
Difficulties with the Residual Sum of Squares as a Selection Criterion   252
The Prediction Sum of Squares (PRESS)   254
Finger Lakes Data   257
Summary   259

Unit 21 Coping with Large Standard Errors and Unrealistic Estimates   260

The Barley Response Data   261
Data Augmentation to Achieve a Specified v_c 263
Ridge Regression   266
Firefly Data   267
Examination of Results   272
Summary   273

Unit 22 Split Experimental Units and Repeated Measures   276

Drug and Alcohol Data   278
The Reasons for Different Residual Terms   282
Missing Data in the Split-Unit Trial   285
More on the Interpretation of Interaction   288
Missing Data in the Repeated-Measures Trial   289
Comments on Using Statistical Package Programs   292
Summary   292

Unit 23 Nonlinear Models   294

Where Nonlinear Models Are Used   295
How Nonlinear Models are Fitted   296
Drug Assimilation Data   299
Nonlinear Least Squares Algorithm   302
Constructing the Approximating Linear Model   306
Summary   309

Unit 24	A Regression Program for Microcomputers	310
Appendix A	Exercises	326
Appendix B	Annotated Computer Output	344
Appendix C	Suggested Reading	378
Appendix D	Answers to Selected Exercises	382
Index		393

List of Tables

Table 2.1 Data for New York Rivers Study   11
Table 2.2 Data for Thoroughbred Auction Study   12
Table 2.3 Data for Study of Hypomagnesaemia in Dairy Cattle   14
Table 4.1 Data for Arsenic Content Study   33
Table 6.1 Analysis of Variance of Arsenic Data for Evaluating the Mean Model as the Reduced Model   52
Table 6.2 Decomposition and Analysis of Variance of Arsenic Data for Evaluating the Slope-Through-the-Origin Model as the Reduced Model   53
Table 7.1 Sums of Squares and Cross Products for Four-Variable Problems   60
Table 7.2 Abbreviated Doolittle Algorithm for the New York Rivers Data   61
Table 7.3 Sums of Squares Table Emphasizing Sequencing of Fitted Models   63
Table 7.4 ANOVA When Model Does Not Include $R(0)$ 64
Table 7.5 Sums of Squares and Cross Products Incorporating Weights   66
Table 8.1 Vital Statistics Data   71
Table 8.2 Common Curves Found in Practice and Transformations for Straightening   79
Table 9.1 Blood Pressure Data   82
Table 10.1 Data for Firefly Study   92
Table 10.2 Interpretation of Estimated Regression Coefficients   102
Table 11.1 Protein Content of Soymilk (g/100 ml) Versus Boiling Time (h)   111
Table 11.2 ABDO Calculations for Fitting the Quadratic Polynomial Model with the Soymilk Data   112

Table 11.3	ANOVA Table for a Candidate Model Sequence for the Soymilk Data	113
Table 12.1	ELIM Calculations for Quadratic Polynomial Example	127
Table 12.2	Linear Combinations of the Observations	128
Table 12.3	Summary of Results for Firefly Data	131
Table 13.1	Analysis of Variance of the Firefly Data for the Model Sequence $x_0\|x_1\|x_2$	137
Table 13.2	Electricity Load Data and Residuals	143
Table 13.3	Six-Variable Model for Electricity Load Data	145
Table 13.4	ANOVA for Electricity Load Data	146
Table 13.5	Sequential and Partial Sums of Squares	147
Table 14.1	South Dakota Unimproved Agricultural Land Data (1967–1969)	151
Table 14.2	Summary of Preliminary Analysis for the Land Evaluation Data	154
Table 14.3	Sequential and Partial Sums of Squares from Four-Variable General Model	154
Table 15.1	Calculating Sequential Residuals	171
Table 15.2	Estimated Regression Coefficients in Terms of Means	174
Table 16.1	Potato Leafhopper Survival Data	179
Table 16.2	ANOVA for General Means Model	180
Table 16.3	Results for Contrasts 2 and 3	182
Table 16.4	ANOVA for Testing Equality of Means	183
Table 16.5	Predicted Values for Four-Treatment Model	185
Table 16.6	Interpretation of Sequential and Partial Coefficients	190
Table 17.1	Lymphocyte Treatments and Data	194
Table 17.2	ANOVA for Lymphocyte Data	194
Table 17.3	Cross-Classification Format for Lymphocyte Data	195
Table 17.4	ANOVA with Breakdown of Model Sum of Squares	197
Table 17.5	Fat Digestibility Data	200
Table 17.6	Calculations for Fat Digestibility Data	201
Table 17.7	Estimated Contrasts from a Regression Program	203
Table 17.8	Calculations for Lymphocyte Data	206
Table 18.1	Final Weights (Grams) of Chicks at Six Weeks in Nutrition Study	216
Table 18.2	ANOVA for Nutrition Data	216
Table 18.3	Results for Two Contrasts	217
Table 18.4	ANOVA for ORTHO Approach	220

Table 18.5	pH Readings for Swamp Soil Samples	221
Table 18.6	Indicator Variables and **X′X**|**X′Y**	222
Table 18.7	ABDO Calculations	223
Table 18.8	Even Steps in ABDO When Columns Precede Rows in Model Sequence	225
Table 18.9	Guidelines for Selecting a Model	230
Table 19.1	Seed Yields (Grams) and Initial Plant Heights (Centimeters) for Soybean Study	234
Table 19.2	ANOVA for Soybean Study	234
Table 19.3	ABDO Calculations for the Soybean Covariance Model	235
Table 19.4	Analysis of Covariance	238
Table 19.5	Contrast Coefficients (**c**) for Contrasts Among Means	240
Table 19.6	ANOVA for the Sequence of Models	244
Table 19.7	Comparison of Factorial Data Analysis Approaches with Diet (D) Qualitative and PCB Quantitative	244
Table 20.1	Sequential Output of SWEEP Applied to the Sums of Squares and Cross Products Matrix of the Firefly Data	249
Table 20.2	Variance of Predicted Value When Model Includes Only x_0	253
Table 20.3	Variance of Predicted Value When Model Includes x_0 and x_1	253
Table 20.4	Variance of Predicted Value When Model Includes x_0, x_1, and $x_2 = x_1^2$	253
Table 20.5	Finger Lakes Phosphorus Pollution Data	256
Table 20.6	Variables Identified for the Finger Lakes Data by the SELECT Procedure	258
Table 21.1	Barley Yield Data	262
Table 21.2	Summary Statistics for the Barley Data for Augmented and Nonaugmented Models	265
Table 21.3	Selection of Three **c**'s	267
Table 21.4	Summary of Results for Three Linear Combinations	272
Table 22.1	Alcohol-Drug Data	280
Table 22.2	ANOVA for Data When y Is Regression on x_0 Through x_{23}	280
Table 22.3	Results from Fitting the General Means Model	281
Table 22.4	Data from a Drug Trial	282
Table 22.5	Data from a Drug Trial with Three Missing Values	287

Table 22.6	Data from a Drug Trial with Three Missing Values; Repeated-Measures Trial	290
Table 22.7	Successive Differences for Data in Table 22.6	290
Table 22.8	Successive Differences for Means	291
Table 22.9	Data When There is No Interaction	291
Table 22.10	Successive Differences When There Is No Interaction	292
Table 23.1	Drug Assimilation Data	300

Preface

FOR WHOM

This book is for:

— the practicing researcher who engages in data analysis using regression and for the graduate student who expects to become such a researcher.
— those who use computer programs for regression and are curious about how and why their data are transformed into the program's output.
— someone who wants to learn a single methodology that can replace a large assortment of specialized formulas and techniques.

THE BOOK'S PHILOSOPHY

There are two major premises of this book. The first premise is that data analysis should be approached with an objective well in mind, but there is seldom a rigid sequence of steps that can be applied. The data analyst should allow the findings at each stage to influence the direction through subsequent stages.

The second premise is that regression analysis should be used for classical regression, experimental design, and covariance situations. This results in a unified methodology rather than a collection of specialized techniques. This unified approach is possible because of today's computers and statistical packages.

FEATURES

Features of the book include:

— An integrated data analysis approach which can be easily adapted to modern statistical packages.
— Emphasis on model formulation, fitting, and selection under a variety of conditions.

— Use of small but real data sets to illustrate concepts in practice. The data sets are from a variety of fields including Agriculture, Biology, Environment, Management, Medicine, and Pharmaceutics.
— Three case studies.
— Strong emphasis on computer packages and their use in data analysis.

REQUIRED BACKGROUND

Ideally, the reader should have some data analysis "maturity," gained from experience in experimentation or from use of statistical computer packages. Required statistical background is modest, assuming an awareness of basic statistical concepts such as mean, variance, sample, population, and inference. Matrix notation and basic addition and multiplication operations are developed, but additional matrix calculations, such as inverses, are not utilized.

ORGANIZATION

This book is organized in a number of compact Units, instead of the conventional and more lengthy chapters. The purpose is to focus more sharply on a given data analysis situation and thereby help the reader hone his or her analytic tools. The early Units discuss some philosophy of experimentation, statistics and inference. Means and slopes are very important concepts throughout regression and experimental design. These are discussed in their simplest context in Units 4 and 5. Units 6 through 19 deal with techniques, specific experimental situations, and case studies. Units 20 through 23 are more advanced and employ more specialized techniques. Unit 24 describes a computer program for microcomputers that can perform nearly all of the techniques presented in this book. Appendix B contains annotated runs of widely distributed statistical systems to assist in preparing input and interpreting the output.

David M. Allen
Foster B. Cady

Analyzing Experimental Data by Regression

Unit 1

Strategy of Data Analysis

Objectives
- To define *data analysis*.
- To focus on the role of a data analyst.
- To describe the iterative nature of research.
- To develop a general strategy for data analysis.

Research is dynamic. The process can be envisioned as a stepwise progression that builds upon exploratory efforts and becomes refined and more specific as the evidence accumulates. Data analysis is also a dynamic and sequential process that serves to illuminate and interpret the collected data. While providing support to the ongoing investigation, the analysis may demonstrate the need for further research.

As data analysts we cannot restrict ourselves to a static approach. We need to be innovative and flexible so that data can speak to us, revealing facets of the study that were not known to exist. Our thoughts should be influenced but not restricted by traditional approaches; we must be dynamic too.

Formally stated, *data analysis* is the application of one or more techniques to a set of data as guided by the subject matter of the problem. The end product should be a meaningful and quantitative evaluation of the data. Data analysis is an essential step in the scientific process, and data analysts are persons within the scientific community who have special interests and abilities in data analysis. Our group is diverse, including full-time data analysts, researchers from many disciplines who have sufficient expertise to handle their own data analysis, and specialists in statistics and computing.

The roles of data analysts vary, as do our academic training, our familiarity with computing, and our understanding of concepts and methodology of statistical inference. Despite the diversity of backgrounds and experience, however, persons involved in scientific research have the unifying framework of the scientific learning process to guide their search. Steps in the process include those shown in Figure 1.1.

Let us consider the links of this chain. The researcher who collects the data is basically concerned with formulating hypotheses, working out an appropriate research design, and interpreting the data. Applied statisticians focus on developing and refining methodology, thus giving the researcher a continually improved battery of techniques.

After the first steps in the process, including the development of a sound research design, the data analyst enters the process when these

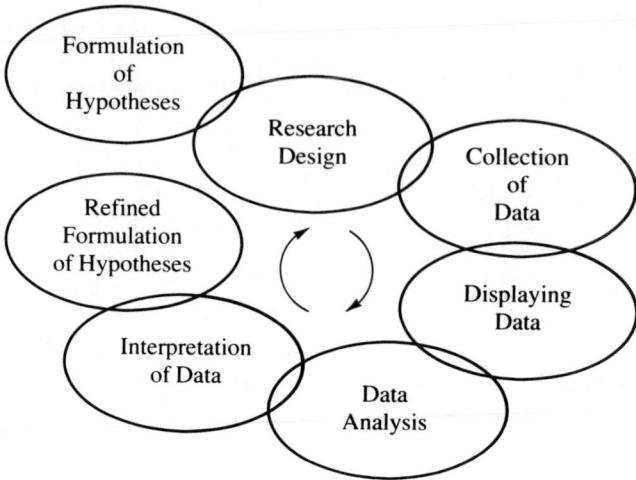

Figure 1.1 Steps in the Scientific Learning Process

actions become necessary:

- To display and describe findings.
- To create meaning from a vast amount of data.
- To negotiate an understanding of the researcher's data so that others can appreciate the results.

The data analyst holds the key to illuminating the research results, of creating sense and order so that others may find meaning in what has been accomplished. This step involves time and money, and the data analyst, to be cost-effective, needs to approach his or her task by using the most efficient techniques available.

This book emphasizes a practical and efficient approach to data analysis using regression methodology. The book has two major goals:

1. Application of specific techniques with practical utility in analyzing data sets.
2. Development of a general and unified approach to data analysis.

A general outline of a data analysis strategy, starting with the concept of a model, will get us started in the right direction.

The term *model* has different meanings among scientists. *Regression models* quantitatively describe the variability among the observations, or responses, by partitioning an observation into two parts. The first part of this decomposition incorporates everything known about the magnitude of a response. For example, if the observations were tumor weights, the first part would be the values we would predict for each of the tumors based on all our known information. The remaining portion, called the *residual*, is the difference between the observed value and the predicted value and has to be ascribed to unknown sources. The residual should be small relative to the predicted value and should display characteristics of random variability.

In the simplest model the predicted portion depends on only one component; that one component is a characteristic that can be ascribed to all the observations considered as a group. For example, the arithmetic mean, the average of all the observed tumor weights, will usually account for a portion of each observation. Without other information we would use the mean tumor weight as the systematic part of the model; that is, the predicted weight for each tumor would be the mean tumor weight. The residuals would be the difference between the observed and mean tumor weights. The model is written as follows:

$$\text{observation} = \text{prediction} + \text{residual} = \text{group mean} + \text{residual}$$

In general, observations are numerical values measured on a group of units. (The generic term *unit* usually will be used rather than the terms *subjects* or *plots*.) Usually we have more information on the units than indicated above. A common example is gender, a characteristic that divides the units into subgroups, each with its own mean. Within each subgroup the model is

$$\text{observation} = \text{subgroup mean} + \text{subgroup residual}$$

Note that two means have now been calculated, compared with only one mean in the single-group-mean model. The residuals of the subgroup model would, one hopes, be smaller than those of the single-group model.

In designed experiments a set of rules governs the allocation of the treatments to the experimental units. In a completely randomized design each treatment is allocated at random to a certain number of units. The first part of the model will include the fitting of a mean for each of the treatment subgroups, and the second part will again include the residuals, which are called *experimental errors*. Another experimental design is used if part of the variability among the observations is due to known differences among the experimental units. Then the units can be arranged in a randomized block design; that is, the treatments are allocated at random within each block or subgroup of units. Now the predicted part of each observation includes a second component due to the block associated with each observation. The predicted portion of the model is now based on more information and should account for a larger portion of each observation. Again the residuals should be smaller.

In either experimental or observational studies, information other than meaningful subdivisions of the units might be known, information that will give additional explanation about the variability among the observations. So far only the response variable has been measured. The next step in decreasing the residual portion of the observation is to actually measure other variables on each unit. These additional measurements are known or hypothesized to affect the response variable. All the known information will be related to the predicted portion, so that the residuals should be smaller.

Where no questions exist on the nature of the model and the true model can be written without any conjecture, the objective of the data analysis is to fit the known model. The residual is nothing more than random error. But usually the true model is not known. In fact, much of the research is directed toward investigations exploring several feasible models. The data analysis will provide insight into the problem and, specifically, will give an indication of the complexity of the model needed to adequately fit the data. When the true model is not known, several models can be hypothesized and different predicted values can be calculated, depending on the formulation of the model. These potentially feasible models will be called *candidate models*. Now the data analysis can become complex, and thus a general approach or strategy needs to be developed.

The general strategy is to examine a sequence of candidate models. One specific strategy might be to start with the most general model consistent with the data and subject matter. Then the relative adequacy of fitting a sequence of simpler models could be examined. For example, in the consideration of premature birth weights, length of gestation is known to account for variability among birth weights. The mother's smoking behavior may also be related to birth weight. If these two factors are the only ones perceived as having an effect on birth weight, then the most general model formulation would include components for both gestation time and smoking behavior. Simpler or reduced models would delete the hypothesized smoking component or delete both factors.

As already indicated, judgments are made on the adequacy of a candidate model to describe the data. The basis for these decisions is the residual part of the model. The magnitude and pattern of the residuals when plotted in various displays will give clues to the nature of an inadequate model or faulty assumptions in fitting the model. With an adequate fit to the model, the residuals will exhibit characteristics of random variability. In the choice between two candidate models, both giving a satisfactory residual pattern, the relative magnitude of the residuals will be one deciding factor.

After one candidate model has been selected, the residuals are used again in evaluating the fitted model, since they are part of the estimated standard errors of various statistics. The verification of the selected candidate model is provided by the performance of the fitted model when used in the future. For instance, a model fitted from one set of data can give poor prediction when used in slightly different circumstances; thus this model would have to be refined or eliminated altogether. Even with a single set of data and before a postevaluation on future data sets is available, the predictive ability of fitted candidate models should be evaluated.

The analysis of the data can also give information on designing future investigations that will better discriminate among feasible models. This iterative process between analysis and investigation continues until sufficient model resolution is reached. In practice, time and cost considerations may restrict us to one study, and thus we must fully utilize the available methodology. In fact, we try to do more than model discrimination with each data set. Whenever possible, we will want to make predictions and inferences to other situations by confidence interval estimation and hypothesis testing.

In summary, the strategy of data analysis presented here is to find the simplest model adequate for data description and inference. The general flow of data analysis can be outlined as follows:

- An eyeball study of the raw data, assisted by plotting.
- Conjecture and fitting of several candidate models.
- An examination of the relative adequacy, or goodness of fit, of the candidate models.
- A check on the validity of the assumptions.
- The selection and validation of one of the fitted models.
- Inference based on the fitted model and associated standard errors.
- Storage of the data and summary statistics for later retrieval and communication.

Unit 2

Data Link Between Research Objectives and Analysis

Objectives

- To show how research objectives can be integrated with data collection and analysis.
- To explore actual data sets that exemplify the data link between research objectives and analysis.
- To outline a general approach to data analysis.

In a research proposal general objectives are set forth as clearly as possible. The variables to be measured or the treatments to be compared are usually described in detail. Less often, however, is the proposed analysis of the data given clearly or in detail. Sometimes the total elaboration is "a statistical analysis of the data will be done," or "the data will be subjected to a regression analysis." Today, though, this situation is changing. Research administrators and granting agencies are becoming increasingly partial toward proposals including a clear outline of the data analysis strategy. This trend is an indication that researchers must know just what data and analysis are needed to support their hypotheses.

In textbooks and the literature, examples encompassing both an outline of the research objectives and an approach to the data analysis are difficult to find. Subject matter journals concentrate on objectives, experimental techniques, and summarizations of data. On the other hand, statistical journals and statistical methodology textbooks start with the data as given —for example, "suppose we have a random sample of n observations, $y_1, y_2, \ldots, y_n$." But research is not the process of collecting data and then making a decision about how to effect analysis. Rather, researchers need to envision, at least conceptually, an actual data set and formulate a strategy for the analysis concurrently with initial research planning. Thus this unit stresses the flow of research objectives, data, and analysis. Examples of data sets are presented with the following purposes:

- To indicate continuity between the general objectives of a study and the outline of a data analysis strategy. The connector is the data set.

- To provide real data examples as background for the general development in subsequent units.

- To present real data sets in a particular format—namely, columnar, with one column for each variable called an explanatory, predictor, input, or independent variable and one column for the response, output, or dependent variable. The *explanatory variables* will be denoted as x variables and the *response variable* by y. The x variables can be numerical measurements on the sampled units or values constructed by the data analyst from known information (e.g., by categorizing the units into subgroups, as indicated in Unit 1).

NEW YORK RIVERS DATA

In a 1976 study on land use and water quality in New York rivers, Haith used total nitrogen content as a measure of water quality in 20 river basins. The general objectives included the following:

1. Do quantitative relationships exist between water quality (nitrogen) and land use?
2. What land uses have the most significant impact on water quality among various categories such as agricultural, forest, residential, and commercial?
3. Can the relative importance of the categories of land use be identified?

The data are shown in Table 2.1.

Possible steps in the data analysis would include the following:

- A plot of y against each of the x variables.
- Formulation of a model relating the response variable to the x variables.
- A fit of the model.
- Calculation of residual plots to indicate adequacy of the fitted model.
- Formulation of simpler models, with certain x variables deleted, based on the fitted model, subject matter knowledge, and plots of the original data.
- Comparison of the candidate models by analysis of variance methodology and residual plotting.
- Estimation of the model, the residual mean square, and the standard errors, if a simpler model adequately describes the data.

This study could be considered a data description and summarization problem, since emphasis is on estimating and understanding relationships. Plotting will show qualitatively possible relationships and will give leads to the nature of the model to be assumed. Fitting the formulated model will give a quantitative evaluation of the assumed relationships and will point to the variables that have strong impact on water quality. Along with consideration of various candidate models, an ordering of the relative importance of the four x variables can become evident.

THOROUGHBRED AUCTION DATA

One of the well-known yearly horse auctions is held at Keeneland, near Lexington, Kentucky, each summer. Within the thoroughbred horse industry it is felt that certain factors are important in determining the auction price. In a study of this issue 37 yearlings were selected from the 1979

Table 2.1 Data for New York Rivers Study

River Basin	y	x_1	x_2	x_3	x_4
Olean	1.10	26	63	1.2	0.29
Cassadaga	1.01	29	57	0.7	0.09
Oatka	1.90	54	26	1.8	0.58
Neversink	1.00	2	84	1.9	1.98
Hackensack	1.99	3	27	29.4	3.11
Wappinger	1.42	19	61	3.4	0.56
Fishkill	2.04	16	60	5.6	1.11
Honeoye	1.65	40	43	1.3	0.24
Susquehanna	1.01	28	62	1.1	0.15
Chenango	1.21	26	60	0.9	0.23
Tioughnioga	1.33	26	53	0.9	0.18
West Canada	0.75	15	75	0.7	0.16
East Canada	0.73	6	84	0.5	0.12
Saranac	0.80	3	81	0.8	0.35
Ausable	0.76	2	89	0.7	0.35
Black	0.87	6	82	0.5	0.15
Schoharie	0.80	22	70	0.9	0.22
Raquette	0.87	4	75	0.4	0.18
Oswegatchie	0.66	21	56	0.5	0.13
Cohocton	1.25	40	49	1.1	0.13

Definition of Variables

y = total nitrogen; mean concentration, in milligrams per liter (mg/l), based on samples taken at regular intervals during the spring, summer, and fall months

x_1 = active agriculture; percentage of land area currently in agricultural use (cropland, pasture, orchards, horticulture, floriculture, vegetable and truck farms, etc.)

x_2 = forest; percentage of land area in forest, forest brushland, and plantations

x_3 = residential; percentage of land area in residential use [including urban, surburban, and rural communities and strip developments with more than four residences per 1000 feet (ft)]

x_4 = commercial and/or industrial; percentage of land area in either commercial or manufacturing use

Source: Douglas A. Haith, "Land Use and Water Quality in New York Rivers," *Journal of the Environmental Engineering Division*, *ASCE* 102 (no. EE1, Proc. Paper 11902, February 1976): 1–15.

Table 2.2 Data for Thoroughbred Auction Study

y	x_1	x_2	x_3	x_4	x_5
62	112	8	20	11	41
200	78	73	25	13	243
225	1317	233	13	25	226
125	576	119	24	7	55
45	414	124	6	13	30
50	46	88	5	55	85
50	405	7	17	1	68
57	355	4	20	10	64
110	304	14	15	14	23
50	224	7	7	4	53
40	538	131	6	14	110
185	1462	29	15	17	86
80	76	4	26	12	133
95	224	6	15	23	53
38	46	20	5	21	85
105	1317	63	20	9	226
72	1317	169	21	21	226
145	576	201	6	33	55
100	796	45	8	13	109
155	271	29	6	19	32
70	123	254	4	20	41
100	604	41	19	10	39
57	110	105	9	40	34
225	76	291	10	19	133
130	576	28	6	12	55
125	581	22	7	7	304
275	169	38	6	13	44
120	637	138	7	10	46
50	256	53	10	11	39
55	160	37	21	9	52
35	1111	91	4	10	52
80	279	118	6	13	34
300	796	17	13	23	57
85	671	27	15	17	73
260	1750	5	15	16	53
250	538	53	8	12	110
72	500	92	19	16	42

Definition of Variables

y = sales price ($\times \$1000$)

x_1 = total money won by the sire ($\times \$1000$)

x_2 = total money won by the dam ($\times \$1000$)

x_3 = total number of stakes winners and stakes placed horses listed under the first three dams

x_4 = total number of wins for the first three dams

x_5 = median yearling price the previous year for the designated sire ($\times \$1000$)

Source: David B. Foye, unpublished data (The Jockey Club, Lexington, Ky).

summer sale. The objective of the study was to determine an equation for predicting future sales price (y) by using several predictor variables. The data are shown in Table 2.2.

For an analysis where prediction is the major and perhaps sole objective, a general outline would include the following steps:

- Evaluation of several candidate models and selection of the simplest adequate model by using these techniques:

   Analysis of variance for comparison of the candidate models.

   A plot of the residuals.

   Independent validation, if possible.

- Calculation of the prediction equation and standard errors for the selected model.

DAIRY HYPOMAGNESAEMIA DATA

A nervous disorder of dairy cattle is grass tetany or subclinical hypomagnesaemia, which is symptomatically revealed in an abnormal lowering of blood serum magnesium levels. A 1979 New Zealand study integrated results of field plot fertilizer work with animal-grazing trials. Sixty cows, including 12 Friesians, 24 Jerseys, and 24 Jersey-Friesian crosses, were assigned to grazing paddocks to which either a control (no applied magnesium) or a treatment [110 kilograms of magnesium per hectare (kg/ha)] had been allocated.

The response variable y is blood serum magnesium (milligrams per 100 milliliters, mg/100 ml), specifically, the difference between measurements before and after the experiment. Two other variables, which were believed to influence the y variable and were measured on each cow before the experiment, were DCALF, the number of days until calving (as estimated from estrous evidence), and SCORE, the overall physical condition of each animal (an assigned score determined largely by fat cover). These two measured variables, commonly called *covariates* (uncontrolled variables not affected by the treatments), will be two of the x variables in the analysis. The data are shown in Table 2.3.

In a regression analysis of the data each of the six breed and magnesium level combinations will enter the analysis as an x variable. The details of constructing six indicator x variables when no measured values exist are developed in later units, starting with Unit 15. Briefly, each indicator variable is equal to one if the observation is associated with a particular breed and magnesium level combination and is equal to zero if not. In a regression format the dairy hypomagnesaemia data consist of a y variable and eight x variables, the indicator variables $x_1, x_2, \ldots, x_6$ and the two measured x variables x_7 and x_8.

Table 2.3 Data for Study of Hypomagnesaemia in Dairy Cattle

\multicolumn{3}{c}{JERSEY AND CONTROL COMBINATION}	\multicolumn{3}{c}{JERSEY AND TREATMENT COMBINATION}
y	DCALF	SCORE	y	DCALF	SCORE
−0.26	79	4.0	0.19	52	4.0
−0.05	61	5.0	0.01	63	3.5
0.04	82	4.0	0.09	70	4.0
−0.21	56	4.5	0.19	69	4.5
−0.18	59	5.0	0.09	60	4.0
−0.15	70	5.5	0.00	70	6.0
0.08	72	5.0	0.19	62	4.5
−0.31	66	4.5	0.10	71	4.0
−0.28	67	4.0	0.32	69	4.5
0.03	56	5.0	−0.03	76	3.5
−0.25	74	4.5	0.08	52	4.5
−0.04	56	3.0	0.24	59	3.5

JERSEY-FRIESIAN AND CONTROL COMBINATION			JERSEY-FRIESIAN AND TREATMENT COMBINATION		
y	DCALF	SCORE	y	DCALF	SCORE
−0.15	55	5.0	0.02	56	5.0
−0.35	58	4.5	0.08	69	4.5
0.00	63	4.0	−0.08	69	4.0
−0.17	61	4.0	−0.08	55	5.0
−0.15	55	4.0	0.10	68	3.5
−0.34	66	4.0	0.38	61	5.5
−0.37	66	4.5	0.00	66	4.0
−0.22	54	4.5	0.01	70	4.5
−0.15	59	4.5	0.20	56	4.5
−0.38	56	5.5	−0.05	60	5.5
−0.36	64	6.0	0.09	55	4.0
−0.63	69	4.0	0.23	55	3.5

\multicolumn{3}{c}{FRIESIAN AND CONTROL COMBINATION}	\multicolumn{3}{c}{FRIESIAN AND TREATMENT COMBINATION}
y	DCALF	SCORE	y	DCALF	SCORE
−0.19	58	4.0	−0.11	79	3.0
−0.15	73	6.0	0.04	68	5.0
−0.07	59	3.0	0.24	56	5.0
−0.19	52	5.0	0.12	70	3.5
−0.18	70	4.5	0.23	55	5.0
−0.09	53	5.0	0.04	57	5.0

Definition of Variables

y = blood serum magnesium; the difference between measurements before and after the experiment, in milligrams per 100 milliliters

DCALF = number of days until calving (estimated from estrous evidence)

SCORE = overall physical condition of animal (an assigned score determined largely by fat cover)

Source: C. B. Dyson, unpublished data (New Zealand Ministry of Agriculture and Fisheries).

Under the assumption that the difference between the control and the treatment is the same for each breed (no interaction), the primary objective of the experiment involves the difference between the control group and the treatment group. Since this difference could be influenced by the measured covariates, part of the analysis will examine the importance of the covariates. If the interaction is not important but the covariates are important, the last part of the analysis will estimate the mean difference between the control and treatment groups adjusted for the covariates. Specifically, the flow of the data analysis follows these steps (to be discussed in detail in Units 17 and 19):

- Estimation of the model between y and all the x variables.
- Consideration of reduced models, deleting one or both covariates.
- Evaluation of the importance of the interaction between breeds and magnesium levels.
- Calculation of the difference in the means of the treatment and control groups and the associated standard error for the selected model.

SUMMARY

Our data analysis strategy starts with an evaluation of a full or general model. We then look for a reduced or simpler model within a sequence of candidate models that will also be adequate for describing the data and for making inferences. During the planning stage of a study, research design and data analysis aspects need to be integrated so that a correspondence between subobjectives and specific aspects of the data analysis is clear. Specifically, the subobjectives and the data analysis approach are modified iteratively until the correspondence is obtained. Then the required data are collected, and we can interpret the data meaningfully.

Unit 3

Model Formulation and Inference

Objectives

- To explore the nature of the relationships between the response (dependent) and explanatory (independent) variables, specifically, the mathematical form of the relationship, including the residual, denoted ε.

- To explore the concepts of statistical inference and, in particular the influence of sampling variability. A notation for describing and handling data sets is also developed.

- To discuss the basic computational aspects of solving a set of equations.

Each data set in Unit 2 has one response variable, y, and several explanatory or predictor variables, the x variables. If y is a function of the several x variables, the value of y depends on the values of the x's, but the x values are free to vary (within limits). Thus the y variable has historically been called the *dependent variable* and the x variables are called the *independent variables*.

RESPONSE AND EXPLANATORY VARIABLES

In statistical terminology the measured response variable y and the residual ε of our statistical model are random variables with a probability distribution, usually assumed to be a normal distribution. The explanatory x variables are strictly mathematical variables with no associated probability distribution. These x variables can be qualitative (categorical) or quantitative.

Categorical variables do not usually involve measurement. Each unit or subject is classified—for example, type of occupation or type of treatment allocated to experimental units in designed studies. Then a label—for instance, lawyer or control treatment—is assigned to each category. Each category will be a variable in the regression format, constructed by assigning a code of zero to indicate that the unit has not been given that particular label and a code of one to indicate that it has. The dairy hypomagnesaemia data set in Unit 2 provides an example of categorical variables. The basic interest in the analysis usually is to estimate linear contrasts among the mean responses for each category (these ideas are developed in Units 15–19).

For now, measured *quantitative variables*, either discrete or continuous, will be emphasized. Measurements on *discrete variables* are limited to certain values, usually integer values, such as counts—for example, the number of days in a hospital stay or the number of cells in an organism. (The number of single cells is discrete, but if it is in the thousands, the distinction between discrete and continuous can become blurred.) Examples of *continuous variables* are hours of artificial light in an egg-laying house or age of patients in a geriatric survey. Theoretically, hours or ages can take on any value within a defined range. Now our interest includes not only the particular values of the variables recorded in the study but also values between the recorded ones. Consequently, a fitted smooth curve between the response variable and a continuous x variable will allow us to make predictions at nonrecorded values of the x variable. Note that a functional relationship between the response variable and a discrete x variable can also be envisioned and a smooth curve fitted, but prediction is reasonable only at the possible discrete values.

In Unit 2 explanatory variables are regression-formulated as columns, with numerical values of x variables measured directly from the sampled units (as in observational studies) or preselected to have specific values (as in designed experiments). Treatments in a designed study could also be categorical x variables with assigned zeros and ones for each category. For example, in a soil fertility experiment relating crop yield to explanatory factors, the treatments could be preselected levels of an applied nutrient (quantitative) or different chemical formulations of an applied nutrient (categorical). In addition, uncontrollable environmental variables affecting yield, such as drought, can be measured. Depending on the objectives of a study, some variables, like insect infestation, can be either controlled at predetermined levels or not controlled, with the indigenous population levels measured. These examples indicate the diversity and flexibility of explanatory x variables.

FORMULATION OF A REGRESSION MODEL

The responses are related to the x variables through a model. As indicated in Unit 1, a regression model is a decomposition of each observation into two parts, the predicted portion, which can be assigned to sources known to affect the response variable, and the residual portion, which is assumed to be due to random variability. The known sources are the explanatory variables.

The simplest approach to explicitly incorporating the x variables into a model is to assume that a good approximation to the relationship between y and each x is a straight line. The *slope* of the straight line is denoted by β. If k explanatory variables are identified, β_1 is the partial slope between y and x_1; β_2 is the partial slope between y and x_2, and, in general, β_j is the partial slope between y and x_j, where $j=1,2,\ldots,k$. The product $\beta_j x_j$ is the contribution to the magnitude of y that is associated with x_j. If we also assume that the value of y is the sum of these $\beta_j x_j$ contributions plus the random component ε, then the regression model we assume is given by

$$y=\beta_0+\beta_1 x_1+\beta_2 x_2+\cdots+\beta_k x_k+\varepsilon$$

The first beta, β_0, is called the *intercept*, the explainable part of y when all the x's are equal to zero. When the value of all the x's is one, then y is the sum of the betas and epsilon. If x_j changes by one unit while the other x's are held constant, then y is assumed to change by β_j. Hence the partial slopes are commonly called the *partial regression coefficients*, and each β_j gives the change in y per unit change of the x variable, assuming that the other x variables stay constant. The betas are called the *unknown parameters* of the assumed regression model.

The right-hand side of the regression model is a sum, the last part being the random component ε. The residuals are simple differences between the y's and the x variable contributions. We would expect the average of these simple differences to be zero, an acceptable assumption if the sum of the $\beta_j x_j$'s (including the intercept) accounts for all the explainable variability among the observations. It is also assumed that (1) the variability among the residual components is constant—that is, the variability does not depend on the values of the x's—and (2) the value of ε associated with each observation does not depend on the value of any other residual component.

The sum of all the terms on the right-hand side of the regression model, except for epsilon, can be expressed by one summary parameter:

$$y = \mu(x_1, x_2, \ldots, x_k) + \varepsilon$$

where

$$\mu(x_1, x_2, \ldots, x_k) = \beta_0 + \beta_1 x_1 + \beta_2 x_2 + \cdots + \beta_k x_k$$

The $\mu(x_1, x_2, \ldots, x_k)$ is read, "the parameter μ depends on (or is a function of) the k explanatory x variables." The beta parameters indicate the exact nature of the relationship.

The job of the data analyst is to calculate an estimated value of μ from the data, called the *predicted value*, by fitting the regression model. Fitting refers to calculating values of the parameters from a set of data. In operational jargon the data analyst refers to $\mu(x_1, x_2, \ldots, x_k)$ as "the model to be fitted"; then the difference between the observation and the predicted or fitted value is the residual, a term also used for ε in the regression model. Procedures for fitting models start in the next unit.

In Unit 1 a regression model was defined as a decomposition of an observation into two parts, where the first part could involve several components. Let us relate Unit 1 with this unit. The several components of Unit 1 are the several x variables, $x_1, x_2, \ldots, x_k$, of this unit. The regression model of Unit 1 was stated in terms of the data, that is, it was actually a fitted model,

$$\text{observation} = \text{prediction} + \text{residual}$$

or
$$y = \hat{y} + (y - \hat{y})$$

where $\hat{y}$ (y hat) is the notation for a predicted value. Actually $\hat{y}$ should be written as $\hat{y}(x_1, x_2, \ldots, x_k)$ to show that $\hat{y}$ depends on values of the x variables. This fitted model can be written as an assumed model by replacing the predicted part by $\mu(x_1, x_2, \ldots, x_k)$ and the residual part by ε:

$$y = \mu(x_1, x_2, \ldots, x_k) + \varepsilon$$

In this unit the concept of an assumed regression model has been developed from basic principles and assumptions. Note that an assumed model can be written without having any data. Also, in this unit the regression model is explicit about the contribution of each x variable going into the μ parameter. The explicit form of the regression model assumed in this unit is $\beta_0 + \beta_1 x_1 + \cdots + \beta_k x_k$, but other forms will be assumed in Unit 23.

REGRESSION INFERENCE

An assumed regression model conceptually relates the response variable to the x variables through the parameters. A fitted model summarizes a data set by giving numerical estimates of the parameters, which are few relative to the number of observations in a data set. Such a data reduction is one objective of data analysis commonly called descriptive statistics. However, when we try to generalize and say something about an unknown parameter, we are making inferences to units not sampled and included in the data set. Before proceeding further, we should outline the basic concepts of the inference problem in a regression frame.

1. A clear, concise statement is made of the target population, that is, those units to which we want to make inferences. Of special importance is the definition of membership in the target population. It must be clear to the researcher and everybody else whether or not a given unit is a member of the target population.

2. For each unit in the population it is assumed that a response variable of interest conceptually can be measured along with certain x variables. An assumed model is written, indicating the relationship between the response variable and the x variables, either measured or constructed. The beta parameters of the model are unknown constants.

3. A random sample of n units is selected from the population of N units. At times all the units in the population will not be accessible to the researcher due to practical, legal, or other considerations. The sample is then selected from the redefined accessible population. Strictly speaking,

any inferences are limited to the accessible population. Statements about the target population involve a certain degree of hand waving in the sense of indicating that the nonaccessible units are just like the accessible units.

4. The response variable and the x variables are measured on the sample units, and the data set is recorded.
5. The partial regression coefficients are estimated from the data set. The formulas for calculating the estimates of the unknown parameters are called *statistics* or *estimators*. The *parameters* characterize or describe the population, while the statistics describe the sample. The numerical values of the statistics or estimators are the *estimates*.
6. The estimators of the partial regression coefficients are used as descriptive statistics for the data in the sample or as inferential statistics in making an inference concerning the betas of the assumed regression model. Use of the estimators will yield point estimates of the beta values. But usually an inference is made to the betas by use of a confidence interval, the point estimate plus and minus a multiple of the estimated standard error of the estimator. The estimators and calculated standard errors are also used in testing null hypotheses concerning the beta parameters.

The inference problem can be summarized in diagram form, as shown in Figure 3.1.

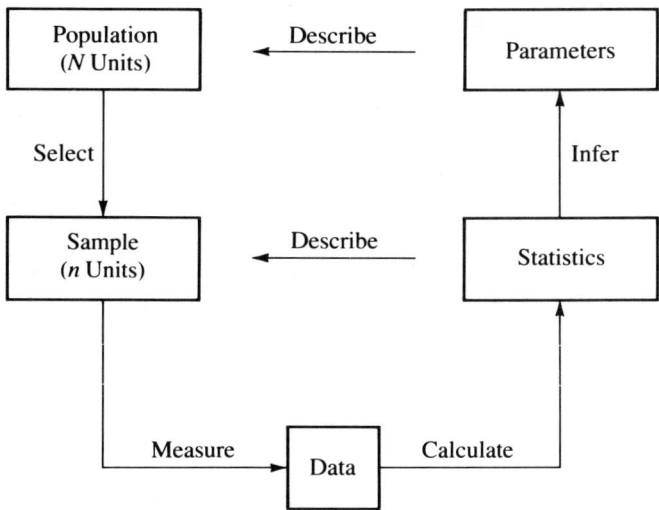

Figure 3.1 Summary of the Inference Problem

NOTATION FOR REGRESSION DATA

In practice we will find it convenient to use a simplified notation for a row or column of numbers (a vector) or for a *matrix*, defined to be a rectangular array of numbers. The numbers in a matrix are referred to as its elements. The numbers in a horizontal line constitute a row of the matrix, those in a vertical line, a column. Boldface letters are used to denote matrices (or vectors), and italicized letters are used for elements; thus **A** is a matrix with elements a_{ij}. The first subscript position in a_{ij} indicates the row location, and the second subscript position indicates the column location of the elements under discussion. For example,

$$\mathbf{A} = \begin{bmatrix} 1 & 0 & 2 \\ 3 & 4 & 5 \end{bmatrix} \quad \text{implies that} \quad \begin{matrix} a_{11}=1 & a_{12}=0 & a_{13}=2 \\ a_{21}=3 & a_{22}=4 & a_{23}=5 \end{matrix}$$

A matrix with n rows and p columns is said to have dimension $n \times p$; thus **A** has dimension 2×3. The matrix sum $\mathbf{A} + \mathbf{C}$, where **A** and **C** are of the same dimension, is defined to be a matrix containing the sums of the corresponding elements of **A** and **C**; that is, $\mathbf{F} = \mathbf{A} + \mathbf{C}$ means that $f_{ij} = a_{ij} + c_{ij}$. For example,

$$\begin{bmatrix} 1 & 0 & 2 \\ 3 & 4 & 5 \end{bmatrix} + \begin{bmatrix} 1 & 1 & 7 \\ 0 & 2 & 8 \end{bmatrix} = \begin{bmatrix} 2 & 1 & 9 \\ 3 & 6 & 13 \end{bmatrix}$$

Similarly, one matrix or vector may be subtracted from another matrix with the same dimensions.

The product of a $1 \times m$ row vector with an $m \times 1$ column vector is given by

$$[a_1 a_2 \cdots a_m] \begin{bmatrix} c_1 \\ c_2 \\ \vdots \\ c_m \end{bmatrix} = a_1 c_1 + a_2 c_2 + \cdots + a_m c_m = \sum_{q=1}^{m} a_q c_q$$

This leads to the general definition of matrix multiplication, namely, $\mathbf{F} = \mathbf{AC}$, where

$$f_{ij} = \sum_{q=1}^{m} a_{iq} c_{qj}$$

A little reflection will show that for this definition to work, the number of columns of **A** must equal the number of rows of **C**. In fact, if **A** is $r \times s$ and **C** is $s \times t$, then $\mathbf{F} = \mathbf{AC}$ is $r \times t$.

$$\begin{bmatrix} 1 & 2 & 3 \\ 4 & 4 & 7 \end{bmatrix} \begin{bmatrix} 1 & 4 & 2 \\ 1 & 0 & -1 \\ 1 & 6 & 3 \end{bmatrix} = \begin{bmatrix} 6 & 22 & 9 \\ 15 & 58 & 25 \end{bmatrix}$$

$$(2 \times 3) \qquad (3 \times 3) \qquad (2 \times 3)$$

where $6 = (1)(1) + (2)(1) + (3)(1)$, $22 = (1)(4) + (2)(0) + (3)(6), \ldots,$ and $25 = (4)(2) + (4)(-1) + (7)(3)$.

Within this framework for the multiple regression model, we define

$$\mathbf{Y} = \begin{bmatrix} y_1 \\ y_2 \\ \vdots \\ y_i \\ \vdots \\ y_n \end{bmatrix} \qquad \mathbf{X} = \begin{bmatrix} x_{10} & x_{11} & \cdots & x_{1k} \\ x_{20} & x_{21} & \cdots & x_{2k} \\ \vdots & \vdots & & \vdots \\ x_{i0} & x_{i1} & \cdots & x_{ik} \\ \vdots & \vdots & & \vdots \\ x_{n0} & x_{n1} & \cdots & x_{nk} \end{bmatrix} \qquad \boldsymbol{\beta} = \begin{bmatrix} \beta_0 \\ \beta_1 \\ \vdots \\ \beta_k \end{bmatrix} \qquad \boldsymbol{\varepsilon} = \begin{bmatrix} \varepsilon_1 \\ \varepsilon_2 \\ \vdots \\ \varepsilon_i \\ \vdots \\ \varepsilon_n \end{bmatrix}$$

Now we write the model for all our observations as a matrix equation,

$$\mathbf{Y} = \mathbf{X}\boldsymbol{\beta} + \boldsymbol{\varepsilon}$$
$$(n \times 1) \quad (n \times p)(p \times 1) \quad (n \times 1)$$

where **Y** is a column vector of observed responses, **X** is a known matrix of explanatory or predictor variables, $\boldsymbol{\beta}$ is an unknown vector of parameters relating the data in the **Y** column to the variables in the **X** matrix ($p = k + 1$), and $\boldsymbol{\varepsilon}$ is a column vector of unknown random variables. Usually it is assumed that each of these random variables can be conceptualized as coming from a continuous distribution with a mean of zero and a constant variance of σ^2, and that the ε of one observation is uncorrelated with the ε of any other observation. (Roughly speaking, two errors are uncorrelated if they bear no linear relationship to each other.) In addition, the errors are assumed to be normally distributed in inference situations involving the t and F distributions.

The **X** matrix includes the data on the x variables in the last k columns, but it also has an x_0 column, the first column. Each element in the first column is equal to one for the intercept. When **X** and $\boldsymbol{\beta}$ are multiplied, the observational equations will be reproduced. For example, for the New

York rivers data (Table 2.1) of Unit 2, we have

$$Y = \begin{bmatrix} 1.10 \\ 1.01 \\ 1.90 \\ 1.00 \\ 1.99 \\ 1.42 \\ 2.04 \\ 1.65 \\ 1.01 \\ 1.21 \\ 1.33 \\ 0.75 \\ 0.73 \\ 0.80 \\ 0.76 \\ 0.87 \\ 0.80 \\ 0.87 \\ 0.66 \\ 1.25 \end{bmatrix} = \begin{bmatrix} 1 & 26 & 63 & 1.2 & 0.29 \\ 1 & 29 & 57 & 0.7 & 0.09 \\ 1 & 54 & 26 & 1.8 & 0.58 \\ 1 & 2 & 84 & 1.9 & 1.98 \\ 1 & 3 & 27 & 29.4 & 3.11 \\ 1 & 19 & 61 & 3.4 & 0.56 \\ 1 & 16 & 60 & 5.6 & 1.11 \\ 1 & 40 & 43 & 1.3 & 0.24 \\ 1 & 28 & 62 & 1.1 & 0.15 \\ 1 & 26 & 60 & 0.9 & 0.23 \\ 1 & 26 & 53 & 0.9 & 0.18 \\ 1 & 15 & 75 & 0.7 & 0.16 \\ 1 & 6 & 84 & 0.5 & 0.12 \\ 1 & 3 & 81 & 0.8 & 0.35 \\ 1 & 2 & 89 & 0.7 & 0.35 \\ 1 & 6 & 82 & 0.5 & 0.15 \\ 1 & 22 & 70 & 0.9 & 0.22 \\ 1 & 4 & 75 & 0.4 & 0.18 \\ 1 & 21 & 56 & 0.5 & 0.13 \\ 1 & 40 & 49 & 1.1 & 0.13 \end{bmatrix} \begin{bmatrix} \beta_0 \\ \beta_1 \\ \beta_2 \\ \beta_3 \\ \beta_4 \end{bmatrix} + \begin{bmatrix} \varepsilon_1 \\ \varepsilon_2 \\ \varepsilon_3 \\ \varepsilon_4 \\ \varepsilon_5 \\ \varepsilon_6 \\ \varepsilon_7 \\ \varepsilon_8 \\ \varepsilon_9 \\ \varepsilon_{10} \\ \varepsilon_{11} \\ \varepsilon_{12} \\ \varepsilon_{13} \\ \varepsilon_{14} \\ \varepsilon_{15} \\ \varepsilon_{16} \\ \varepsilon_{17} \\ \varepsilon_{18} \\ \varepsilon_{19} \\ \varepsilon_{20} \end{bmatrix}$$

Reproducing the observational equations, we obtain

$$1.10 = \beta_0 + 26\beta_1 + 63\beta_2 + 1.2\beta_3 + 0.29\beta_4 + \varepsilon_1$$
$$1.01 = \beta_0 + 29\beta_1 + 57\beta_2 + 0.7\beta_3 + 0.09\beta_4 + \varepsilon_2$$
$$\vdots \quad \vdots \quad \vdots \quad \vdots \quad \vdots \quad \vdots \quad \vdots$$
$$1.25 = \beta_0 + 40\beta_1 + 49\beta_2 + 1.1\beta_3 + 0.13\beta_4 + \varepsilon_{20}$$

Expressed in either the matrix equation or observational equation format, our immediate need is to estimate the parameters of $\boldsymbol{\beta}$ from the data. The calculations will involve sums of squares of x variables and the sums of cross products among the x variables and between y and each x variable. These quantities can be systematically calculated by forming the matrix products $\mathbf{X'X}$ and $\mathbf{X'Y}$, where $\mathbf{X'}$, the *transpose* of $\mathbf{X}$, is defined as the matrix obtained by transposing the rows and columns of $\mathbf{X}$. Thus if $\mathbf{X}$ is a

20×5 matrix, $\mathbf{X}'$ is a 5×20 matrix, and each column of $\mathbf{X}$ is the corresponding row in $\mathbf{X}'$.

We will expend considerable effort in the upcoming units developing the methodology for computing the needed statistics, rather than going directly to the computer. General awareness of the computing approach, especially with simple regression models, is not only important in understanding the output from regression computer programs but is also the key to achieving a complete analysis and interpretation of complex data sets by regression. The methodology will be slowly and systematically developed, starting with the simplest regression model in the next unit. But first let us discuss some general computational aspects.

COMPUTATIONAL CONSIDERATIONS

As we will see unfolding in Units 4, 5, and 7, the core of regression calculations involves solving a set of p equations with p unknowns. Before launching into regression models, let's review our algebraic basics by solving the equations

$$3b_0 - b_1 + b_2 = 10$$
$$-b_0 + 5b_1 - b_2 = -10$$
$$b_0 - b_1 + 3b_2 = 12$$

where b_0, b_1, and b_2 are the unknowns. The rationale is to systematically eliminate the unknowns until the equations can be readily solved. We begin by dividing the first equation by 3. Then in order to eliminate b_0 from the remaining equations, we must add the new first equation to the second equation and subtract it from the third. Thus we obtain

$$b_0 - (1/3)b_1 + (1/3)b_2 = 10/3$$
$$(14/3)b_1 - (2/3)b_2 = -20/3$$
$$- (2/3)b_1 + (8/3)b_2 = 26/3$$

We continue in the same pattern; b_1 is eliminated from the third equation by multiplying the second equation by $1/7$ (dividing by $14/3$ and multiplying by $2/3$) and adding the new second equation to the third. Now we have

$$b_0 - (1/3)b_1 + (1/3)b_2 = 10/3$$
$$b_1 - (1/7)b_2 = -10/7$$
$$(18/7)b_2 = 54/7$$

If we work backward from the last equation, the triangular system of equations is easy to solve. Numerically, $b_2 = 3$, $b_1 = (-10/7) + (1/7)(3) = -1$, and $b_0 = (10/3) + (1/3)(-1) - (1/3)(3) = 2$.

The original three equations and three unknowns can be expressed in terms of matrices:

$$\mathbf{A} = \begin{bmatrix} 3 & -1 & 1 \\ -1 & 5 & -1 \\ 1 & -1 & 3 \end{bmatrix} \quad \mathbf{b} = \begin{bmatrix} b_0 \\ b_1 \\ b_2 \end{bmatrix} \quad \mathbf{g} = \begin{bmatrix} 10 \\ -10 \\ 12 \end{bmatrix}$$

The matrix equation $\mathbf{Ab} = \mathbf{g}$ is equivalent to the original set of equations. For instance, the product of the second row of $\mathbf{A}$ and $\mathbf{b}$, $-b_0 + 5b_1 - b_2$, equated to the second element in $\mathbf{g}$ is the second equation of the set of equations to be solved.

In our example note that $a_{12} = a_{21} = -1$, $a_{13} = a_{31} = 1$, and $a_{23} = a_{32} = -1$; that is, $\mathbf{A}$ is symmetric. An alternative procedure for eliminating unknowns, which takes advantage of a symmetric $\mathbf{A}$, is the *Abbreviated Doolittle Method*. The operations of the Abbreviated Doolittle are patterned and repetitive and can be easily carried out on a hand or desk calculator. By writing out the calculation steps for our example, we can see an emerging pattern that will be useful with other sets of equations.

The starting point is the upper triangular part of $\mathbf{A}$ and $\mathbf{g}$ written side by side and denoted as $\mathbf{A}|\mathbf{g}$:

$$\begin{array}{rrrr} 3 & -1 & 1 & 10 \\ & 5 & -1 & -10 \\ & & 3 & 12 \end{array}$$

Step 1: Rewrite the first row of $\mathbf{A}|\mathbf{g}$:

$$3 \quad -1 \quad 1 \quad 10$$

Step 2: Divide each number in step 1 by the first number in step 1:

$$1 \quad -1/3 \quad 1/3 \quad 10/3$$

The value $-1/3$ is a *pivotal number* in the next step.

Step 3: Subtract from each number in row 2 of $\mathbf{A}|\mathbf{g}$ the product of $-1/3$ and the corresponding number in step 1:

$$14/3 \quad -2/3 \quad -20/3$$

Step 4: Divide each number in step 3 by the first number in step 3:

$$1 \quad -1/7 \quad -10/7$$

Now the third value in step 2, $1/3$, and the second value in step 4, $-1/7$, are the pivotal numbers in the next step.

Step 5: Subtract from each number in row 3 of $A|g$ both the product of $1/3$ and the corresponding number in step 1 and the product of $-1/7$ and the corresponding number in step 3:

$$18/7 \quad 54/7$$

Step 6: Divide each number in step 5 by the first number in step 5:

$$1 \quad 3$$

Repeating $A|g$ and collecting the numerical results from the six steps, we have

	3	−1	1	10
		5	−1	−10
			3	12
Step 1.	3	−1	1	10
Step 2.	1	−1/3	1/3	10/3
Step 3.		14/3	−2/3	−20/3
Step 4.		1	−1/7	−10/7
Step 5.			18/7	54/7
Step 6.			1	3

Even though the unknowns haven't been explicitly carried along, the same triangular set of equations calculated previously can be identified in the even steps above, namely,

$$b_0 - (1/3)b_1 + (1/3)b_2 = 10/3$$
$$b_1 - (1/7)b_2 = -10/7$$
$$b_2 = 3$$

and the solutions are found as before.

A systematic procedure for handling the necessary calculations with **A** and **g** in a prescribed stepwise order is called a computing algorithm. The Abbreviated Doolittle is a time-honored algorithm for solving a set of equations. It was developed well before the advent of electronic calculators and computers—in fact, before the turn of the century while Doolittle was with the U.S. Coast and Geodetic Survey. The method is designed to solve equations, and to perform other useful functions, with the minimum amount of computation and recording of intermediate results. In this regard it remains a superior technique, and basic regression computations can be done most efficiently on a hand or desk calculator by using the Abbreviated Doolittle algorithm.

Regression programs in today's statistical packages use the sweep operator, which is similar to the Abbreviated Doolittle but the steps are in a somewhat different order. Adopted by the packages because inventory recording of intermediate results is not a problem to computers, the sweep operator is also more adaptable to automatic variable selection techniques, a topic discussed in Unit 20.

When using the Abbreviated Doolittle or the sweep operator algorithm, certain precautions should be taken. All significant digits carried by your calculator must be recorded at the intermediate stages. Also, significant digits must not be confused with decimal places. For instance, the number 0.0045 has four decimal places but only two significant digits. You might say, "I would be satisfied to have an answer accurate to four significant digits. Why should I carry ten digits?" To understand the answer to that question, suppose you carry six digits and compute $648,321. - 648,217.$ The solution is 104. and has only three significant digits. Hence because of the cancellation of the significant digits during intermediate computations, we must carry more significant digits than are needed for the final result. In later units numerical results have been rounded to a fixed number of decimal places, which may introduce minor inconsistencies in the calculations. If the computations are done by using a good statistical library program, the computations will automatically use a large number of significant digits or an algorithm that minimizes cancellation of the most significant digits. These alternative algorithms are not practical for hand and desk calculators but are incorporated in a microcomputer program for regression models called Statistical Analysis (STAN) and presented in Unit 24.

For the analysis of a data set a statistical package should be used, if available. If a computer is available but it does not support a statistical package, then STAN should be adopted or mimicked. Otherwise, the Abbreviated Doolittle and related methods presented in this book should be used. Regardless of the choice, the development of this book embraces the premise that an understanding of data analysis concepts and practice is enhanced by doing some computations with a hand or desk calculator. Through insight into computer data manipulations, we can understand ambiguous output statistics, better utilize the program's capabilities, and supplement any limitations of the program. A more complete interpretation of the data will invariably result.

Unit 4

Fitting the Mean Model

Objectives

- To describe the method of least squares estimation for fitting regression models.
- To develop the Abbreviated Doolittle algorithm for fitting the Mean Model and to show the equivalence between the estimated regression coefficient and the arithmetic mean of sample observations.

One common step in the data analysis strategies outlined for each data set in Unit 2 is the fitting of candidate models. In the terminology of Unit 3 we need to estimate the partial regression coefficients of a multiple regression model between y and the x variables. Involved in these calculations are sums of squares and cross products among the x variables and sums of cross products between y and the x variables. Thus a computing algorithm that will handle the necessary calculations for any data set is needed. The Abbreviated Doolittle algorithm (ABDO) is chosen here for its characteristics as presented in Unit 3 and for its usefulness in interpreting the estimated partial regression coefficients. But before the details of ABDO for the model with several x variables is given (Unit 7), the general method of least squares model fitting or estimation is described. This presentation is followed by a development of ABDO for three simple models, the Mean Model (no measured x variables), and then, in Unit 5, the fitting of two straight-line models.

LEAST SQUARES MODEL FITTING

Before carrying out the calculation for estimating mean and straight-line models, we will discuss the *least squares method* of estimating $\mathbf{X}\boldsymbol{\beta}$ models in general. In Unit 3 our general regression model was expressed as a matrix equation,

$$\mathbf{Y} = \mathbf{X}\boldsymbol{\beta} + \boldsymbol{\varepsilon}$$

We denote the vector of estimated regression coefficients as $\mathbf{b}$. The predicted values are then calculated by multiplying each row in the $\mathbf{X}$ matrix by the $\mathbf{b}$ column, that is $\hat{\mathbf{Y}} = \mathbf{X}\mathbf{b}$. The estimated residuals are then written as

$$\mathbf{Y} - \hat{\mathbf{Y}} = \mathbf{Y} - \mathbf{X}\mathbf{b}$$

We want to choose a method of calculating $\mathbf{b}$ so that the elements or components of the residual vector, $\mathbf{Y} - \mathbf{X}\mathbf{b}$, are as small as possible. The specific criterion of smallness will be the sum of squares of the residuals. If $\mathbf{b}$ is the solution to the set of equations $\mathbf{X}'\mathbf{X}\mathbf{b} = \mathbf{X}'\mathbf{Y}$, then the residual sum of squares is minimized. Consequently, one important phase of regression calculations is solving the set of equations $\mathbf{X}'\mathbf{X}\mathbf{b} = \mathbf{X}'\mathbf{Y}$.

A direct correspondence between the terms in the set of equations $\mathbf{X'Xb}=\mathbf{X'Y}$ and the general notation of $\mathbf{Ab}=\mathbf{g}$ used in the section "Computational Considerations" of Unit 3 can be made. With our $\mathbf{X}\boldsymbol{\beta}$ models the $\mathbf{A}$ matrix is the symmetric $\mathbf{X'X}$ matrix of sums of squares and cross products among the x variables, and $\mathbf{g}$ is the $\mathbf{X'Y}$ vector of sums of cross products between each x and the y variable. The Abbreviated Doolittle computing algorithm developed in Unit 3 will be utilized for solving the set of equations for the estimated regression coefficients $\mathbf{b}$, starting with the upper triangular part of $\mathbf{X'X}$ and $\mathbf{X'Y}$ written side by side and denoted as $\mathbf{X'X}|\mathbf{X'Y}$.

The simplest model with no measured x variable is the Mean Model,

$$y_i = \mu + \varepsilon_i, \qquad i=1,2,\ldots,n$$

Implicitly assumed is that n units have been randomly sampled from a population and that an observation y_i has been observed and recorded on each unit. The parameter μ is defined as the arithmetic mean of the population. The variance of the population is $\sigma^2 = \Sigma(y_i-\mu)^2/N$, where $i=1,2,\ldots,N$ is the number of experimental or sampling units in the population.

It seems reasonable to use the arithmetic mean of the sample, $\bar{y}$, as an estimator of μ. The estimator $\bar{y}$ also results from an approach that starts with the specific criterion of choosing an estimator of μ to minimize the analogous sample sum of squared deviations from μ, $\Sigma(y_i-\mu)^2, i=1,2,\ldots,n$. Algebraically,

$$\Sigma(y_i-\mu)^2 = \Sigma(y_i-\bar{y})^2 + n(\bar{y}-\mu)^2$$

and this expression is minimized when $\mu=\bar{y}$. The estimator or statistic $\bar{y}$ is known as a *least squares estimator* because it minimizes the sum of squared deviations.

ARSENIC DATA

In this section we examine the environmental study of Bencko and Symon in which they reported arsenic determinations on hair samples from people residing near a power plant that burned coal with a high arsenic content. Groups of 10-year-olds, each with 20 to 27 boys, were selected from ten communities southwest of the plant. Only part of the data given by the authors will be used here, and for now we will work with only one response value for each community, the average concentration of arsenic (in parts per million, ppm) for the individual boys within each community. For the moment we will assume that we have no other information—specifically, no measured x variables. Then the model we would fit with the data would be the Mean Model,

$$y_i = \mu + \varepsilon_i, \quad i = 1, 2, \ldots, 10$$

Fitting this model involves estimating the parameter μ by the sample mean $\bar{y} = \Sigma y/n$ and estimating the residuals by $y_i - \bar{y}$. The data and estimated residuals are shown in Table 4.1.

Table 4.1 Data for Arsenic Content Study

y_i, Average Concentration of Arsenic (ppm)	$\bar{y}$, Sample Mean	$y_i - \bar{y}$
3.19	1.63	1.56
3.26	1.63	1.63
1.82	1.63	0.19
1.02	1.63	−0.61
1.85	1.63	0.22
2.05	1.63	0.42
1.34	1.63	−0.29
0.79	1.63	−0.84
0.66	1.63	−0.97
0.30	1.63	−1.33

Estimated variance $= s^2 = \Sigma(y_i - \bar{y})^2/(n-1) = 1.025$

Source: Vladimir Bencko and Karel Symon, "Health Aspects of Burning Coal with a High Arsenic Content. Part I: Arsenic in Hair, Urine and Blood in Children Residing in a Polluted Area," *Environmental Research* 13 (1977): 378–385.

In an examination of a sequence of candidate models, as outlined in the earlier units, the Mean Model will usually be the simplest model. Despite the simplicity it will be beneficial to us in later units if we now go through the steps needed to fit the Mean Model by utilizing the ABDO computing algorithm. Reformulating the Mean Model as an $\mathbf{X}\boldsymbol{\beta}$ regression model, we can calculate $\bar{y}$ by using our general regression notation and terminology. Now for the details.

FITTING THE MEAN MODEL

Writing out the observational equations of the Mean Model (realizing that a value of one is associated with μ) for the arsenic data, we have

$$y_1 = 3.19 = 1\mu + \varepsilon_1$$
$$y_2 = 3.26 = 1\mu + \varepsilon_2$$
$$\vdots$$
$$y_{10} = 0.30 = 1\mu + \varepsilon_{10}$$

Then the Mean Model can be expressed in a regression format as follows:

Mean Model: $\quad y_i = \mu x_{i0} + \varepsilon_i, \quad i = 1, 2, \ldots, 10$

Remember that no x variable has been measured. However, the observational equations give a clue that we could construct a variable, denoted as x_0, by setting each value of x_{i0} equal to one.

The general notation for a regression parameter is β_j, but since μ is the commonly accepted notation for the overall mean, it is also used here in the regression format. This simplest of regression models, the Mean Model, can be written in matrix notation as

$$\mathbf{Y} = \mathbf{X}\mu + \boldsymbol{\varepsilon}$$

where $\mathbf{Y}$ is a column of the observations and $\mathbf{X}$ is a column of ones. Thus

$$\mathbf{Y} = \begin{bmatrix} y_1 \\ y_2 \\ \vdots \\ y_{10} \end{bmatrix} = \begin{bmatrix} 3.19 \\ 3.26 \\ \vdots \\ 0.30 \end{bmatrix} \quad \text{and} \quad \mathbf{X} = \begin{bmatrix} x_{10} \\ x_{20} \\ \vdots \\ x_{10,0} \end{bmatrix} = \begin{bmatrix} 1 \\ 1 \\ \vdots \\ 1 \end{bmatrix}$$

Fitting the Mean Model with n observations is nothing more than calculating $\bar{y} = \Sigma y/n$. The number of observations, n, can be expressed as the sum of squares of x_0, and the sum of the observations can be expressed as the sum of cross products between x_0 and y. Specifically,

$$n = \sum_{i=1}^{n} x_{i0}^2 \quad \text{and} \quad \Sigma y = \sum_{i=1}^{n} x_{i0} y_i$$

Here the summations are clear, so we drop the i subscripts and write $n = \Sigma x_0^2$ and $\Sigma y = \Sigma x_0 y$.

The sum of squares of x_0 and the sum of cross products between x_0 and y can be expressed as matrices. Remember that for multiplication of matrices (or in this case vectors), the number of columns in the first matrix has to equal the number of rows in the second matrix. In order to calculate the sum of squares of x_0 and the sum of cross products between x_0 and y, we need the transpose of $\mathbf{X}$, denoted as $\mathbf{X}'$ and defined as the matrix obtained by transposing the rows and columns of $\mathbf{X}$. Thus if $\mathbf{X}$ is a column vector of ones, then the first column (and only column in this case) becomes the first (and only) row of $\mathbf{X}'$. Then

$$\mathbf{X}'\mathbf{X} = [x_{10}\, x_{20} \cdots x_{n0}] \begin{bmatrix} x_{10} \\ x_{20} \\ \vdots \\ x_{n0} \end{bmatrix} = [1\ 1\ \cdots\ 1] \begin{bmatrix} 1 \\ 1 \\ \vdots \\ 1 \end{bmatrix}$$

which is nothing more than Σx_0^2 or n. Similarly,

$$\mathbf{X}'\mathbf{Y} = [x_{10}\, x_{20} \cdots x_{n0}] \begin{bmatrix} y_1 \\ y_2 \\ \vdots \\ y_n \end{bmatrix} = [1\ 1\ \cdots\ 1] \begin{bmatrix} y_1 \\ y_2 \\ \vdots \\ y_n \end{bmatrix}$$

is the sum of cross products between x_0 and y.

In general, fitting the model is the process of solving the system of equations $\mathbf{X}'\mathbf{X}\mathbf{b} = \mathbf{X}'\mathbf{Y}$. For the Mean Model the $\mathbf{b}$ vector includes only $\bar{y}$; that is, there is only one equation to be solved. The ABDO algorithm can now be invoked, starting with

$$\mathbf{X}'\mathbf{X} | \mathbf{X}'\mathbf{Y}: \qquad \Sigma x_0^2 \qquad \Sigma x_0 y$$

and followed by two steps.

Step 1: Rewrite the first and only row of $\mathbf{X}'\mathbf{X} | \mathbf{X}'\mathbf{Y}$:

$$\Sigma x_0^2 \qquad \Sigma x_0 y$$

Step 2: Divide each element in step 1 by the first element in step 1:

$$1 \qquad \Sigma x_0 y / \Sigma x_0^2 = \bar{y}$$

For the arsenic data $\mathbf{X}'\mathbf{X} | \mathbf{X}'\mathbf{Y} = 10 | 16.28$, and the steps are as follows:

Step 1:   10   16.28
Step 2:   1 $1.628 = \bar{y}$

The prediction equation is $\hat{y}_i = \bar{y}$, and each estimated residual, $y_i - \hat{y}_i$, is $y_i - \bar{y}$. Consequently, $s^2 = 1.025$.

RESIDUAL PLOTTING

After a model has been fit, a good practice in data analysis is to plot the estimated residuals against any reasonable variables, for instance, the order of the observations. In this example the observations happen to be recorded by geographic location, $y_1 = 3.19$ being the observation of the community closest to the plant. Even when no natural reason is evident for plotting the residuals against the order in which the observations were recorded, the data analyst might want to do this plot in order to detect any unsuspected influences. For example, a time-of-day bias introduced in a series of laboratory determinations can give a systematic pattern to the residuals.

If the model and assumptions are correct, the residual pattern should reflect random variability around zero. An indicator of an inadequate model is a systematic pattern among the residuals, such as the pattern shown in Figure 4.1. In the plot the residuals show a pattern of decreasing value with order. Those communities near the plant are being underestimated by using $\bar{y}$ as the predicted value, while those far away are overestimated, as shown by the negative values of $y_i - \hat{y}$. If the observations had not been recorded by geographic location, the pattern of Figure 4.1 would not have been obvious. However, the magnitudes of the residuals are sufficiently large to be disturbing, and additional inquiry would have revealed the information about the distance from the plant. In the reported study the investigators had measured both the arsenic concentrations and the distances and had conjectured the dependency. A tentative hypothesis is that arsenic con-

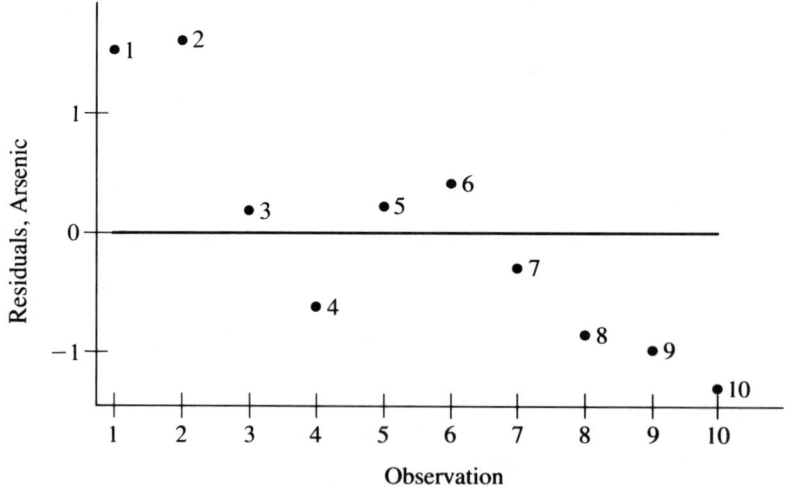

Figure 4.1 Estimated Residuals Versus Order of Observation for the Arsenic Data

centration in the hair depends on the distance of the community from the plant; that is, the further away the community is from the station, the less will be the concentration of arsenic in the hair. This conjecture will be evaluated in the next unit.

SUMMARY

In this unit a rationale has been given for using $\bar{y}$ as the estimator of μ in the simple regression model $y_i = \mu x_{i0} + \varepsilon_i$. In our regression format a systematic two-step procedure was developed by using the sums of squares and cross products of $\mathbf{X'X}$ and $\mathbf{X'Y}$. The regression estimator is equal to the sum of cross products between x_0 and y divided by the sum of squares for x_0. Since each value of x_{i0} equals one, the estimated regression coefficient and the arithmetic mean of a sample of n observations are equivalent formulations of $\bar{y}$; that is,

$$\frac{\Sigma x_0 y}{\Sigma x_0^2} = \frac{\Sigma y}{n} = \bar{y}$$

Unit 5

Fitting a Straight-Line Model

Objectives

- To extend the Abbreviated Doolittle algorithm for fitting two straight-line models.
- To explore and compare the fitting of three models: the Mean Model, the Mean and Slope Model, and the Intercept and Slope Model.

The arsenic residual plot of Figure 4.1 showed a trend among the residuals, indicating that a more general model should be evaluated. Specifically, a model including a measured x variable, in addition to the constructed x_0 variable, is indicated by the residual plot. For the arsenic concentration study the distance between each community and the power station was measured by the investigators, and the data are plotted in Figure 5.1, with y = arsenic concentration (ppm) and x_1 = distance (kilometers, km) from the source of emission. The simplest expression for the relationship between y and x_1 is a straight line. Further, the straight-line trend of the residuals in Figure 4.1 is additional evidence that a more general model, including the distance information, would give an improved fit to the data compared with the model containing x_0 alone.

The fulcrum ($\triangle$) in the data plot is at the coordinate point $(\bar{y}, \bar{x}_1)$. If we want to predict y when x_1 is equal to $\bar{x}_1$, then $\bar{y}$ would be a reasonable predictor. Consequently, any line we fit to the data should go through the point $(\bar{y}, \bar{x}_1)$. Picture this point as the support under a seesaw board, with the slope of the fitted line as the angle of the board. For values of x_1 less than $\bar{x}_1$ we want our predicted value of y to be greater than $\bar{y}$; for values of x_1 greater than $\bar{x}_1$ the predicted value should be less than $\bar{y}$.

The need for another variable, allowing a negative slope to the straight line, is evident. The associated parameter β_1, when multiplied by $(x_1 - \bar{x}_1)$, should give predicted values closer to the observed values. Writing our two models together, we have the following:

Mean Model: $\quad y_i = \mu x_{i0} + \varepsilon_i \quad$ (Model 1)

Mean and Slope Model: $\quad y_i = \mu x_{i0} + \beta_1(x_{i1} - \bar{x}_1) + \varepsilon_i \quad$ (Model 2)

The y_i for both models are the same, of course, but the ε_i may differ. Adding the distance variable should explain an additional part of the variability among the y_i. The ε_i from Model 2 would be expected to be smaller, in general, than the ε_i from Model 1.

FITTING THE MEAN AND SLOPE MODEL

Alternatively, we could plot $(x_{i1} - \bar{x}_1)$ instead of x_{i1} on the x axis of Figure 5.1. The relative spacing of the points along the x axis would remain the same; that is, subtracting the constant $\bar{x}_1$ from each x_{i1} only shifts the x axis

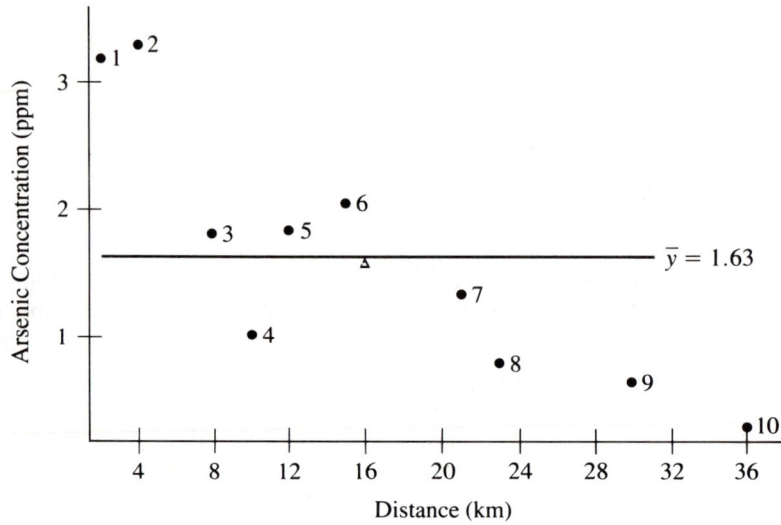

Figure 5.1   Arsenic Concentration Versus Distance from the Power Station

values. The estimated slope of the fitted straight line will be the same, and we can focus on fitting the straight line as formulated by Model 2.

The coefficient associated with x_0, the mean, was estimated in Unit 4 as the sum of cross products between x_{i0} and y divided by the sum of squares of x_{i0}, or $\bar{y} = \Sigma x_{i0} y / \Sigma x_{i0}^2$. For reasons discussed in the next section, it is also appropriate to estimate the slope by simply calculating the sum of cross products between $x_{i1} - \bar{x}_1$ and y and dividing by the sum of squares of $x_{i1} - \bar{x}_1$. For the arsenic data displayed in Figure 5.1 and given in the next section, β_1 is estimated as

$$b_1 = \sum_{i=1}^{10}(x_{i1}-\bar{x}_1)y_i \Big/ \sum_{i=1}^{10}(x_{i1}-\bar{x}_1)^2 = -88.068/1126.90$$
$$= -0.07815$$

From Unit 4, $\bar{y} = 1.628$. The prediction equation for Model 2 is then written as

$$\hat{y} = 1.628 - 0.07815(x_{i1} - \bar{x}_1)$$

Using the prediction equation, we can calculate $\hat{y}$ for values of $x_{i1} - \bar{x}_1$ in the study and the $y_i - \hat{y}_i$ residuals (rounded to two decimals) are found to be 0.46, 0.69, −0.44, −1.08, −0.10, 0.34, 0.09, −0.30, 0.12, and 0.23 for the ten data points. These residuals are plotted against the x_1 variable as shown in Figure 5.2. Alternatively, the residuals could be plotted against the

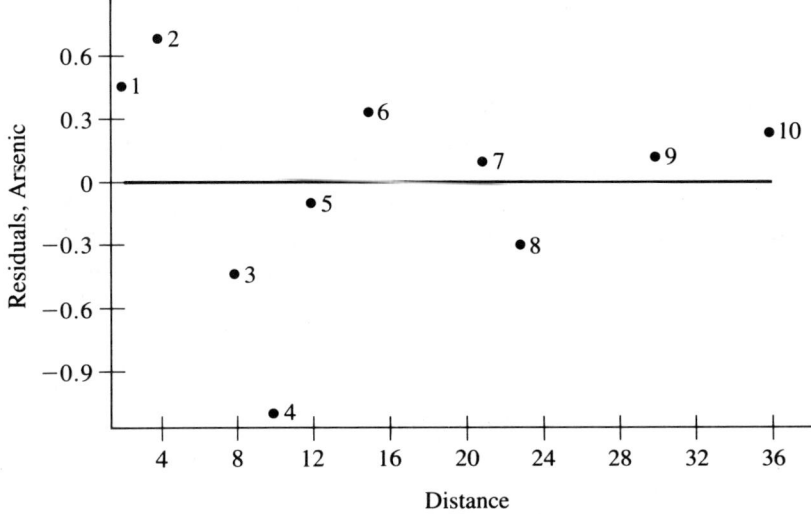

Figure 5.2 Residuals Versus Distance for the Arsenic Data

$x_1 - \bar{x}_1$ variable. Compared with the first residual plot, Figure 4.1, two features are visually noticeable:

1. In general, the residuals are smaller in Figure 5.2. For example, only two residuals have absolute values greater than 0.5, whereas six were greater in Figure 4.2.
2. No particular trend is evident in Figure 5.2, as contrasted with the obvious trend of Figure 4.2.

The second residual plot does show the four largest residuals to be the first four, an indication that the straight line is not fitting the communities closest to the power plant. Of particular concern is the community (4) at a distance of 10 km. In an effort to refine the model an extra variable might be added to the model or a transformation considered, for example, the logarithm of arsenic concentration. It might also be conjectured that the variance of the responses does not remain constant with increased distance from the plant. However, none of these considerations will be pursued now.

The next step in the data analysis is to estimate the variance σ^2 by calculating the residual mean square. In general, the *residual mean square* is the sum of squares of the $y_i - \hat{y}_i$ deviations divided by the degrees of freedom. The number of degrees of freedom will be equal to the number of observations minus the number of x variables, including x_0, assuming that no linear combination of the x variables equals zero. In our example we have

$$\text{Residual Mean Square} = s^2 = \Sigma(y_i - \hat{y}_i)^2/(10-2) = 0.2925$$

Note that s^2 has dropped from 1.025 when the Mean Model in Unit 4 was fitted to 0.2925 with the straight-line model fit. This magnitude of reduction would usually be of importance.

Using the Mean and Slope Model allowed us to find estimates of both the mean and the slope by dividing a sum of cross products by a sum of squares. In general, we cannot be so computationally naive, and the ABDO computing algorithm will be needed to solve the set of equations $\mathbf{X'Xb} = \mathbf{X'Y}$.

Before formulating and fitting another straight-line model, however, let's consider why subtracting $\bar{x}_1$ from each x_{i1} permitted our naive sequential approach to work. Stated briefly, x_{i0} and $x_{i1} - \bar{x}_1$ explained different parts of the variability among the y_i observations. The key to the Mean and Slope Model was to formulate or adjust the x_1 variable by subtracting that part of the distance measurements that could account for variability among the y_i's already accounted for by the x_0 variable. Subsequent units will develop the theme that whenever a new x variable is added to a regression equation, its coefficient can be obtained by a single-variable regression of y on the additional x variable, properly adjusted for the previous x variables. For now we will return to the fitting of the straight-line model.

FITTING THE INTERCEPT AND SLOPE MODEL

An algebraically equivalent version of the Mean and Slope Model, with x_0 and $x_1 - \bar{x}_1$ as the x variables, is the following:

Intercept and Slope Model: $\quad y_i = \beta_0 x_{i0} + \beta_1 x_{i1} + \varepsilon_i \quad$ (Model 3)

where β_0 is the intercept. In other words, the straight-line model always involves two parameters, including the slope. However, the second parameter can be defined as the intercept, the average value of y when x_1 is equal to zero, or as the mean, the average y when x_1 is equal to $\bar{x}_1$. Model 3 is sometimes written in the notation of $y = \alpha + \beta x + \varepsilon$.

From the development of Unit 3 we see that two x variables are now in the $\mathbf{X}$ matrix, the first column being $x_{i0} = 1$, $i = 1, 2, \ldots, n$. The second-column elements of the $\mathbf{X}$ matrix are the measured values of x_1, not the $(x_1 - \bar{x}_1)$ deviations as in Model 2. Then

$$\mathbf{Y} = \begin{bmatrix} y_1 \\ y_2 \\ \vdots \\ y_n \end{bmatrix} \quad \mathbf{X} = \begin{bmatrix} 1 & x_{11} \\ 1 & x_{21} \\ \vdots & \vdots \\ 1 & x_{n1} \end{bmatrix} \quad \boldsymbol{\beta} = \begin{bmatrix} \beta_0 \\ \beta_1 \end{bmatrix} \quad \boldsymbol{\varepsilon} = \begin{bmatrix} \varepsilon_1 \\ \varepsilon_2 \\ \vdots \\ \varepsilon_n \end{bmatrix}$$

Multiplying both **X** and **Y** by **X'**, we obtain

$$\mathbf{X'X} = \begin{bmatrix} 1 & 1 & \cdots & 1 \\ x_{11} & x_{21} & \cdots & x_{n1} \end{bmatrix} \begin{bmatrix} 1 & x_{11} \\ 1 & x_{21} \\ \vdots & \vdots \\ 1 & x_{n1} \end{bmatrix} = \begin{bmatrix} \Sigma x_{i0}^2 & \Sigma x_{i0}x_{i1} \\ \Sigma x_{i0}x_{i1} & \Sigma x_{i1}^2 \end{bmatrix}$$

and

$$\mathbf{X'Y} = \begin{bmatrix} 1 & 1 & \cdots & 1 \\ x_{11} & x_{21} & \cdots & x_{n1} \end{bmatrix} \begin{bmatrix} y_1 \\ y_2 \\ \vdots \\ y_n \end{bmatrix} = \begin{bmatrix} \Sigma x_{i0} y_i \\ \Sigma x_{i1} y_i \end{bmatrix}$$

When **X'** and **X** are multiplied to form **X'X**, the sums of squares of the two x variables, calculated by multiplying each row of **X'** by the corresponding column of **X**, are in the main diagonal of the (2×2) matrix, while the one sum of cross products appears in the off-diagonal position.

For fitting the Intercept and Slope Model, we need the ABDO computing algorithm, starting with **X'X|X'Y**:

Row 1 of X'X|X'Y: Σx_0^2 $\Sigma x_0 x_1$ $\Sigma x_0 y$
Row 2 of X'X|X'Y: Σx_1^2 $\Sigma x_1 y$

Note that $\Sigma x_0 x_1$ in the second row of **X'X** has been deleted and for notational convenience the i subscript has been dropped, since each summation adds n values, starting at i equal to one and going through $i = n$ ($i = 1, 2, \ldots, n$). Also note that $\Sigma x_0 x_1 = \Sigma x_1$ and $\Sigma x_0 y = \Sigma y$.

Applying ABDO is straightforward, following the development in Unit 3 and remembering the correspondence between **A** and **X'X** and between **g** and **X'Y**.

Step 1: Rewrite row 1 of **X'X|X'Y**:

Σx_0^2 $\Sigma x_0 x_1$ $\Sigma x_0 y$

Step 2: Divide each element in step 1 by the first element in step 1:

1   $\bar{x}_1$ $\bar{y}$

The value of $\bar{x}_1$ is a pivotal number in the next step.

Step 3: Subtract from each element in row 2 of $\mathbf{X'X}|\mathbf{X'Y}$ the product of $\bar{x}_1$ and the corresponding element in step 1:

$$\Sigma x_1^2 - \bar{x}_1 \Sigma x_0 x_1 \qquad \Sigma x_1 y - \bar{x}_1 \Sigma x_0 y$$
$$= \Sigma(x_1 - \bar{x}_1)^2 \qquad = \Sigma(x_1 - \bar{x}_1)y$$

Step 4: Divide each element in step 3 by the first element in step 3:

$$1 \qquad \frac{\Sigma(x_1 - \bar{x}_1)y}{\Sigma(x_1 - \bar{x}_1)^2}$$

As shown in Unit 3, the solutions haven't been explicitly carried along but the necessary triangular set of equations needed in solving for b_0 and b_1, the estimators of β_0 and β_1 in the Intercept and Slope Model, can be identified in the even steps, namely,

$$b_0 + \bar{x}_1 b_1 = \bar{y}$$
$$b_1 = \Sigma(x_1 - \bar{x}_1)y / \Sigma(x_1 - \bar{x}_1)^2$$

Setting up ABDO for the arsenic data, we obtain the following:

$$\mathbf{Y} = \begin{bmatrix} 3.19 \\ 3.26 \\ 1.82 \\ 1.02 \\ 1.85 \\ 2.05 \\ 1.34 \\ 0.79 \\ 0.66 \\ 0.30 \end{bmatrix} \qquad \mathbf{X} = \begin{bmatrix} 1 & 2 \\ 1 & 4 \\ 1 & 8 \\ 1 & 10 \\ 1 & 12 \\ 1 & 15 \\ 1 & 21 \\ 1 & 23 \\ 1 & 30 \\ 1 & 36 \end{bmatrix}$$

Row 1 of $\mathbf{X'X}\|\mathbf{X'Y}$:	10	161	16.28
Row 2 of $\mathbf{X'X}\|\mathbf{X'Y}$:		3719	174.04
Step 1:	10	161	16.28
Step 2:	1	16.1	1.628
Step 3:		3719 − (16.1)(161)	174.04 − (16.1)(16.28)
		= 1126.9	= −88.068
Step 4:		1	−0.07815

Then

$$b_1 = -0.07815 \quad \text{and} \quad b_0 = 1.628 - (16.1)(-0.07815) = 2.89$$

Interestingly, the right-hand elements of steps 2 and 4 are $\bar{y}$ and b_1, the estimators for the Mean and Slope Model (Model 2). Applying ABDO with x_0 and $(x_1 - \bar{x}_1)$ of Model 2, we obtain

Row 1 of X'X|X'Y: $\quad \Sigma x_0^2 \quad \Sigma x_0(x_1 - \bar{x}_1) \quad \Sigma x_0 y$

Row 2 of X'X|X'Y: $\quad\quad\quad \Sigma(x_1 - \bar{x}_1)^2 \quad \Sigma(x_1 - \bar{x}_1)y$

For Model 2, $\Sigma x_0(x_1 - \bar{x}_1)$ is equal to zero since $\Sigma x_0(x_1 - \bar{x}_1) = \Sigma(x_1 - \bar{x}_1) = 0$.

The ABDO calculations are especially straightforward when the pivotal element in step 2 is zero and no subtraction is needed in step 3. Then the solutions can be read directly from steps 2 and 4. For the arsenic data we have

Step 1: $\Sigma x_0^2 = 10 \quad\quad 0 \quad\quad \Sigma x_0 y = 16.28$

Step 2: $\quad\quad\quad 1 \quad\quad 0 \quad\quad 1.628$

Step 3: $\quad\quad\quad\quad\quad \Sigma(x_1 - \bar{x}_1)^2 \quad \Sigma(x_1 - \bar{x}_1)y$

$\quad\quad\quad\quad\quad\quad = 1126.9 \quad\quad = -88.068$

Step 4: $\quad\quad\quad\quad\quad 1 \quad\quad\quad\quad -0.07815$

Solving the implicit set of equations from steps 2 and 4, we obtain

$$b_1 = -0.07815 \quad \text{and} \quad \bar{y} + (0)(b_1) = 1.628$$

It is now possible to see the connection in the ABDO calculations between Model 2 and Model 3. ABDO is sequenced in a series of stages, each consisting of two steps. The number of stages is equal to the number of rows in X'X|X'Y or the number of x variables in the model. The stages can be numbered, starting with zero, so that stage 0 (steps 1 and 2) corresponds with x_0 operations, and stage 1 (steps 3 and 4) with x_1 or $x_1 - \bar{x}_1$. Model 2 starts with $x_{i1} - \bar{x}_1$ as the explanatory variable; that is, each x_{i1} has already been corrected for the mean of x_1. The only nonzero calculation in stage 0 was the calculation involved in solving for $\bar{y}$. Stage 1 then solved for b_1, using the already corrected sum of cross products and sum of squares from the second row of X'X|X'Y. However, the explanatory variable in Model 3 was x_1, not $x_1 - \bar{x}_1$ as in Model 2. The ABDO calculations for Model 3 solved for $\bar{x}_1$ in stage 0 and then made the proper corrections in step 3, resulting in the same b_1 in step 4 for either model. Specifically, the uncorrected sum of squares, Σx_1^2, has to be corrected for the mean $\bar{x}_1$ by subtracting $\bar{x}_1(\Sigma x_0 x_1) = (\Sigma x_1)^2/n = n\bar{x}_1^2$, and the uncorrected sum of cross products $\Sigma x_1 y$ has to be corrected by subtracting $\bar{x}_1(\Sigma x_0 y) = (\Sigma x_1)(\Sigma y)/n$. The corrected sum of cross products is usually written as $\Sigma(x_1 - \bar{x}_1)(y - \bar{y})$, which is algebraically equivalent to $\Sigma(x_1 - \bar{x}_1)y$ since $\bar{y}\Sigma(x_1 - \bar{x}_1)$ equals zero. The three quantities needed for the two correction factors in step 3, $\bar{x}_1$, $\Sigma x_0 x_1$, and $\Sigma x_0 y$, are systematically picked up from stage 0.

SUMMARY

In Units 4 and 5 a systematic calculation procedure, the ABDO algorithm, has been developed for handling three models:

Model 1: $y_i = \mu x_{i0} + \varepsilon_i$

Model 2: $y_i = \mu x_{i0} + \beta_1(x_{i1} - \bar{x}_1) + \varepsilon_i$

Model 3: $y_i = \beta_0 x_{i0} + \beta_1 x_{i1} + \varepsilon_i$

In Model 1 the mean is estimated. In the other models two parameters are estimated: the mean and the slope for Model 2, and the intercept and the slope for Model 3. The fitted straight line has the same slope for Models 2 and 3 and will give the same predicted values. Graphically, the two straight-line models are as shown in Figure 5.3.

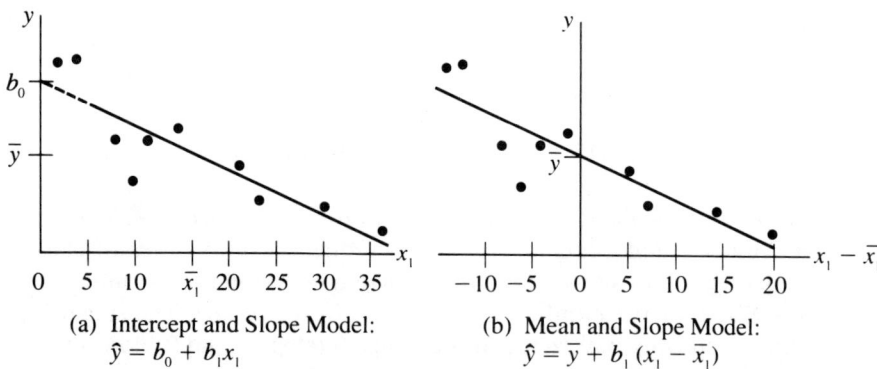

(a) Intercept and Slope Model:
$\hat{y} = b_0 + b_1 x_1$

(b) Mean and Slope Model:
$\hat{y} = \bar{y} + b_1(x_1 - \bar{x}_1)$

Figure 5.3 Plots of Arsenic Concentration (y) Versus Distance for the Intercept and Slope and the Mean and Slope Models

In practice, the ABDO calculations would be carried out by computer, even though this unit indicates the ease of using a hand or desk calculator. The program used by a computer probably will not be the ABDO algorithm exactly as developed in this unit, but the general flow of calculations will be similar. The data will be in a format of n rows and three columns, x_0, x_1, and y, and the objective is the reduction of the data from the n pairs (x_1 and y) of data to two summary statistics of straight-line regression, the mean and slope (or the intercept and slope). If the residuals $y - \hat{y}$ are acceptably small in magnitude and no pattern is evident when plotted against x_1, a good data analysis strategy then considers the possibility that a reduced model will also adequately explain the variability in the y observations.

Unit 6

Analysis of Variance

Objectives

- To develop an analysis of variance (ANOVA) methodology for the decomposition of the response variable.
- To examine a sequence of candidate models for the purpose of selecting the simplest model that adequately explains variability in the response variable.
- To fit a straight-line model with zero intercept.

We have previously compared two models, starting with the Mean Model of Unit 4. An unacceptable residual plot motivated the fitting of a straight-line model in Unit 5. The major reason for this particular sequencing of the two candidate models was the development of the computational procedure ABDO. Usually the analysis of a straight-line model problem would start with fitting a straight line and checking the residual plot. Then the sequence would continue with fitting a mean model and comparing the two candidate models. Indeed, the general data analysis strategy developed in Unit 1 is to start with the most general model and then consider other candidate models with fewer variables. The pattern and magnitude of the residuals for each candidate model is examined with the objective of finding a model that is as reduced as possible while adequately describing the data. Even though residual plots, as used in Units 4 and 5, are helpful in comparing candidate plots, a more systematic methodology using significance tests is also needed. The *analysis of variance* (ANOVA), a quantitative technique useful in comparing candidate models, is developed in this unit.

DECOMPOSITION OF REGRESSION DATA

The ANOVA is an extension of the concept of partitioning y into a predicted value and a residual. In general, we begin with a model that contains all the x variables known to be important and perhaps several variables we think might be important. Such a model is called the *general* or *full model*. A model that contains a subset of the variables in the full model is called a *simpler* or *reduced model*. If $\hat{y}$ is the predicted value for the full model and $\hat{y}_r$ is the predicted value for the reduced model, then y may be decomposed into components by the algebraic identity

$$y = \hat{y}_r + (\hat{y} - \hat{y}_r) + (y - \hat{y})$$

The component $(y - \hat{y})$ is the residual from fitting the full model; it tends to be large if the inherent variability of the data is large. The residual mean square is the usual estimate of the variance, σ^2. The component $(\hat{y} - \hat{y}_r)$ is the difference between the predicted value of the full model and the predicted value of the reduced model. It tends to be large if the inherent variability is large or if the reduced model does not adequately fit the data.

The sum of squares of this component divided by its degrees of freedom is called the mean square for the difference between the full model and the reduced model. A mean square for the difference that is considerably larger than the residual mean square indicates that the reduced model does not adequately fit the data. Formal testing for deciding between the two models is deferred until Unit 13. The remainder of this unit illustrates the concepts of data decomposition and shows how to display the pertinent results in an analysis of variance table.

The example uses the arsenic data of the previous unit, where the full model consists of x_0 and x_1, the distance from the plant. Saying the reduced model is adequate is the same as saying that distance from the plant has no predictive ability. In other words, we might as well predict every observation by $\bar{y}$. The decomposition is

$$y = \hat{y}_r + (\hat{y} - \hat{y}_r) + (y - \hat{y}) = \bar{y} + (\hat{y} - \bar{y}) + (y - \hat{y})$$
$$= \hat{y}(0) + [\hat{y}(0,1) - \hat{y}(0)] + [y - \hat{y}(0,1)]$$

where $\hat{y}(0) = \bar{y}x_0 = \bar{y}$ and $\hat{y}(0,1) = \bar{y}x_0 + b_1(x_1 - \bar{x}_1) = b_0 x_0 + b_1 x_1$, where b_0 and b_1 are the estimated parameters of the Intercept and Slope Model. For the arsenic data set, with values rounded to two decimal places, we have

	y	$\hat{y}(0)$	$\hat{y}(0,1) - \hat{y}(0)$	$y - \hat{y}(0,1)$
	3.19	1.63	1.10	0.46
	3.26	1.63	0.95	0.69
	1.82	1.63	0.63	−0.44
	1.02	1.63	0.48	−1.08
=	1.85	1.63	0.32	−0.10
	2.05	1.63	0.09	0.34
	1.34	1.63	−0.38	0.09
	0.79	1.63	−0.54	−0.30
	0.66	1.63	−1.09	0.12
	0.30	1.63	−1.56	0.23

The residuals in the last column were plotted in Unit 5. In general, a residual plot from the full model should be routinely examined for adequacy of the full model. In Figure 6.1 the observations are plotted as in Figure 5.1. The horizontal line is $\hat{y}(0) = \bar{y}$, and the sloping line is formed by the predicted values of the full model. Also displayed are the three components of the data decomposition for two observations, the second and eighth. Since the two lines differ, the $\hat{y}(0,1) - \hat{y}(0)$ components vary, from zero at $x = \bar{x}$, the point where the two lines cross, to relatively large values at the two extremes. In general, the $\hat{y}(0,1) - \hat{y}(0)$ components are larger than the $y - \hat{y}(0,1)$ components.

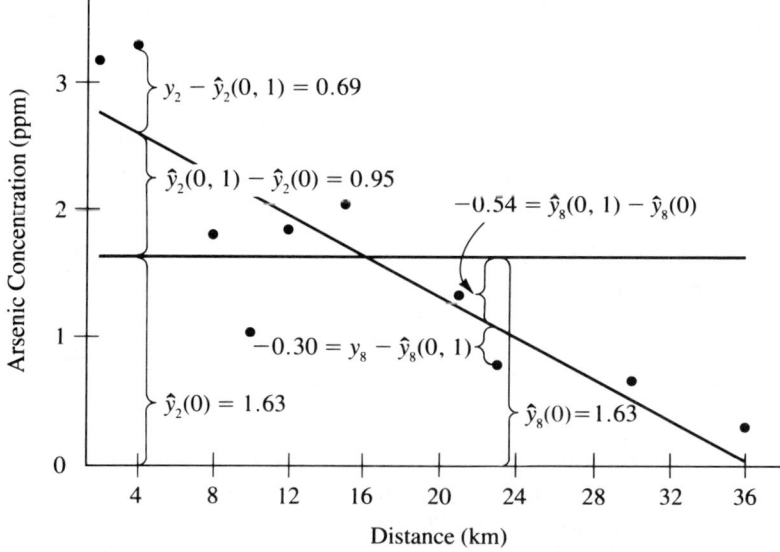

Figure 6.1 Arsenic Concentration Versus Distance from the Power Station, Showing the Decomposition Components of y_2 and y_8

SUMS OF SQUARES AND ANALYSIS OF VARIANCE TABLE

The display of the decomposition of every observation can be very space consuming and is not usually done by data analysts. It is done here to illustrate a principle. The essential information is contained in the sums of squares of the three terms and their respective degrees of freedom. The ANOVA table is a systematic record of the various sums of squares and degrees of freedom for each component of the decomposition. In addition, columns for a description of each source of the variability and the mean squares are included. For completeness the first entry, as shown for the arsenic data in Table 6.1, is the total sum of squares.

Note from Table 6.1 that the mean square for the difference is considerably larger than the residual mean square. The conclusion is that the reduced model does not provide an adequate fit to the data. This conclusion is supported by Figure 6.1. Each of the two model sources is associated with one degree of freedom since one variable is associated with both the reduced model and the additional variable of the general model. In general, the degrees of freedom associated with the difference sum of squares is equal to the difference in the number of x variables between the full and reduced models, assuming no linear combination of the x variables equals zero. The number of residual degrees of freedom is the remainder of the total.

Table 6.1 Analysis of Variance of Arsenic Data for Evaluating the Mean Model as the Reduced Model

Source	Sum of Squares	Degrees of Freedom	Mean Square
Total	35.72	10	
Reduced model (x_0 alone)	26.50	1	26.50
Difference between full and reduced models	6.88	1	6.88
Residual	2.34	8	0.29

The decomposition of y, although helpful in understanding analysis of variance, is not needed to compute the sums of squares. These may be computed easily by using ABDO. Arranging the **X** matrix so that the variable of the reduced model is first, we find the necessary values for computing the sums of squares located in the right-hand column of the ABDO calculations in Unit 5; the four steps are repeated here.

Step 1: 10 0 16.280

Step 2: 1 0 1.628

Step 3: 1126.90 −88.068

Step 4: 1 −0.07815

The following steps can be used to calculate the sums of squares for Table 6.1:

1. The sum of squares for the reduced model is the product of the last element in step 1, the sum of the responses (16.28), and the last element of step 2, the mean (1.628), which is equal to $(\Sigma x_0 y)\bar{y} = 26.50$.
2. The sum of squares for the difference between the full and reduced models is the product of the last element in step 3, the corrected sum of cross products (−88.068), and the estimated slope (−0.07815), which is equal to $[\Sigma(x_1 - \bar{x}_1)y]b = 6.88$. In other words, the difference sum of squares is the additional sum of squares that can be associated with the slope parameter of the full model. The residual sum of squares is found by evaluating $\Sigma[y_i - \hat{y}_i(0, 1)]^2$ or by subtracting 26.50 plus 6.88 from the total sum of squares.

FITTING A STRAIGHT LINE THROUGH THE ORIGIN

Sometimes we may encounter situations where it is known in advance that the response variable must be zero if the explanatory variable is zero. Suppose that y is the amount of a substance that has passed through a filter and x is the time that the filtering apparatus has been in operation. In this case when x is equal to zero, then y must also be zero. The line describing the relationship between x and y is said to have a zero intercept, or to pass through the origin. Then the straight-line regression model will not include x_0 and is written as follows:

Slope-Through-the-Origin Model: $\quad y_i = \beta_1 x_{i1} + \varepsilon_i$

For this situation the **X** matrix contains only the x_1 variable. In the ABDO algorithm the estimator of β_1 is simply $\Sigma x_1 y / \Sigma x_1^2$. Then, since x_1 is one of the two x variables of the straight-line model with x_0 and x_1, the Slope-Through-the-Origin Model is a candidate reduced model. An alternative model sequence for the arsenic data starts with a candidate reduced model of x_1 alone. For this reduced model to be adequate, the data would

Table 6.2 Decomposition and Analysis of Variance of Arsenic Data for Evaluating the Slope-Through-the-Origin Model as the Reduced Model

$$y = \begin{bmatrix} 3.19 \\ 3.26 \\ 1.82 \\ 1.02 \\ 1.85 \\ 2.05 \\ 1.34 \\ 0.79 \\ 0.66 \\ 0.30 \end{bmatrix} = \hat{y}(1) = \begin{bmatrix} 0.09 \\ 0.19 \\ 0.37 \\ 0.47 \\ 0.56 \\ 0.70 \\ 0.98 \\ 1.08 \\ 1.40 \\ 1.68 \end{bmatrix} + \hat{y}(0,1) - \hat{y}(1) = \begin{bmatrix} 2.64 \\ 2.39 \\ 1.89 \\ 1.64 \\ 1.39 \\ 1.01 \\ 0.26 \\ 0.01 \\ -0.86 \\ -1.61 \end{bmatrix} + y - \hat{y}(0,1) = \begin{bmatrix} 0.46 \\ 0.69 \\ -0.44 \\ -1.08 \\ -0.10 \\ 0.34 \\ 0.09 \\ -0.30 \\ 0.12 \\ 0.23 \end{bmatrix}$$

Source	Sum of Squares	Degrees of Freedom	Mean Square
Total	35.72	10	
Reduced model (x_1 alone)	8.14	1	8.14
Difference between full and reduced models	25.24	1	25.24
Residual	2.34	8	0.29

have to lie near a line that goes through the origin. Figure 6.1 shows this not to be the case, and the data decomposition and ANOVA of Table 6.2 both confirm that the reduced model consisting of x_1 alone is inadequate.

SUMMARY

The purposes of the ANOVA technique developed in this unit include the calculation of the residual mean square and the comparison of candidate models, specifically, the general or full model with a reduced or simpler model. This unit compared a general model, the mean and slope models of Unit 5, with a reduced model, the Mean Model of Unit 4. Evaluating the adequacy of the Mean Model is equivalent to evaluating the need of the second variable in the Mean and Slope Model. With one additional variable in the full model, one degree of freedom is associated with the difference sum of squares between the two models. The difference mean square is compared with the residual mean square from the general model for evaluation of the adequacy of the reduced model. A formal significance test of the comparison is deferred to Unit 13.

Each observation can be decomposed into three components, and a sum of squares for each component is associated with a row in the ANOVA table. The various component parts are additive. Similarly, the sums of squares in the ANOVA table add to the total sum of squares. These sums of squares may be calculated from the observational decomposition or directly from ABDO. Then the two mean squares for the sequence of reduced and full models are compared to evaluate the adequacy of the reduced model. In general, the sequence of candidate models will include more than one reduced model. The methodology of this unit is extended in Unit 13.

Unit 7

Computations for Multivariable Models

Objectives

- To extend the Abbreviated Doolittle calculations to include several x variables.
- To extend the ANOVA technique for the evaluation of a candidate model sequence.
- To estimate and compare sequential and partial regression coefficients.

The Abbreviated Doolittle (ABDO) algorithm has been used in previous units for giving us a systematic procedure for estimating regression models. Estimates of $\bar{y}$ and b_1 were calculated for the Mean and Slope Model. Similarly, b_0 and b_1 estimates of the Intercept and Slope Model have also been found. The nature of the adjustment needed for calculating the slope—namely, correcting x_1 for its mean and working with the $x_1 - \bar{x}_1$ values—was observed to be systematically built into the ABDO algorithm for fitting the Intercept and Slope Model.

In this unit ABDO will be generalized and formally presented so that any number of x variables can be handled. As a first step in the extension of ABDO to handling more than the x_0 and x_1 variables of the previous units, a brief summary of fitting the straight-line one-variable model is given below.

	Model:	$y = \beta_0 x_0 + \beta_1 x_1 + \varepsilon$		
X'X|X'Y				
	Row 1:	Σx_0^2	$\Sigma x_0 x_1$	$\Sigma x_0 y$
	Row 2:		Σx_1^2	$\Sigma x_1 y$

Stage 0 of ABDO

	Step 1:	Σx_0^2	$\Sigma x_0 x_1$	$\Sigma x_0 y$
	Step 2:	1	$\bar{x}_1$	$\bar{y}$

Stage 1 of ABDO

	Step 3:		$\Sigma(x_1 - \bar{x}_1)^2$	$\Sigma(x_1 - \bar{x}_1)y$
	Step 4:		1	$\dfrac{\Sigma(x_1 - \bar{x}_1)y}{\Sigma(x_1 - \bar{x}_1)^2}$

Five summary features of ABDO are recorded in the listing:

1. The means of y and x_1 are found in stage 0.
2. The corrected sum of squares for x_1 and the corrected sum of cross products are found in stage 1 (step 3), derived from the uncorrected

quantities in the second row of **X'X|X'Y** and the correction factors from stage 0. The product $\bar{x}_1(\Sigma x_0 x_1) = (\Sigma x_1)^2/n$ is the correction factor for x_1, and $\bar{x}_1 \Sigma x_0 y = \bar{x}_1 \Sigma y$ adjusts the uncorrected sum of cross products.

3. The $\bar{y}$ and b_1 estimators of the Mean and Slope Model can be found from the right-most elements of the sequential steps 2 and 4. With the slope estimate from step 4, the intercept can be found algebraically from step 2 by $b_0 = \bar{y} - b_1 \bar{x}_1$. Then $\hat{y} = \bar{y} + b_1(x_1 - \bar{x}_1) = b_0 x_0 + b_1 x_1$.

4. The sum of squares for fitting the two sequential models can be calculated from the products of the right-most elements of the two stages. Specifically, the sum of squares for the reduced model, the Mean Model, is equal to $(\Sigma x_0 y)(\bar{y}) = n\bar{y}^2$, and the sum of squares for the difference between full (mean and slope) and reduced models is equal to $[\Sigma(x_1 - \bar{x}_1)y]^2/\Sigma(x_1 - \bar{x}_1)^2$.

5. In addition to the regression of y on x_0, another regression is actually taking place in stage 0, the regression of x_1 on x_0. We can conceptually envision the model of $x_1 = \mu x_0 + \varepsilon$, where μ is the mean of the response variable x_1. The estimator will be $\bar{x}_1$. We realize, of course, that x_1 is not the real response variable, but viewing the algorithm from this perspective allows us to envision the residuals of $x_1 - \bar{x}_1$. Then stage 1 is seen as the regression of y on $x_1 - \bar{x}_1$, with b_1 being the slope of the regression line.

ABBREVIATED DOOLITTLE CALCULATIONS WITH SEVERAL x VARIABLES

From the ABDO algorithm for the Intercept and Slope Model, certain generalizations to more than one variable may be made.

1. The necessary calculations will utilize sums of squares and sums of cross products as arranged in **X'X|X'Y**. The sums of cross products below the sum of squares in the main diagonal of **X'X** are deleted. Specifically, the first element in each row is the sum of squares of one of the x variables, followed by the sums of cross products with higher-numbered x's and y. For example, with four x variables the second row of **X'X|X'Y** would be

$$\Sigma x_1^2 \quad \Sigma x_1 x_2 \quad \Sigma x_1 x_3 \quad \Sigma x_1 x_4 \quad \Sigma x_1 y$$

2. The number of variables in the assumed model to be fitted is equal to the number of rows in **X'X** and **X'Y**.

3. The number of stages in the calculations is equal to the number of variables to be fitted. With an x_0 in the model the stages will be numbered from 0 to k to correspond with variables $x_0, x_1, \ldots, x_k$.

4. The calculations for each stage are organized in two steps. Each odd step starts with one of the rows in **X'X|X'Y**, and the sum of squares of that

variable and the cross products are corrected or adjusted for the previous stages. Dividing through each odd step by the leading element gives the even steps. No corrections are done in stage 0. With four x variables the means for the four x's and y are calculated as follows:

$$\begin{array}{cccccc} \Sigma x_0^2 & \Sigma x_0 x_1 & \Sigma x_0 x_2 & \Sigma x_0 x_3 & \Sigma x_0 x_4 & \Sigma x_0 y \\ 1 & \bar{x}_1 & \bar{x}_2 & \bar{x}_3 & \bar{x}_4 & \bar{y} \end{array}$$

5. Stage 1 starts with the uncorrected sum of squares of x_1 and uncorrected sums of cross products between x_1 and the other x's and y from the second row of $\mathbf{X'X|X'Y}$ and subtracts the correction factors from stage 0. Systematically, we have the following:

Uncorrected:	Σx_1^2	$\Sigma x_1 x_2$	$\cdots$	$\Sigma x_1 y$
Correction factor:	$\bar{x}_1(\Sigma x_0 x_1)$	$\bar{x}_1(\Sigma x_0 x_2)$	$\cdots$	$\bar{x}_1(\Sigma x_0 y)$
Step 3:	$\Sigma(x_1-\bar{x}_1)^2$	$\Sigma(x_1-\bar{x}_1)x_2$	$\cdots$	$\Sigma(x_1-\bar{x}_1)y$

6. Each succeeding stage starts with the uncorrected sum of squares and sums of cross products from a succeeding line of $\mathbf{X'X|X'Y}$ and subtracts correction factors from the previous stages. The necessary corrections increase by one for each succeeding stage. Whereas stage 1 calculations involved corrections from stage 0, stage 2 calculations will involve corrections from stage 0 and stage 1. With four variables stage 4 will involve corrections from the previous four stages.

7. The stages continue until an even-numbered step contains two elements, a value of one and another value that is the partial regression coefficient for the last variable x_k. With four x variables plus x_0, stage 4 gives the estimated partial regression coefficient for x_4.

8. Except for the last x variable, the other partial regression coefficients are calculated from the even-numbered steps in reverse order. The right-most elements of each even step are called *estimated sequential regression coefficients*. For the last x variable the partial and sequential regression coefficients are the same.

We are now ready to develop ABDO for a multivariable model, using the four-variable New York rivers data set from Unit 2 (Table 2.1) for illustration. The first step is to form the $\mathbf{X'X}$ matrix augmented by $\mathbf{X'Y}$. The necessary sums of squares and cross products are algebraically depicted in Table 7.1. Only the main diagonal elements (sums of squares) and the off-diagonal elements above the diagonal (sums of cross products) of $\mathbf{X'X}$ are needed.

Note that there is an extra entry in Table 7.1, which is not present in $\mathbf{X'X|X'Y}$. The lower right-most element, Σy^2, the uncorrected sum of squares of the responses, has been added to complete the main diagonal. The inclusion of this additional element will be helpful in systematically calculating the residual sum of squares.

Table 7.1 Sums of Squares and Cross Products for Four-Variable Problems

Row	Column					
	1	2	3	4	5	6
1	Σx_0^2	$\Sigma x_0 x_1$	$\Sigma x_0 x_2$	$\Sigma x_0 x_3$	$\Sigma x_0 x_4$	$\Sigma x_0 y$
2		Σx_1^2	$\Sigma x_1 x_2$	$\Sigma x_1 x_3$	$\Sigma x_1 x_4$	$\Sigma x_1 y$
3			Σx_2^2	$\Sigma x_2 x_3$	$\Sigma x_2 x_4$	$\Sigma x_2 y$
4				Σx_3^2	$\Sigma x_3 x_4$	$\Sigma x_3 y$
5					Σx_4^2	$\Sigma x_4 y$
6						Σy^2

The sums of squares and cross products of Table 7.1 can be calculated for the river data and are given in the upper part of Table 7.2. For the identification of a particular element in Table 7.2, the notation (i, j) is used to designate the number in the ith line and jth column. For example, $(2,4)$ designates the value 620.700. From the earlier generalizations the ABDO algorithm applied to the data is defined by the following steps, where rows 7 and 8 are steps 1 and 2 of stage 0, rows 9 and 10 are steps 3 and 4 of stage 1,..., and rows 17 and 18 are steps 11 and 12 of stage 6 of the ABDO calculations in the lower part of Table 7.2.

General expressions for the first three stages are as follows:

Stage 0

Step 1: $(7, j)=(1, j)$, for $j=1,\ldots,6$, defines line 7.

Step 2: $(8, j)=(7, j)/(7,1)$, for $j=1,\ldots,6$, defines line 8.

Stage 1

Step 3: $(9, j)=(2, j)-(8,2)\times(7, j)$, for $j=2,\ldots,6$, defines line 9.

Step 4: $(10, j)=(9, j)/(9,2)$, for $j=2,\ldots,6$, defines line 10.

Stage 2

Step 5: $(11, j)=(3, j)-(8,3)\times(7, j)-(10,3)\times(9, j)$, for $j=3,\ldots,6$, defines line 11.

Step 6: $(12, j)=(11, j)/(11,3)$, for $j=3,\ldots,6$, defines line 12.

Stage 3

Step 7: $(13, j) = (4, j) - (8, 4) \times (7, j) - (10, 4) \times (9, j) - (12, 4) \times (11, j)$, for $j = 4, 5, 6$, defines line 13.

Step 8: $(14, j) = (13, j)/(13, 4)$, for $j = 4, 5, 6$, defines line 14.

Table 7.2 Abbreviated Doolittle Algorithm for the New York Rivers Data

Row	\multicolumn{6}{c}{Column}
	1	2	3	4	5	6
1	20.0000	388.000	1,257.00	54.3000	10.3100	23.1500
2		11,650.0	20,975.0	620.700	126.880	498.110
3			85,051.0	2322.40	568.920	1340.49
4				925.130	106.867	92.9170
5					16.1543	15.2706
6						30.4187
7	20.0000	388.000	1,257.00	54.3000	10.3100	23.1500
8	1.00000	19.4000	62.8500	2.71500	0.515500	1.15750
9		4,122.80	−3,410.80	−432.720	−73.1340	49.0000
10		1.00000	−0.827302	−0.104958	−0.0177389	0.0118851
11			3,226.79	−1448.34	−139.567	−73.9497
12			1.00000	−0.448850	−0.0432527	−0.0229174
13				82.1981	8.55450	2.01534
14				1.00000	0.104072	0.0245181
15					2.61522	0.797716
16					1.00000	0.305028
17						1.05273
18						1.00000

MODEL SEQUENCING WITH SEVERAL x VARIABLES

In the previous analysis of variance unit the total amount of the variability among the responses was divided into the sum of squares associated with fitting a sequence of two candidate models and the residual sum of squares. The model sequence sum of squares was further decomposed into an additive sequential set of two sums of squares. Specifically:

- $R(0)$ is the sum of squares associated with x_0 by fitting the Mean Model.

- $R(1|0)$ is the additional sum of squares associated with x_1 by fitting the straight-line model. In other words, this is the extra amount of variability associated with the sloping straight line that couldn't be accounted for by the horizontal straight line.

The $R(\ )$ notation is introduced here for convenience. The $R(1|0)$ is to be read as "the additional sum of squares associated with the x_1 variable given that the x_0 variable model had already been fitted." This additional sum of squares is equivalent to the sum of squares for the difference between full and reduced models of Table 6.1 of the previous unit.

If we generalize the one-variable situation to the present four-variable model, we can calculate an analogous analysis of variance table directly from the right-hand column of the ABDO calculations in Table 7.2:

Source of Variation	Sum of Squares
$R(0)$	$(23.15)(1.1575) = 26.7961$
$R(1\|0)$	$(49)(0.0118851) = 0.5824$
$R(2\|0,1)$	$(-73.9497)(-0.0229174) = 1.6947$
$R(3\|0,1,2)$	$(2.01534)(0.0245181) = 0.0495$
$R(4\|0,1,2,3)$	$(0.797716)(0.305028) = 0.2433$
Residual	$(1.05273)(1) = 1.0527$

To emphasize the sequential nature of the sums of squares calculated from Table 7.2, we could formally envision the process as a sequence of the following fitted models, where the values in the parentheses after $\hat{y}$ indicate the x variables used in each fitted model:

Model 0: $\quad \hat{y}(0) = \bar{y}x_0$

Model 1: $\quad \hat{y}(0,1) = b_0 x_0 + b_1 x_1$

Model 2: $\quad \hat{y}(0,1,2) = b_0 x_0 + b_1 x_1 + b_2 x_2$

Model 3: $\quad \hat{y}(0,1,2,3) = b_0 x_0 + b_1 x_1 + b_2 x_2 + b_3 x_3$

Model 4: $\quad \hat{y}(0,1,2,3,4) = b_0 x_0 + b_1 x_1 + b_2 x_2 + b_3 x_3 + b_4 x_4$

In general, the numerical values of the intercept b_0 and the other estimated partial regression coefficients b_1, b_2, and b_3 will be different for each of the models, as we will show in Unit 10. Also, the particular sequence of the sums of squares in general depends on the particular order of the explanatory variables. If the sums of squares table is rewritten to emphasize the sequencing of the fitted models, it will have the format shown in Table 7.3.

Table 7.3 Sums of Squares Table Emphasizing Sequencing of Fitted Models

Source of Variation	Sum of Squares	Degrees of Freedom
Total	30.4187	20
Fitting model 0, $R(0)$	26.7961	1
Residual (1)	3.6226	19
Fitting model 1, $R(1\|0)$	0.5824	1
Residual (2)	3.0402	18
Fitting model 2, $R(2\|0,1)$	1.6947	1
Residual (3)	1.3455	17
Fitting model 3, $R(3\|0,1,2)$	0.0495	1
Residual (4)	1.2960	16
Fitting model 4, $R(4\|0,1,2,3)$	0.2433	1
Residual (5)	1.0527	15

Note that the | symbol has different meanings. In our sum of squares notation the symbol is read as "given that"; for example, $R(2|0,1)$ indicates the additional sum of squares associated with x_2 given that a model with x_0 and x_1 had already been fitted. We have also seen the symbol with an *augmented* interpretation, for example, $\mathbf{X'X|X'Y}$. In this sense later units will employ the symbol for a sequence of several models. Model 0, followed by Model 1 and Model 2, is denoted as $x_0|x_0\ x_1|x_0\ x_1\ x_2$.

If the researcher is primarily interested in the variability among the responses after the mean has been subtracted from each response, the model sum of squares does not include $R(0)$, and the analysis of variance table has the format shown in Table 7.4.

The model sum of squares is the sum of four individual sums of squares, one for each measured x variable. The sums of squares are sequential; that is, each one is the additional sum of squares that can be associated with the variability in the response variable remaining after fitting the previous variables. Excluding the sum of squares associated with the mean, approximately 70% of the total corrected sum of squares is associated with the fitted model. The proportion of the total corrected sum of squares associated with the model sum of squares is defined as the *coefficient of multiple determination*, the square of the coefficient of multiple

Table 7.4 ANOVA When Model Does Not Include $R(0)$

Source of Variation	Sum of Squares
Total (uncorrected)	30.4187
$R(0)$	26.7961
Total (corrected)	3.6226
Model	2.5699
$R(1\|0)$	0.5824
$R(2\|0,1)$	1.6947
$R(3\|0,1,2)$	0.0495
$R(4\|0,1,2,3)$	0.2433
Residual	1.0527

correlation R. In our example we have

$$R^2 = \frac{\text{model sum of squares }[\text{excluding }R(0)]}{\text{total corrected sum of squares}} = \frac{2.5699}{3.6226} = 0.71$$

The descriptive statistic R^2 is one summary measure of the overall goodness of the fitted model.

SEQUENTIAL AND PARTIAL REGRESSION COEFFICIENTS

In Units 5 and 6 a prediction equation for the one-variable model was calculated in terms of the estimated mean and slope or estimated intercept and slope:

$$\hat{y} = \bar{y}x_0 + b_1(x_1 - \bar{x}_1) \quad \text{or} \quad \hat{y} = b_0 x_0 + b_1 x_1$$

As developed in Unit 3, the intercept and slope are called partial regression coefficients. The mean of the Mean and Slope Model is called a sequential regression coefficient, since it is calculated from an intermediate step in the fitting of a sequence of two models. In the spirit of sequential fitting the slope could also be called a sequential coefficient.

We can now generalize about several features of the ABDO development in this unit.

Each calculated sequential or partial regression coefficient is a sum of cross products divided by a sum of squares, which can be expressed as

$$b_j = \sum_{i=1}^{n}(x_{ij} - \hat{x}_{ij})y_i \bigg/ \sum_{i=1}^{n}(x_{ij} - \hat{x}_{ij})^2$$

where j is the index for the explanatory variable, $j=0,1,2,\ldots,k$. The estimated regression coefficient, b_j, can be either a sequential or partial regression coefficient depending on the set of x variables used to predict x_{ij}. The numerator is a weighted sum of the response variable observations, the $(x_{ij}-\hat{x}_{ij})$ values being the weights. The $\hat{x}_{ij}$ represents the magnitude of x_{ij} that can be predicted by other x variables. For sequential regression coefficients these weighted sums or linear combinations of the observations are the right-most elements in the odd steps of ABDO. For partial regression coefficients the x_{ij} values are corrected for all the other x variables in the model, not just for the preceding x variables in the ABDO calculations. The last sequential coefficient from ABDO, b_k, is also the partial coefficient.

In stage 0 of ABDO no corrections have been made, the x_{i0} values are all equal to one, and the numerator is a simple unweighted sum, $(1)y_1 + (1)y_2 + \cdots + (1)y_n$.

In stage 1 of ABDO each variable has been corrected for its mean, for example, $\hat{x}_{i1} = \bar{x}_1$, and the numerator of the sequential regression coefficient is the weighted sum,

$$(x_{11}-\bar{x}_1)y_1 + (x_{21}-\bar{x}_1)y_2 + \cdots + (x_{n1}-\bar{x}_1)y_n$$

Consequently, the straight-line slope between y and x_1 is a linear combination of the observations, where the weights depend on the distance of the associated x value from $\bar{x}$. Returning to the even step of stage 0, we can calculate the partial regression coefficient, the intercept of the Intercept and Slope Model.

The right-most element in the odd step of the subsequent stages, the numerator of the sequential regression coefficients, continues to be a weighted sum. The weights $(x_{ij}-\hat{x}_{ij})$ are the remainders of each value of the corresponding x variable that cannot be predicted by the previous x variables. Naively, the partial regression coefficients for the full model are calculated by running $p=(k+1)$ ABDO's since the partials are based on x's that are adjusted for all the other x's in the model, for example, by fitting each variable last. However, by starting with the last stage for one sequence of models (one ABDO), the partial regression coefficients can be calculated by working backward through the even steps of ABDO. These details are developed in Unit 10.

WEIGHTED REGRESSION

On occasion we may have data for which the observations were measured with different levels of precision. A particular situation is when the y's are sample means based on different numbers of observations. Sample means based on a large number of observations deserve more weight than those based on fewer observations, because the variance of a mean depends on the sample size. In fact, the appropriate weights are the numbers of observations that went into the means.

To incorporate different weights in the computation procedure, we have to modify $\mathbf{X'X}$ and $\mathbf{X'Y}$. With four x variables the sums of squares and cross products in Table 7.5 are used instead of those in Table 7.1. The weights are denoted by the w's. All subsequent computations are unchanged.

Table 7.5 Sums of Squares and Cross Products Incorporating Weights

Row	Column					
	1	2	3	4	5	6
1	$\Sigma w_i x_{i0}^2$	$\Sigma x_{i0} w_i x_{i1}$	$\Sigma x_{i0} w_i x_{i2}$	$\Sigma x_{i0} w_i x_{i3}$	$\Sigma x_{i0} w_i x_{i4}$	$\Sigma x_{i0} w_i y_i$
2		$\Sigma w_i x_{i1}^2$	$\Sigma x_{i1} w_i x_{i2}$	$\Sigma x_{i1} w_i x_{i3}$	$\Sigma x_{i1} w_i x_{i4}$	$\Sigma x_{i1} w_i y_i$
3			$\Sigma w_i x_{i2}^2$	$\Sigma x_{i2} w_i x_{i3}$	$\Sigma x_{i2} w_i x_{i4}$	$\Sigma x_{i2} w_i y_i$
4				$\Sigma w_i x_{i3}^2$	$\Sigma x_{i3} w_i x_{i4}$	$\Sigma x_{i3} w_i y_i$
5					$\Sigma w_i x_{i4}^2$	$\Sigma x_{i4} w_i y_i$
6						$\Sigma w_i y_i^2$

As it happens, the arsenic data were actually means of several children in each town. The respective numbers of children are 25, 23, 24, 27, 25, 22, 26, 20, 22, and 23. These weights were ignored in the previous analysis because they are so nearly equal. Their use would make no practical difference, and the unweighted analysis is simpler. They are introduced here simply to illustrate the computations. The numerical values of the weighted $\mathbf{X'X}$ and $\mathbf{X'Y}$ matrices needed for a weighted regression of the arsenic data in Unit 5 are

$$\Sigma w_i x_{i0}^2 = 237 \qquad \Sigma x_{i0} w_i x_{i1} = 3{,}728 \qquad \Sigma x_{i0} w_i y_i = 389.36$$

$$\Sigma w_i x_{i1}^2 = 84{,}908 \qquad \Sigma x_{i1} w_i y_i = 4094.80$$

$$\Sigma w_i y_i^2 = 855.26$$

For example, the weighted sum of cross products between x_1 and y is

$$\Sigma x_{i1} w_i y_i = (2)(25)(3.19) + (4)(23)(3.26) + \cdots + (36)(23)(0.30)$$
$$= 4094.80$$

The estimated coefficients are $b_0 = 2.86$ and $b_1 = -0.077$, which are virtually the same as those previously obtained. Weighted regression can be used in several contexts, as will be developed in later units.

SUMMARY

The computing algorithm ABDO was generalized in this unit from the one-variable model of earlier units. The calculations follow an easily recognized pattern, allowing generalization to k variables. Envisioning a sequence of models, we can associate a sequential sum of squares with each additional x variable (or group of variables). These additive sums of squares can be listed in an ANOVA table showing the magnitude of the remaining variability of the response variable accounted for by each additional x variable in the sequence. Like the estimated sequential regression coefficients, which are available at each intermediate stage of the calculations, the sequential sums of squares are order-dependent. The partial regression coefficients can also be calculated from ABDO. Both types of coefficients can be expressed as a sum of cross products between an adjusted explanatory variable and a response variable divided by a sum of squares of the adjusted explanatory variable. In other words, any sequential or partial regression coefficient in multivariable models can be calculated from a single-variable model with an appropriate construction of the x variable. Algebraically, we have

$$b_j = \sum_{i=1}^{n} (x_{ij} - \hat{x}_{ij}) y_i \Big/ \sum_{i=1}^{n} (x_{ij} - \hat{x}_{ij})^2$$

where $\hat{x}_{ij}$ is the predicted value of x_{ij} using other x variables.

Unit 8

Coping with Model Violations

Objectives

- To explore data analysis approaches when we violate regression model assumptions.
- To develop a systematic and data-based transformation methodology that emphasizes residual plotting.
- To straighten curvilinear relationships and to ensure that the variability of the response variable does not depend on the magnitude of an explanatory variable.

In Unit 3 the regression model with k variables,

$$y = \beta_0 x_0 + \beta_1 x_1 + \cdots + \beta_k x_k + \varepsilon$$

was formulated. Ideally, all the important x variables are identified, and the sum of the $\beta_j x_j$ products, $\Sigma \beta_j x_j$, accounts for the explainable (predictable) part of y. Assumptions about the random component $\varepsilon = y - \Sigma \beta_j x_j$ were also made in Unit 3. To ensure that the results of a regression analysis are meaningful, we should routinely examine the data and the estimated residuals for consistency with the assumptions. In practice, two violations of the assumptions in Unit 3 are sufficiently frequent to warrant special attention:

- Violation I: The relationship between y and an x_j is not a straight line (it is curvilinear).
- Violation II: The variability in the random component ε is dependent on the value of x_j.

An alternative statement of Violation I is that the assumed model has not been correctly formulated. Perhaps not all the important x variables have been identified or, if identified, have not been measured. Here we will only worry about doing our best with the k measured independent variables, since it is usually too much to expect that the relationship between y and each x is a straight line. One methodology for handling curvilinear relationships is to add variables to the model that are mathematical functions of the measured variables, for example, polynomials. This additional-variable methodology is introduced in Unit 11. For now emphasis is on reexpressing the y or x variable in different units—for instance, logarithms—so that the relationship is approximately a straight line. Reasons usually exist for the units in which the y and x variables were measured, but we must be flexible and realize that the fitting of the model might be enhanced by using a transformed unit for one or more of the measured variables.

Whereas Violation I concentrates on the basic model formulation between y and x, Violation II is concerned with an important assumption used in fitting the model; namely, each observation is given the same weight in fitting the model. If some observations are more suspect than others—for example, if some observations have a larger variance—then these observations should be given less weight in the ensuing calculations than the more precise observations. One approach is to transform the y variable so that the variability in y does not depend on the x's.

The objective of this unit is to present a systematic and data-based procedure for selecting transformations to correct for one or both of the violations. The emphasis in this unit and the next is on transformations for one-variable models, although the principles are the same for many variables. The main diagnostic tool will be the residual plot. The residuals $y - \hat{y}$ will be plotted against x for one-variable models (against $\hat{y}$ to obtain a single plot for more than one variable). The plots of y against each x will give clues to various transformations. Usually we are looking for a transformation that will adequately straighten a curvilinear relationship. For a given set of data several transformations can result in approximately straight-line relationships. We will want to choose the transformation that ensures us that the variability of y does not depend on x.

In certain situations, usually one-variable situations, subject matter knowledge is sufficient to make the complete model known beyond reasonable doubt. Variability in y is then attributable only to instrument measurement error and sampling error. For instance, suppose biological theory indicates that the relationship between y and x is

$$y = \beta_0 e^{\beta_1 x} \varepsilon = \beta_0 \exp(\beta_1 x) \varepsilon$$

where $\exp(\beta_1 x)$ is another way of writing the exponential term $e^{\beta_1 x}$. In terms of our regression model hardly anything is right. The observations are not expressed as a sum of $\beta_j x_j$ terms, and the model is generally called a *nonlinear* model. The random component ε is not additive; it is multiplicative. And the variability in y depends on the value of x. But taking the logarithm of both sides gives

$$\ln y = \ln \beta_0 + \beta_1 x + \ln \varepsilon$$

Hence by doing our data analysis with $\ln y$ instead of y, we can use the one-variable additive model, where $\ln \beta_0$ is the intercept and $\ln \varepsilon$ is the residual component. In situations where the known subject matter equation can be linearized to give a model in our regression model format, we proceed with the transformation and then check the residual plot for Violation II.

A different situation occurs when the observations are counts, such as the number of colonies growing on a solid medium or the number of accidents in a fixed period of time. In such cases $\sqrt{y}$ will better satisfy the assumptions than will y. Whenever theory or past experience suggests a model, this information should take precedence over the data-oriented techniques of this unit.

VITAL STATISTICS DATA

The data in Table 8.1 come from the United States Department of Health, Education and Welfare.

Table 8.1 Vital Statistics Data

Age	Coded Age, x	Deaths per Thousand Population, y
40	0	4.85
45	1	7.57
50	2	11.79
55	3	18.57
60	4	27.86
65	5	41.18
70	6	58.93
75	7	86.77
80	8	123.87
85 and over	9	178.22

Source: U.S. Department of Health, Education and Welfare, Public Health Service, *Vital Statistics of the United States, 1969, 1970.* Vol. II: *Mortality*, Part A (Washington, D.C.: Government Printing Office, 1974).

A graph of the data and the least squares straight line fit to the data are given in Figure 8.1. It is obvious from this figure that a straight line is not appropriate. The line is too low at its ends and too high in the middle. In fact, the line predicts a negative death rate for the lowest age group, an interesting prospect. Plotting the residuals from the straight-line fit against x will show a U-shaped pattern of the residuals, and most of the residuals will not be close to zero. Both the original y versus x plot and the residual plot indicate that y is related to x through a smooth curve. The first step will be the identification of transformations to straighten the curve. The second step will be to select the best of the straightened curves with regard to the variability pattern of y.

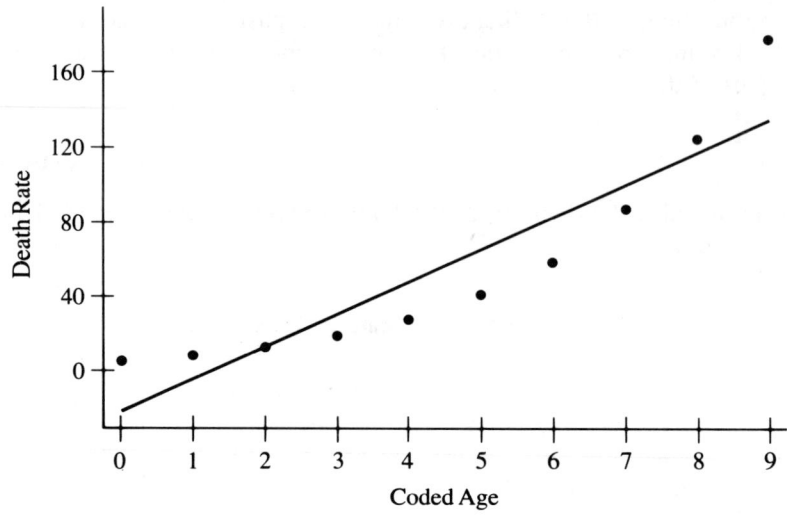

Figure 8.1 Straight Line Fit to the Vital Statistics Data

CURVES THAT ARE ALWAYS UP AND HAVE ONE BEND

Examples of curves that are always up (increasing) and have only one bend are shown in Figure 8.2. These curves are shifted (a constant added) and scaled (a constant multiplied) so that they are equal at zero and one. The constant c will be covered in detail in a later section. For the immediate discussion values of x and y between zero and one are regarded as *small* and values larger than one are regarded as *large*. With actual data other shift and scale constants may be used.

This collection of curves is by no means exhaustive, but it includes most of those commonly used in practice. Of course, the line $y=x$ needs no straightening; it is included for reference.

As an example, consider the curve $y=x^2$ of Figure 8.2. If we let $x^*=x^2$, then y and x^* are linearly related. Another transformation that could be made is $y^*=\sqrt{y}$, and then y^* and x are linearly related.

In general, there are two transformations, involving only one variable, that straighten curves.

1. Form a new x, x^*, equal to the expression for y in terms of x.

2. Form a new y, y^*, equal to the expression for x in terms of y.

The first of these can be viewed graphically as selectively stretching the curve in the horizontal direction. In the case of the curve $y=x^2$ we want to stretch the right-most portion of the x axis so that the curve associated with the large x comes in line with the curve associated with the small x. Since x^2

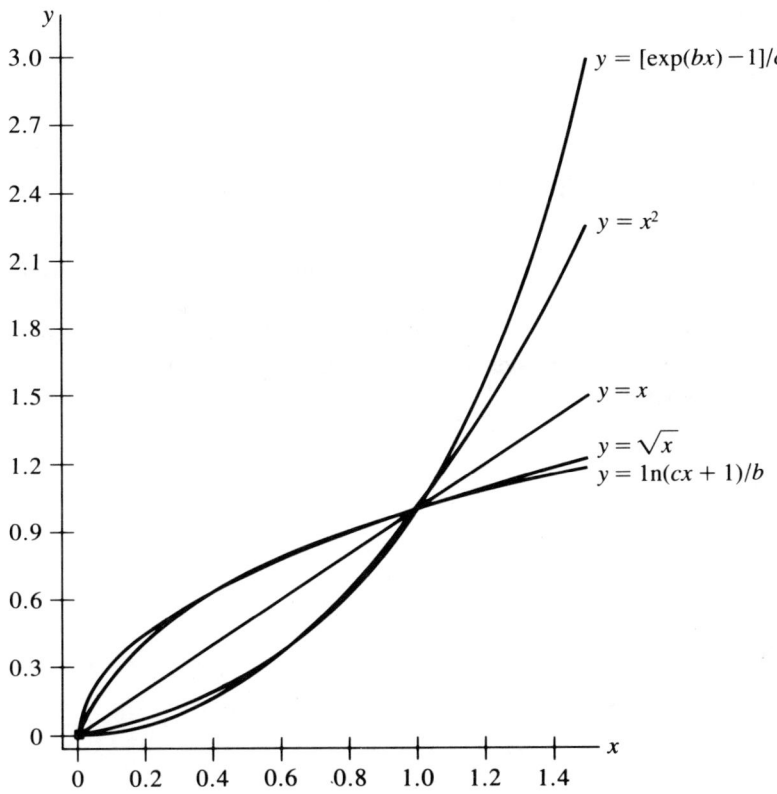

Figure 8.2 Curves That Are Always Up and Have One Bend

is smaller than x when x is small (less than one) and is larger than x when x is large, $x^* = x^2$ provides the appropriate selective stretching. The second transformation can be viewed graphically as selectively stretching the curve in the vertical direction. In the case of the curve $y = x^2$ we want to stretch the lower portion of the vertical axis to bring the curve associated with small y in line with the curve associated with large y. Since $\sqrt{y}$ is larger than y when y is small (less than one) and smaller than y when y is large, $y^* = \sqrt{y}$ provides the appropriate selective stretching.

When dealing with data, we usually don't know the function we are trying to straighten. If it is increasing and concave upward, however, a transformation on x should be a concave upward function of x, such as a square function or an exponential function. A transformation on y should be a concave downward function of y such as a square root function or a logarithmic function. Thus from the standpoint of curve straightening, there are four reasonable transformations to try. Similar reasoning applies when the function is increasing and concave downward. Candidate transformations on x should be concave downward functions of x, such as a square root function or a logarithmic function. A transformation on y should be concave upward, such as a square function or an exponential function.

CURVES THAT ARE ALWAYS DOWN AND HAVE ONE BEND

Examples of curves that are always down (decreasing) and have only one bend are shown in Figure 8.3. Again, this collection of curves is by no means exhaustive, but it includes the commonly used ones. None of the included curves are concave downward since such curves occur less frequently in practice.

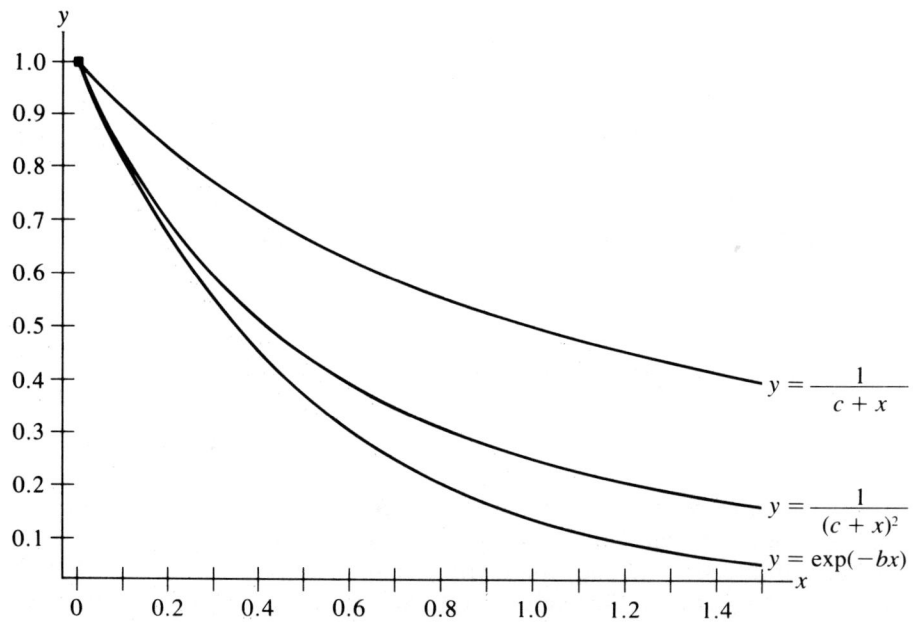

Figure 8.3 Curves That Are Always Down and Have One Bend

Candidate transformations for x are

$$x^* = 1/(c+x) \qquad x^* = 1/(c+x)^2 \qquad x^* = \exp(-bx)$$

Candidate transformations for y are

$$y^* = 1/y \qquad y^* = 1/\sqrt{y} \qquad y^* = \ln y$$

Transformation Selection

As in the preceding section, the particular choice will depend on the variability pattern of y.

SCALING AND SHIFTING

The actual quantities plotted for the exponential and logarithmic functions in Figure 8.2 are

$$y = [\exp(2x) - 1]/1.7183 \quad \text{and} \quad y = \ln(1.7183x + 1)/2$$

The constants 1.7183 and 2 were chosen to make these functions similar to x^2 and $\sqrt{x}$ in the region of discussion. It is sometimes required that constants be chosen so that functions are similar to data in a particular region.

A variable is considered to be *scaled* if it has been multiplied by a constant and to be *shifted* if it has the addition of a constant. For example, $y^* = (y+5)^2$ illustrates the shifting of a variable before application of the square transformation. The expression $x^* = \exp(2x)$ represents a scaling of x before application of the exponential transformation. It is conceivable that we might want to both shift and scale. Since we will do a regression analysis after the transformation, our work is then simplified. For example, the relation

$$y = b_0 + b_1 \ln(c_0 + c_1 x)$$

can be rewritten as

$$y = b_0^* + b_1 \ln(c + x)$$

where $b_0^* = b_0 + b_1 \ln c_1$ and $c = c_0/c_1$. Since regression analysis will determine b_0^* and b_1, only c must be specified before the transformation. It can be shown in a similar manner that only shifts are required in square and square root transformations and only scaling is required in exponential transformations.

TRANSFORMATION SELECTION

Looking at the death rate data (Figure 8.1 and Table 8.1), we see that death rate increases with age and the curve is concave upward. Hence, as mentioned in the previous section, there are four suggested transformations. The first is the logarithmic transformation on y, $y^* = \ln(y + c)$. But what should be used for the numerical value of c? To determine this, choose the y values associated with three equally spaced values of x. For illustration we use the following values:

x	y	$\ln(y+c)$
0	4.85	$\ln(4.85+c)$
4	27.86	$\ln(27.86+c)$
8	123.87	$\ln(123.87+c)$

We want to choose c such that the first and third columns are linearly related. Since the x values are equally spaced, we have a linear relationship if

$$\ln(27.86+c) - \ln(4.85+c) = \ln(123.87+c) - \ln(27.86+c)$$
$$c = 2.40$$

This three-point nonlinear fitting has given us a constant that will straighten better than $\ln y$. The other constants in the relationship are obtained in the regression analysis that will follow.

Next we try a square root transformation on y,

$$y^* = \sqrt{y+c}$$

To determine c, we use the same three points used with the logarithmic transformation. The resulting equation is

$$\sqrt{27.86+c} - \sqrt{4.85+c} = \sqrt{123.87+c} - \sqrt{27.86+c}$$
$$c = -3.60$$

Since this solution involved the squaring of both sides of the equation (in two instances), there is the possibility of an extraneous solution. We must check to see if $c = -3.60$ satisfies the first equation. It does not, so apparently there is no genuine solution. We will not, therefore, pursue this transformation further.

Now two transformations on x are considered, an exponential transformation, $x^* = \exp(bx)$, and a square, $x^* = (x+c)^2$. We have already established the approximate relationship

$$\ln(y+2.4) = b_0 + b_1 x$$

The quantity b_1 in this expression is the same quantity as b in the expression for x^*. Consequently, a two-point fit—say, to the points $(0, 4.85)$ and $(8, 123.87)$—is

$$\ln(123.87+2.40) - \ln(4.85+2.40) = (8-0)b_1$$
$$b_1 = 0.36$$

Another set of two points could, of course, have been selected, but in practice one doesn't expend too much effort trying to find a best value, since small deviations are compensated for in the regression analysis that follows.

For the square transformation three equally spaced values of y do not exist. We may use the largest and smallest values of y, 178.22 and 4.85, and their average, 91.54. The value of x associated with 91.54 is found by interpolating between the next larger and next smaller values of y, giving $x = 7.13$. A three-point fit of $(x-c)^2$ to x gives the equation

$$50.84 + 14.26c = 81.00 - 50.84 + (18.00 - 14.26)c$$

$$c = -1.97$$

which suggests the transformation $x^* = (x - 1.97)^2$. However, this function is not always increasing and so we will not pursue this transformation further.

FINAL SELECTION

From the application of our rules for straightening curves, we have two potential transformations that are increasing and concave upward. To aid in the selection of one of these, we follow these steps:

1. Fit the potential models, one for each selected transformation.
2. For each fitted model plot the predicted and observed values against x.
3. For each fitted model plot the residuals against the predicted values.

The first plot is used primarily to see if a linear relationship was achieved. For the vital statistics data, Figure 8.4 shows a fit for $y^* = \ln(y + 2.4)$ that is similar to the fit in Figure 8.5 for $x^* = \exp(0.36x)$. Both transformations have done an excellent job of straightening the curve.

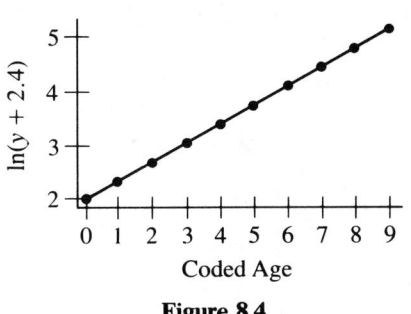

Figure 8.4

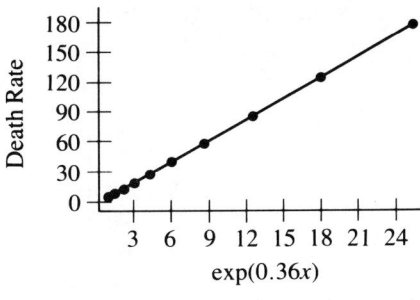

Figure 8.5

Figure 8.4 Plot of Transformed Death Rate Versus Coded Age
Figure 8.5 Plot of Death Rate Versus Transformed Coded Age

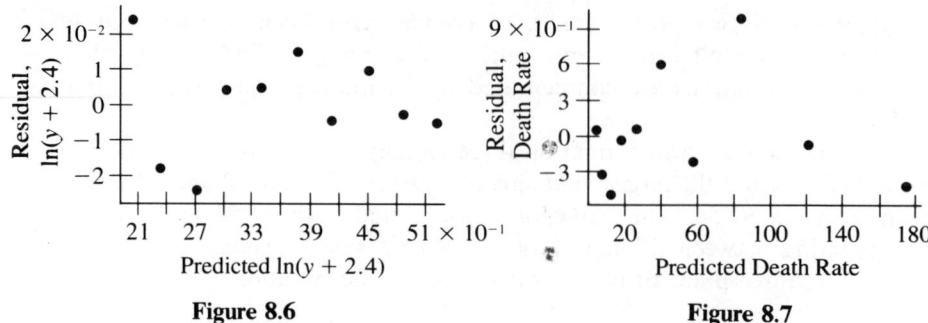

Figure 8.6 Residual Plot of Transformed Death Rate Versus Predicted Transformed Death Rate

Figure 8.7 Residual Plot of Death Rate Versus Predicted Death Rate

The second plot is used primarily to see if the spread in the residuals is similar for all values of x (or, equivalently, for all predicted values). These plots are given in Figures 8.6 and 8.7. No trends are apparent in either residual plot; in fact, the spread is similar. In each, one or two residuals seem large relative to the others. However, the absolute magnitude of these residuals is very small. Both prediction equations seem excellent, and there is little to choose between the two.

The final step is to calculate two prediction equations, one for each transformation. Thus we have

$$\hat{y} = \exp(1.9572 + 0.36055x) - 2.4 = -2.4 + 7.0791 \exp(0.36055x)$$

[the preceding equation was determined by fitting the model $\ln(y+2.4) = b_0 + b_1 x$ and inverting the transformation after finding b_0 and b_1] and

$$\hat{y} = -2.2567 + 7.0845 \exp(0.36x)$$

A plot of this last equation, along with observed values, is given in Figure 8.8. Plots like this one are good for displaying the results of an analysis. Now compare the fitted models of Figures 8.1 and 8.8.

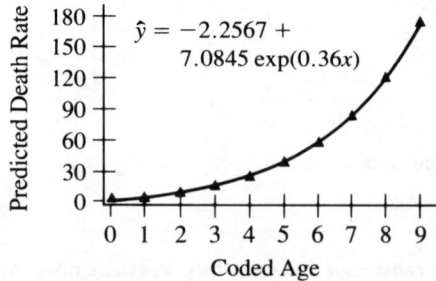

Figure 8.8 Plot of Predicted Death Rate Versus Coded Age

SUMMARY

Emphasized in this unit are two violations of assumptions that occur frequently; namely, (1) the relationship between y and x is not a straight line and (2) the variability of y is dependent on the value of x. These violations may be detected by a plot of the observed and predicted values against x and by a plot of the residuals against the predicted values. Both plots are informative about both assumptions. The first shows what is fit *and* what is not being fit. The second expands the scale on what is not being fit. The first is primarily used with one-variable models or when all x variables are functions of one basic variable. The residual plot can be used regardless of the number and nature of the x variables. Indeed, the residual plot will be our main diagnostic tool in all the units that follow.

Consider the curvilinear problem. Table 8.2 presents curves commonly found in practice and the transformations to be considered for straightening.

Table 8.2 Common Curves Found in Practice and Transformations for Straightening

Shape	Transformation		
⌒ (rising)	$y^* = \ln(y+c)$ $x^* = \exp(bx)$	$y^* = \sqrt{y+c}$ $x^* = (x+c)^2$	
⌒ (concave)	$y^* = \exp(bx)$ $x^* = \ln(x+c)$	$y^* = (y+c)^2$ $x^* = \sqrt{x+c}$	
⌣ (U-shape)	$y^* = 1/y$ $x^* = 1/(c+x)$	$y^* = 1/\sqrt{y}$ $x^* = 1/(c+x)^2$	$y^* = \ln(y+c)$ $x^* = \exp(-bx)$

Transformation methodology, as discussed in this unit, includes the following procedures:

- Shifting or scaling the variable to be transformed in order to achieve the right degree of selective stretching. By selecting three points on the curve so that the values of the variable not being transformed are equally spaced, you can form an equation for the scale or shift parameter.
- Fitting the model for the most promising transformations.
- Graphing two diagnostic plots. If one of the transformed models shows a straight-line linear relationship for the y versus x plot and no patterns for the residual plot, the objective has been achieved. If not, repeat the process by transforming the variable not transformed the first time, or perform a second transformation on the same variable.

Unit 9

Case Study: Dose-Dependent Blood Pressure Data

In a test of the effect of a drug on the blood pressure of an animal, another agent is usually given first to make the blood pressure abnormal. Then the ability of the drug under study to return the blood pressure toward the normal value is examined. The administration of this first agent is called a *challenge*. In the study we examine in this unit the experimental animals are dogs and the challenge agent is isoproterenol, which causes a decrease in blood pressure. The objectives of this study are as follows:

- To determine the effect of the drug on blood pressure of dogs that have been challenged with isoproterenol.
- To determine the dosage that causes a 50% reduction in the response to the isoproterenol challenge.

The experimental protocol for each dog follows seven basic steps:

1. Baseline blood pressure (mmHg) measured.
2. Isoproterenol administered intravenously (0.3 gram per kilogram, g/kg).
3. Blood pressure measured and the control data, denoted as y_c, recorded as percentage change from the baseline.
4. Prescribed dose of the drug administered orally.
5. One hour after administration of the drug, new baseline blood pressure measured.
6. Isoproterenol administered intravenously (0.3 g/kg).
7. Blood pressure measured and the drug data, denoted as y_d, recorded as percentage change from the new baseline.

Other than being preceded by a dose of the drug, steps 5, 6, and 7 are identical to steps 1, 2, and 3.

The control and drug data on each of 20 dogs are given in Table 9.1.

MODELING

Tentatively assumed is a straight-line relationship between response and dose for each dog. Also assumed is that the slope (β_1) of the line quantifies the action of the drug and is the same for each dog. The intercept (β_0) quantifies physical traits that may be different for each dog. For example,

Table 9.1 Blood Pressure Data

Dog	Dose	Control, y_c	Drug, y_d
1	0.033	−27	−25
2	0.330	−15	−21
3	0.100	−28	−15
4	1.000	−41	−15
5	3.330	−29	− 7
6	0.033	−25	−28
7	1.000	−30	−11
8	0.100	−38	−38
9	3.330	−34	− 6
10	0.330	− 5	− 6
11	0.033	−40	−44
12	0.330	−56	−28
13	1.000	−33	−30
14	0.100	−27	−27
15	3.330	−21	−10
16	0.100	−48	−39
17	0.330	− 3	−27
18	3.330	−38	− 8
19	0.033	−27	−25
20	1.000	−18	−12

the observations on the first dog have these tentative models:

$$-27 = \beta_0 + \varepsilon_1 \quad \text{(control)}$$
$$-25 = \beta_0 + \beta_1(0.033) + \varepsilon_2 \quad \text{(drug)}$$

Since the dogs are mongrels and include both sexes, it is anticipated that β_0 will be different for each dog. With major interest in the slopes, the intercepts may be eliminated by taking the differences between observations on the same dog. In this case the model becomes

$$2 = (-25 + 27) = \beta_1(0.033) + (\varepsilon_2 - \varepsilon_1)$$

The same differencing ($y = y_d - y_c$) on each dog-dose combination yields data with the tentative model

$$y = y_d - y_c = \beta_1(\text{dose}) + \varepsilon^*$$

Graphic representation of this model is a straight line through the origin. As a first step in the analysis, the response is plotted against dose in Figure 9.1. From this graph it is seen that the range of the four observations for the 0.33 dose is much larger than for the other doses. Comparing these data with the original case report forms showed that the -3 recorded for the control value of dog 17 should have been -31. This error occurred during the coding or keypunching of the data.

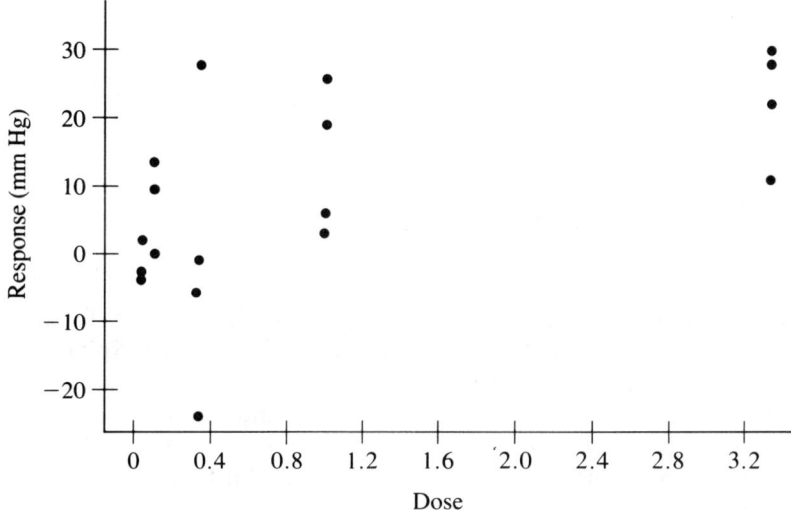

Figure 9.1 First Step: Plotting Response Versus Dose

After a correction is made for this observation, the response-to-dose relationship appears to be curvilinear. For a better assessment of the curvature, the mean for each dose is calculated.

Dose	Mean ($y_d - y_c$)
0.033	−0.75
0.10	5.5
0.33	6.25
1.0	13.5
3.33	22.75

These means are plotted in Figure 9.2. The curve is always increasing and concave downward.

One possible explanation for the curvature is that the action of the drug is multiplicative. A multiplicative model is appropriate when the response

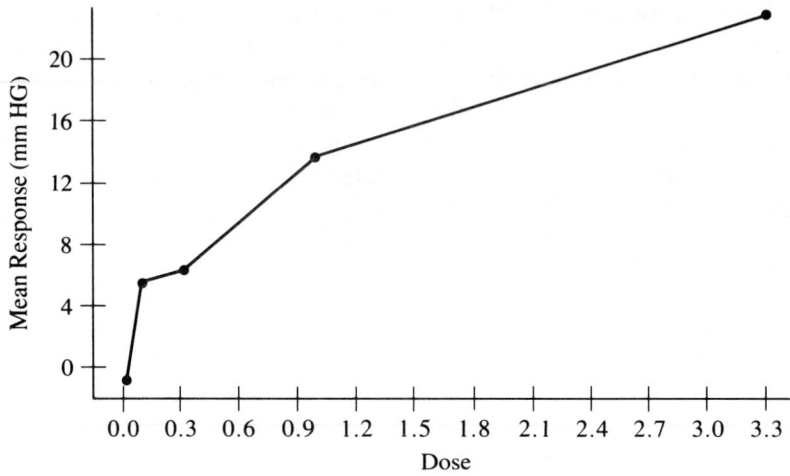

Figure 9.2 Mean ($y_d - y_c$) Versus Dose

changes by a fixed proportion of the response rather than by a fixed absolute amount with a one-unit change in dose. This explanation is plausible since isoproterenol makes the blood pressure abnormally low, and the drug may counteract a proportion of the deviation from normal levels. If the action of the drug were linear, a dose administered to an unchallenged dog would increase blood pressure above the normal level. This is believed not to happen.

For a multiplicative model the observations on the first dog have the models

$$-27 = \beta_0 \varepsilon_1 \quad \text{(control)}$$
$$-25 = \beta_0 \exp(\beta_1 0.033) \varepsilon_2 \quad \text{(drug)}$$

The mean value of $y_d - y_c$ may be expressed as $-\beta_0[1 - \exp(\beta \text{ dose})]$. With β_0 and β_1 both negative, as the data indicate, this curve is increasing and concave downward, which is consistent with Figure 9.2. The appropriate transformation in this case for the first dog is

$$\ln(25) - \ln(27) = \ln(-\beta_0) + \beta_1 0.033 + \ln(\varepsilon_2) - \ln(-\beta_0) - \ln(\varepsilon_1)$$
$$= \beta_1 0.033 + \ln(\varepsilon_2/\varepsilon_1)$$

since the logarithm of a product of two numbers is the sum of the logarithms of the individual numbers. In general, we have the model

$$y = \ln(-y_d) - \ln(-y_c) = \ln(y_d/y_c) = \beta \text{ dose} + \varepsilon^*$$

This model is fit, and the two diagnostic plots proposed in Unit 8 are displayed in Figures 9.3 and 9.4. Discussion of these plots is deferred until we have analogous plots from another transformation.

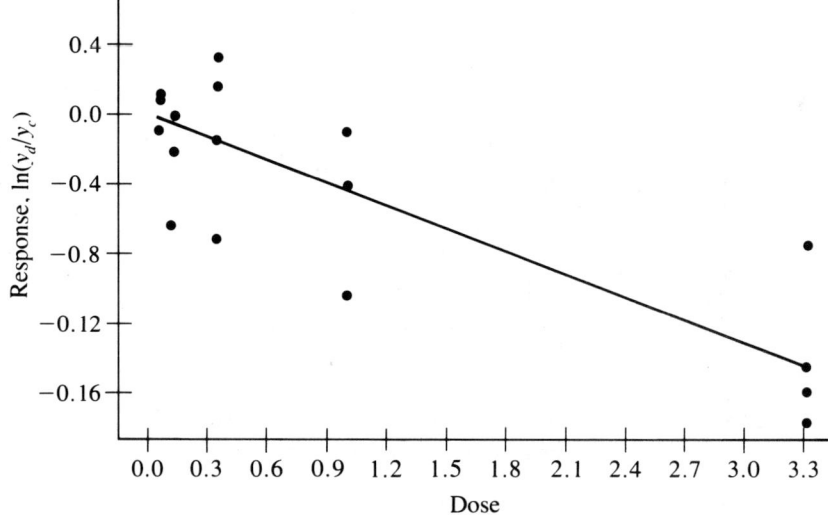

Figure 9.3 Observed and Predicted Values Versus Dose Using the $\ln(y_d/y_c)$ Transformation

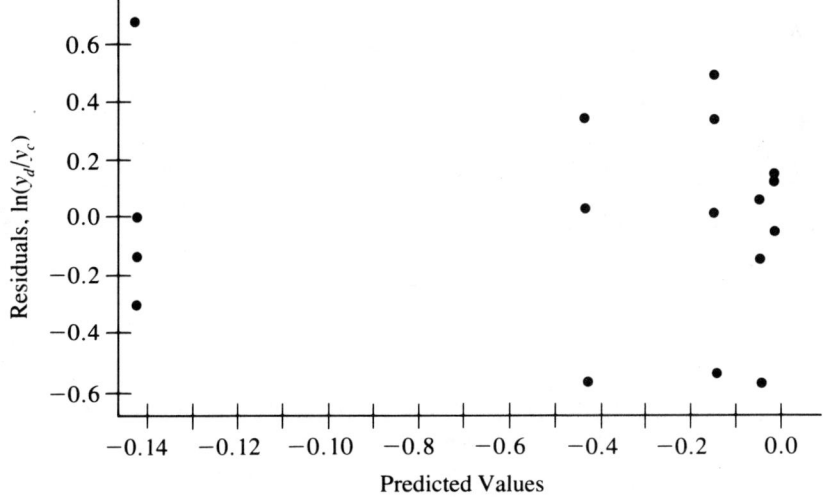

Figure 9.4 Residuals Versus Predicted Values Using the $\ln(y_d/y_c)$ Transformation

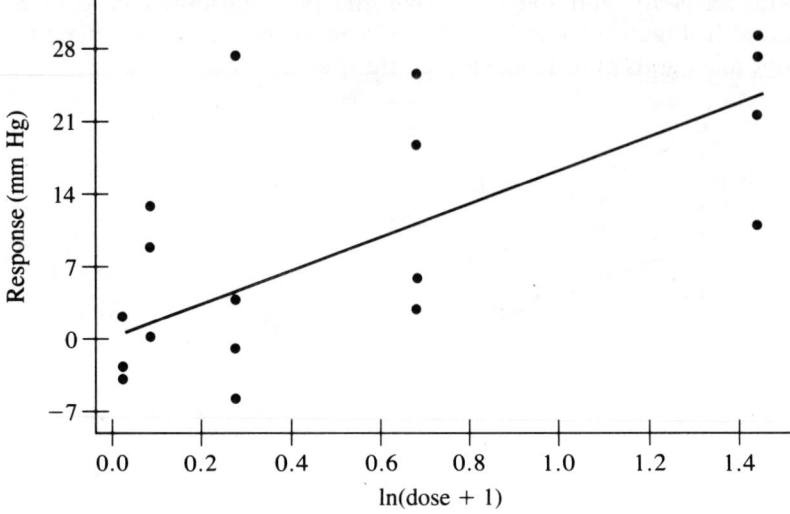

Figure 9.5 Observed and Predicted Values Versus ln(dose+1) Using the ln(dose+1) Transformation

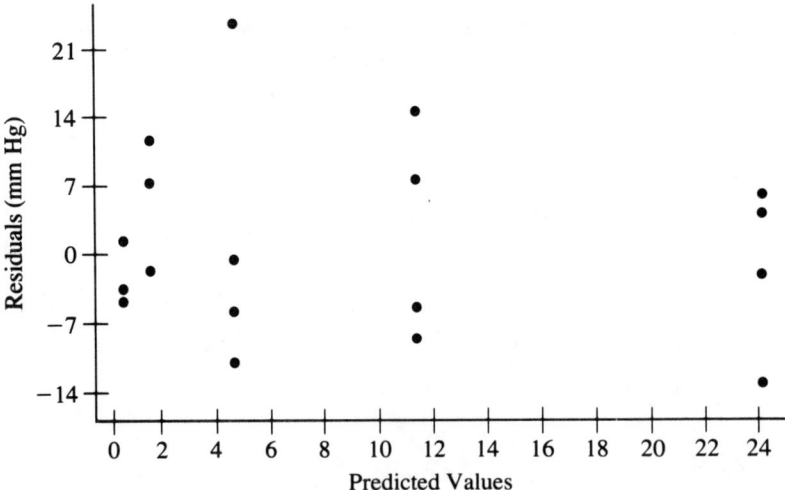

Figure 9.6 Residuals Versus Predicted Values Using the ln(dose+1) Transformation

From the data-oriented guidelines of Unit 8, $x^* = \ln(\text{dose} + c)$ would be one of the suggested transformations, since the curve is increasing and concave downward. This suggestion is consistent with a traditional belief that dose response is often linearly related to $\ln(\text{dose}+c)$. The selection of c is straightforward in this case. When the dose is zero, the mean value of $y_d - y_c$ should be zero. Thus $0 = \ln(0+c)$ implies $c=1$, and a candidate model is

$$y = y_d - y_c = \beta \ln(\text{dose} + 1.0) + \varepsilon$$

This model is fit and the diagnostic plots are displayed in Figures 9.5 and 9.6.

DIAGNOSTIC PLOTS

The guidelines of Unit 8 suggest other possible transformations for straightening an increasing concave downward curve. However, since we have two transformations that have theoretical or historical justification, we will not do other transformations unless both of these prove unsatisfactory.

Examination of Figures 9.3 and 9.5 show that both transformations have straightened the curve. In Figure 9.3, however, the observations are more compact about the line; hence the multiplicative model is preferred based on a comparison of Figures 9.3 and 9.5. Neither Figure 9.4 nor 9.6 shows pronounced patterns in variability with changing dose.

For theoretical reasons, the diagnostic plots, and ease of interpretation of parameters, further analysis will be based on the multiplicative model. You may wonder why the multiplicative model was not chosen initially, since it has theoretical appeal. But hindsight is better than foresight. We may make improper decisions during the course of a data analysis. However, if we allow the data to speak to us, we should return to a reasonable course.

CONCLUSIONS

Having decided on a model, we now address the objectives of the study. Regarding the first objective, the interpretation of β_1 in the multiplicative model is

$$\beta_1 = \frac{\text{change in response per unit change in dose}}{\text{response}}$$

The estimate of β_1 is -0.4272. This number is meaningful when compared to values obtained from other drugs.

Regarding the second objective, we have

$$\beta_0 = \text{mean response for a zero dose}$$
$$\beta_0 \exp(\beta_1 \text{ dose}) = \text{mean response for arbitrary dose}$$

Therefore, the dosage that causes a 50% reduction in the response to the isoproterenol challenge is the dose that satisfies

$$\beta_0 \exp(\beta_1 \text{ dose}) = 0.5\beta_0$$

Solving this expression for dose gives

$$\text{dose} = \ln(0.5)/\beta_1$$

Substituting the estimator for β_1 gives

$$\text{dose} = -0.6931/-0.4272 = 1.6224$$

We have provided specific answers to specific questions. But we would usually want to do more to convey the information in the data. For instance, we might give the dose response equation, which is $y = \beta_0 \exp(-0.4272 \text{ dose})$, where β_0 depends on the particular dog.

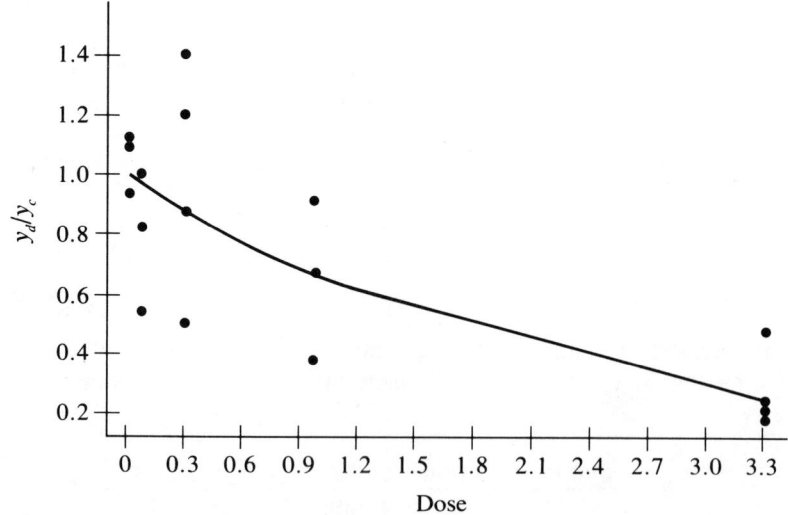

Figure 9.7 Observed and Predicted Values of y_d/y_c Versus Dose Using the $\ln(y_d/y_c)$ Transformation

Conclusions

Having fitted the model on transformed data, we would like to transform back to the original units and plot the results. We cannot do this here since the transformation was such that information about the β_0's was lost. But we can partially back-transform as follows:

$$\ln(\hat{y}_d) - \ln(\hat{y}_c) = -0.4272 \text{ dose}$$
$$\ln(\hat{y}_d/\hat{y}_c) = -0.4272 \text{ dose}$$
$$\hat{y}_d/\hat{y}_c = \exp(-0.4272 \text{ dose})$$

We can, therefore, plot values for y_d/y_c along with the predicted values. This is done in Figure 9.7. Figure 9.7 shows that there is considerable variability in the data. Since we have 20 observations providing information on one parameter, however, the estimate of β_1 is reasonably precise. It should be mentioned that any estimate calculated from data containing random components will have a random component. In Unit 12 procedures for assessing the precision of estimators such as these will be presented.

Unit 10

Interpreting Regression Coefficients in Two-Variable Models

Objectives

- To explore regression calculations with two explanatory variables.
- To express two-variable prediction equations by either sequential or partial regression coefficients.
- To gain insight into interpreting estimated regression coefficients.

Most regression relationships are more complex than a straight line between the response variable and a single explanatory variable, even though in some studies our effort might be restricted to a particular x variable, especially if it were known to be the most important one. Eventually, however, we would want to broaden our horizons to encompass several explanatory variables.

The next reasonable step in our progress, then, is to look at the specific problem of fitting a model with two explanatory variables. Emphasis will be on a better understanding of estimated regression coefficients and sequencing of candidate models.

In previous units the mechanics of computing have been developed. This unit on two-variable models is an extension of the single-variable development of Unit 5, using computation techniques from Unit 7. The two-variable problem is examined in detail for several reasons:

- Two explanatory variables are a commonly occurring data analysis problem.
- The two-variable problem can still be envisioned geometrically without difficulty.
- The basic principles of multiple regression with several x variables can be developed and illustrated with two variables.

Unfortunately, the terminology of independent variables doesn't refer to an effect of each x variable on the response being independent of the other variables. If this were true, the researcher's life would be much easier; each variable could be studied separately and the joint effect obtained by straightforward mathematics. A major weakness with one-variable-at-a-time studies is that the calculated combined effect—for instance, the sum of the individual effects—usually will not equal the measured joint effect of the several variables. If part of the difference is due to interactions among the variables, the multiple regression analysis can be extended to account for this variability. However, at times the major problem is that part of the variability in the response variable can be associated with any of several x variables. Nevertheless, the data analysis, properly carried out and understood, can lead to reasonable statements. But sometimes only the data analysis of a well-designed and controlled experiment can resolve the supposedly conflicting evidence. This was exactly the problem faced by

John Bonner Buck in recording the time of evening when the first flash by fireflies was observed, an experiment we will examine next.

FIREFLY DATA

In trying to clarify the firefly's dependence on external environmental factors rather than on an internal biological clock, Buck measured simultaneously the time of first flash, light intensity, and temperature. The data are shown in Table 10.1 for 17 days.

Table 10.1 Data for Firefly Study

y	x_0	x_1	x_2
45	1	26	21.1
40	1	35	23.9
58	1	40	17.8
50	1	41	22.0
31	1	45	22.3
52	1	55	23.3
54	1	55	20.5
38	1	56	25.5
40	1	70	21.7
28	1	75	26.7
38	1	79	25.0
36	1	87	24.4
36	1	100	22.3
46	1	100	25.5
40	1	110	26.7
31	1	130	25.5
40	1	140	26.7

Definition of Variables

y = time of first flash (number of minutes after 6:30 P.M.)
x_1 = light intensity (in metercandles, mc)
x_2 = temperature (°C)

Source: John Bonner Buck, "Studies on the Firefly. Part I: The Effects of Light and Other Agents on Flashing in *Photinus Pyralic*, with Special Reference to Periodicity and Diurnal Rhythm," *Physiological Zoology* 10 (no. 1, 1937): 45–58.

Preliminary consideration of the data shows that y is negatively related to both x_1 and x_2. If mean and slope straight lines are fitted, one for each independent variable, we have

$$\hat{y}(0,1) = 41.35 - 0.10(x_1 - \bar{x}_1)$$

and

$$\hat{y}(0,2) = 41.35 - 2.12(x_2 - \bar{x}_2)$$

If we assume that the variability in y that is associated with x_2 is uniquely different from that associated with x_1, the $-2.12(x_2 - \bar{x}_2)$ component of the second equation can be added to the first equation to form a prediction equation. However, the simple addition usually doesn't work with measured x variables. Later in this unit the two-variable fitted model will be calculated as

$$\hat{y}(0,1,2) = 41.35 + 0.00069(x_1 - \bar{x}_1) - 2.13(x_2 - \bar{x}_2)$$

The important observation is that the estimated regression coefficients, except for $\bar{y}$, have changed when jointly fitted. The change in b_2 is minor, but the change in b_1 is large and is dramatized by the change in sign from negative to positive. Why the difference between the jointly fitted and individually fitted regression coefficients? To answer that question, we first consider the details of the calculations in model fitting with two explanatory variables.

FITTING A TWO-VARIABLE MODEL WITH ABDO

A reasonable general model is

$$y_i = \mu x_{i0} + \beta_1(x_{i1} - \bar{x}_1) + \beta_2(x_{i2} - \bar{x}_2) + \varepsilon_i = \beta_0 x_{i0} + \beta_1 x_{i1} + \beta_2 x_{i2} + \varepsilon_i$$

From Unit 5 we know that a regression model can be written in terms of the mean and slope or in terms of the intercept and slope. This point can be generalized to more than one independent variable. For two variables, then, $\beta_0 = \mu - \beta_1 \bar{x}_1 - \beta_2 \bar{x}_2$.

Before fitting the general model by ABDO, we can gain some efficiencies in calculations by deciding on the sequence of any candidate reduced models. Suppose we had no previous experience with the type of data. A priori, no particular sequence of candidate models would be favored. Consequently, we would fit the two-variable model as written, realizing that de facto we are examining with ABDO the sequence of

$$y = \mu x_0 + \varepsilon$$
$$y = \beta_0 x_0 + \beta_1 x_1 + \varepsilon$$
$$y = \beta_0 x_0 + \beta_1 x_1 + \beta_2 x_2 + \varepsilon$$

This model sequence will be denoted as

$$x_0|x_0x_1|x_0x_1x_2$$

which is to be read as fitting the mean model first, followed by a one-variable model with x_1, and finally the two-variable model. If we follow a convention of not repeating a variable that is already included in the sequence (a redundant variable), we list the sequence as

$$x_0|x_1|x_2$$

The advantage of this convention is the one-to-one correspondence with the ABDO procedure, assuming that we want each listed variable in the subsequent models in the sequence. For example, x_0 will be retained in both the one- and two-variable models since we want to retain an intercept.

Adopting this convention for the firefly data, we initiate the data analysis with a calculation of the sum of squares and cross products of $X'X|X'Y$ directly from the sequence of candidate models by adding the y variable at the end of the sequence:

$$X'X|X'Y = [x_0|x_1|x_2]'[x_0|x_1|x_2|y]$$

$$= \begin{bmatrix} \Sigma x_0^2 & \Sigma x_0 x_1 & \Sigma x_0 x_2 & \Sigma x_0 y \\ & \Sigma x_1^2 & \Sigma x_1 x_2 & \Sigma x_1 y \\ & & \Sigma x_2^2 & \Sigma x_2 y \end{bmatrix}$$

$$= \begin{bmatrix} 17 & 1{,}244 & 400.9 & 703 \\ & 109{,}328 & 30{,}228.9 & 49{,}557. \\ & & 9{,}555.65 & 16{,}363.1 \end{bmatrix}$$

Then stage 0 is easily carried out.

Step 1:   17   1244   400.9   703

Step 2:   1   73.1765   23.5824 $\boxed{41.3529 = \bar{y}}$

First think about the fate of the y observations in fitting the mean model during stage 0. The mean $\bar{y} = 41.35$ is calculated, the estimated model $\hat{y}(0)$ at this stage is $\bar{y}$, and the response residuals at the end of stage 0 are $y - \hat{y}(0) = y - \bar{y}$. Geometrically, the fitted model is now a plane in the three-dimensional representation of Figure 10.1, where x_1 and x_2 are the horizontal axes and y is the vertical axis. Fitting the Mean Model, $y = \mu x_0 + \varepsilon$, gives a plane without any slope at a height of $\bar{y}$ above the (x_1, x_2) plane. The distance between the observations above and below the plane are the $y - \hat{y}(0)$ residuals.

During stage 0 we are also calculating $\bar{x}_1 = 73.18$ and $\bar{x}_2 = 23.58$. The subsequent "residuals" are $x_1 - \hat{x}_1(0) = x_1 - \bar{x}_1$ and $x_2 - \hat{x}_2(0) = x_2 - \bar{x}_2$.

Fitting a Two-Variable Model with ABDO

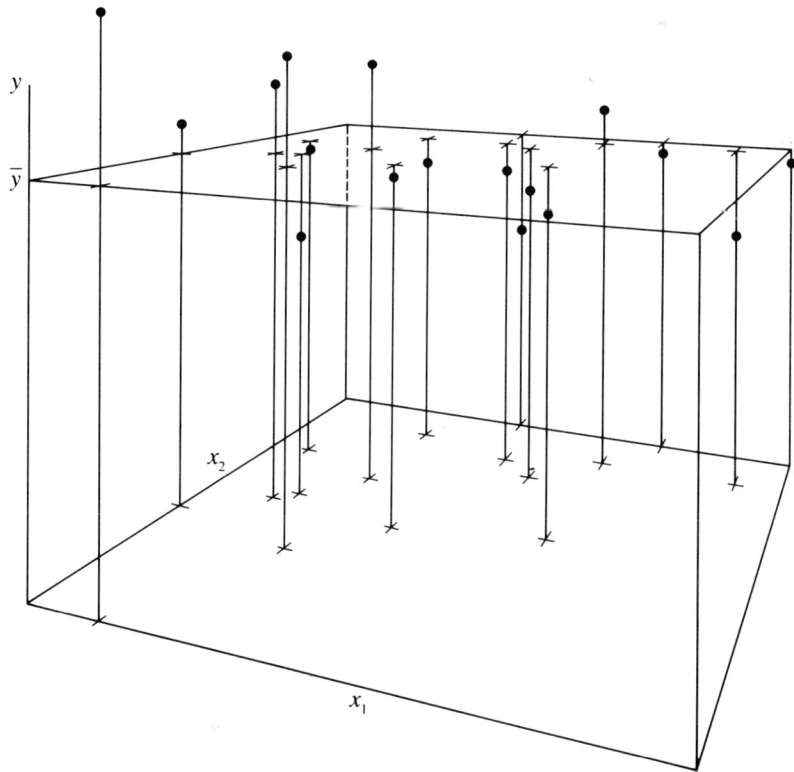

Figure 10.1 Fitted Model of the Firefly Data After Stage 0 of $x_0|x_1|x_2$ Sequence

As we have seen in Units 5 and 7, stage 1 starts with the uncorrected sum of squares for x_1, equal to 109,328, and the uncorrected sums of cross products, $\Sigma x_1 x_2 = 30,228.9$ and $\Sigma x_1 y = 49,557$, and then subtracts the correction factors from stage 0. The three calculations are

$$\Sigma(x_1 - \bar{x}_1)^2 = 109,328 - (73.1765)(1,244) = 18,296.5$$
$$\Sigma(x_1 - \bar{x}_1)x^2 = 30,228.9 - (73.1765)(400.9) = 892.453$$
$$\Sigma(x_1 - \bar{x}_1)y = 49,557 - (73.1765)(703) = -1,886.06$$

Step 3: 18,296.5   892.453 $-1,886.06$

Step 4: 1   0.0487773 $-0.103083 = b_{1 \cdot 0}$

The notation $b_{1 \cdot 0}$ is used now to emphasize that stage 1 is an intermediate stage in the $x_0|x_1|x_2$ sequence of calculations.

Using the sequential coefficients $\bar{y}$ and $b_{1\cdot 0}$, we obtain the following fitted model after stage 1:

$$\hat{y}(0,1) = \bar{y}x_0 + b_{1\cdot 0}(x_1 - \bar{x}_1) = 41.35x_0 - 0.10(x_1 - \bar{x}_1)$$

The flat plane in Figure 10.1 now has a slope in the x_1 direction, as shown in Figure 10.2. The distances between the y observations and the sloping plane are the residuals at the end of stage 1, $y - \hat{y}(0,1)$.

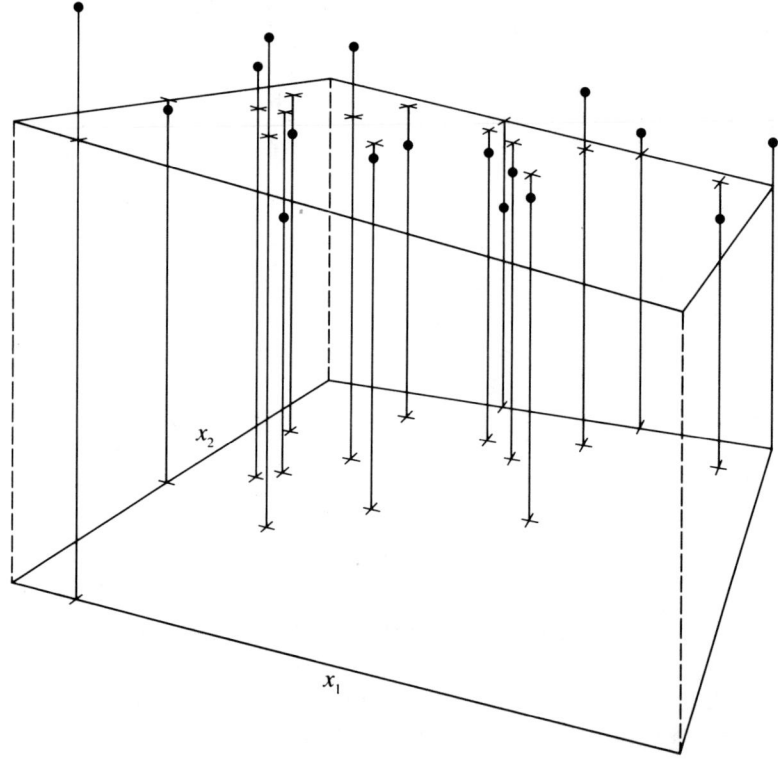

Figure 10.2 Fitted Model of the Firefly Data After Stage 1 of $x_0|x_1|x_2$ Sequence

The emphasis now switches to the fate of the x_2 values. An algebraic representation of the first two stages of ABDO will be helpful.

Stage 0

Step 1: Σx_0^2	$\Sigma x_0 x_1$	$\Sigma x_0 x_2$	$\Sigma x_0 y$
Step 2: 1	$\bar{x}_1$	$\bar{x}_2$	$\bar{y}$

Stage 1:

Step 3: $\quad \Sigma x_1^2 - \bar{x}_1 \Sigma x_1 \qquad \Sigma x_1 x_2 - \bar{x}_1 \Sigma x_2 \qquad \Sigma x_1 y - \bar{x}_1 \Sigma y$

$\qquad\qquad\quad = \Sigma(x_1 - \bar{x}_1)^2 \qquad = \Sigma(x_1 - \bar{x}_1)x_2 \qquad = \Sigma(x_1 - \bar{x}_1)y$

Step 4: $\quad 1 \qquad\qquad\qquad \dfrac{\Sigma(x_1 - \bar{x}_1)x_2}{\Sigma(x_1 - \bar{x}_1)^2} \qquad \dfrac{\Sigma(x_1 - \bar{x}_1)y}{\Sigma(x_1 - \bar{x}_1)^2} = b_{1 \cdot 0}$

Note that the calculations involving x_2 have no influence on the last column (y) of calculations during those first two stages. The value -0.10 is the slope of the line by regressing y on $x_1 - \bar{x}_1$. Also note that the third column (x_2) of calculations is identical to the last column (y) except that y has been replaced by x_2. Momentarily envision x_2 as the last variable. During stage 0 the mean $\bar{x}_1$ is calculated, and the x_1 "residuals" at the end of stage 0 are $x_1 - \hat{x}_1(0) = x_1 - \bar{x}_1$. Stage 1 then solves for the slope of the line from regressing x_2 on $x_1 - \bar{x}_1$. At the end of stage 1 ABDO is working with a new set of "residuals," namely, $x_2 - \hat{x}_2(0, 1)$. The stage 2 estimator is then calculated by regressing y on $x_2 - \hat{x}_2(0, 1)$, that is, by dividing the sum of cross products of the adjusted x_2 and y by the sum of squares of the $x_2 - \hat{x}_2(0, 1)$ residuals. The two necessary calculations for stage 2 are

$$\Sigma[x_2 - \hat{x}_2(0, 1)]^2 = 9555.65 - (23.5824)(400.9) - (0.0487773)(892.453)$$
$$= 57.9352$$
$$\Sigma[x_2 - \hat{x}_2(0, 1)]y = 16{,}363.1 - (23.5824)(703) - (0.0487773)(-1{,}886.06)$$
$$= -123.297$$

Step 5: $\qquad 57.9532 \qquad -123.297$

Step 6: $\qquad 1 \qquad\qquad\quad \boxed{-2.12753 = b_{2 \cdot 0, 1}}$

A definite pattern in the ABDO flow of calculations is now evident. One estimated sequential regression coefficient can be obtained at each stage:

Stage	Estimated Sequential Regression Coefficient	Location
0	$\bar{y}$ (regression on x_0)	Last element of step 2
1	$b_{1 \cdot 0}$ [regression on $x_1 - \hat{x}_1(0)$]	Last element of step 4
2	$b_{2 \cdot 0, 1}$ [regression on $x_2 - \hat{x}_2(0, 1)$]	Last element of step 6

Using these estimators, we can write a prediction equation for the two-variable model:

$$\hat{y}(0, 1, 2) = \bar{y}x_0 + b_{1 \cdot 0}[x_1 - \hat{x}_1(0)] + b_{2 \cdot 0, 1}[x_2 - \hat{x}_2(0, 1)]$$
$$= 41.35x_0 - 0.10(x_1 - \bar{x}_1) - 2.13[x_2 - \hat{x}_2(0, 1)]$$

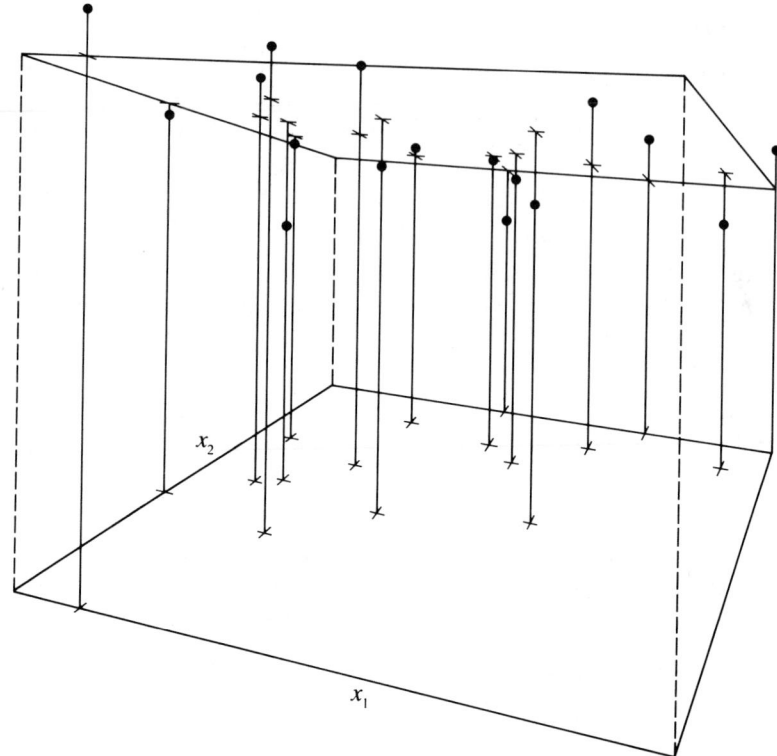

Figure 10.3 Fitted Model of the Firefly Data After Stage 2 of the $x_0|x_1|x_2$ Sequence

Predicted values from this fitted model are shown on the plane in Figure 10.3. This plane has an additional slope in the x_2 direction. Now the distances between the y observations and the tilted plane are the residuals at the end of stage 2, $y - \hat{y}(0, 1, 2)$.

INTERPRETATION OF SEQUENTIAL REGRESSION COEFFICIENTS

The crucial point in interpreting estimated sequential regression coefficients is to ascertain the exact nature of the x variable. The x_2 variable with $b_{2 \cdot 0, 1} = -2.13$ is not actually x_2 but an adjusted x_2, specifically, $x_2 - \hat{x}_2(0, 1)$. As we have just seen, $\hat{x}_2(0, 1)$ is the regression of x_2 on $x_1 - \bar{x}_1$. An important ramification is that the sequential $b_{2 \cdot 0, 1}$ depends on the association between x_1 and x_2. For the fireflies we want to relate the variability in the time of first flash with temperature (x_2). However, temperature is also associated with light intensity (x_1). Accordingly, $b_{2 \cdot 0,1}$ in stage 2 measures the relationship between time of first flash and that part of the temperature variable that is not associated with light intensity. In stage 1 the $b_{1 \cdot 0}$ coefficient estimates the relationship between time of first flash and light intensity without consideration of, or ignoring, temperature. Consequently, $b_{1 \cdot 0}$ would be explaining variability in time of first flash either due to light intensity directly or due to temperature through the association with light intensity.

The $\hat{y}(0,1,2)$ fitted model has been calculated from the sequential stages of ABDO, where the x variables are being corrected for previous stages. Consequently, the fitted model is not written in terms of x_0, x_1, and x_2 but follows the pattern shown below.

Stage	Regression Variables	Estimates
0	x_0	$\bar{y} = 41.35$
1	$x_1 - \hat{x}_1(0) = x_1 - \bar{x}_1 = x_1 - 73.18$	$b_{1 \cdot 0} = -0.10$
2	$x_2 - \hat{x}_2(0,1) = x_2 - [23.58 + 0.048(x_1 - 73.18)]$	$b_{2 \cdot 0,1} = -2.13$

Remember that $\hat{x}_2(0,1)$ is the fitted Mean and Slope Model by regressing x_2 on $(x_1 - \bar{x}_1)$. The mean of x_2 is 23.58 (step 2), the slope is 0.048 (step 4), and the mean of x_1 is 73.18 (step 2). The fitted model can then be written as

$$\hat{y}(0,1,2) = 41.35 x_0 - 0.10(x_1 - 73.18)$$

$$-2.13\{x_2 - [23.58 + 0.048(x_1 - 73.18)]\}$$

or in terms of x_0, $x_1 - \bar{x}_1$, and $x_2 - \bar{x}_2$ as

$$\hat{y}(0,1,2) = 41.35 + 0.00069(x_1 - 73.18) - 2.13(x_2 - 23.58)$$

or in terms of x_0, x_1, and x_2 as

$$\hat{y}(0,1,2) = 91.47 + 0.00069 x_1 - 2.13 x_2$$

Since the last two fitted models were calculated directly from the fitted model in terms of x_0, $x_1 - \hat{x}_1(0)$, and $x_2 - \hat{x}_2(0,1)$, the three forms are algebraically equivalent and any can be used as the equation for the fitted plane in Figure 10.3. The coefficients in the last equation are called partial regression coefficients.

INTERPRETATION OF PARTIAL REGRESSION COEFFICIENTS

The sequentially fitted b values have straightforward interpretations but have the limitation of depending on the order of the x variables in the ABDO calculations. The curious might conjecture about the effect on the estimated coefficients if the model sequence used in ABDO had been $x_0|x_2|x_1$. Carrying out the calculations, we have

$$X'X|X'Y = [x_0|x_2|x_1]'[x_0|x_2|x_1|y]$$

$$= \begin{bmatrix} 17.0000 & 400.900 & 1{,}244.00 & 703.000 \\ & 9{,}555.65 & 30{,}228.9 & 16{,}363.1 \\ & & 109{,}328. & 49{,}557.0 \end{bmatrix}$$

Step 1:	17	400.900	1,244.00	703.000	
Step 2:	1	23.5824	73.1765	41.3529	$= \bar{y}$
Step 3:		101.485	892.453	−215.294	
Step 4:		1	8.79396	−2.12144	$= b_{2 \cdot 0}$
Step 5:			10,448.3	7.23009	
Step 6:			1	0.00069198	$= b_{1 \cdot 0,2}$

From the even-numbered steps of the right-hand column, the sequential regression coefficients are read as follows:

Stages	Regression Variables	Estimates
0	x_0	$\bar{y} = 41.35$
1	$x_2 - \hat{x}_2(0)$	$b_{2 \cdot 0} = -2.12$
2	$x_1 - \hat{x}_1(0,2)$	$b_{1 \cdot 0,2} = 0.00069$

The value of -2.12 for $b_{2 \cdot 0}$ checks with the value reported earlier in the unit when fitting a Mean and Slope Model by regressing y on $x_2 - \bar{x}_2$. Of more interest here is the value 0.00069, which is the same value as that for the b_1 coefficient in the $\hat{y}(0,1,2)$ fitted model in terms of x_0, x_1, and x_2,

$$\hat{y}(0,1,2) = 91.47 + 0.00069 x_1 - 2.13 x_2$$

Now we might conjecture that the 91.47 value could be calculated from ABDO by the sequence of models $x_1|x_2|x_0$. The details are not given here but the conjecture is correct, and $b_{0 \cdot 1,2} = 91.47$ would then be the last element in step 6. The interpretation of $b_{0 \cdot 1,2} = 91.47$, $b_{1 \cdot 0,2} = 0.00069$, and

$b_{2 \cdot 0, 1} = -2.13$, commonly called the *estimated partial regression coefficients*, is now clear. The partials will be denoted simply as b_0, b_1, and b_2. They are indeed the slopes by regressing y on each x variable, where the x variable is corrected for any association with all the other variables.

In practice we need to do only one ABDO to calculate the three partial coefficients. Remember from Unit 5 that we could calculate the intercept b_0 in the one-variable equation by working backward in the ABDO calculations. With two variables the last coefficient, b_2, is read from step 6, the second is calculated from step 4, and the third from step 2. For example, in the ABDO with the sequence $x_0 | x_1 | x_2$, we have calculated

Step 2:	1	73.1765	23.5824	41.3529 $= \bar{y}$
Step 4:		1	0.0487773	$-0.103083 = b_{1 \cdot 0}$
Step 6:			1	$-2.12753 = b_{2 \cdot 0, 1}$

Then we have the following:

Step 6 is read: $\qquad\qquad b_2 = -2.13$

Step 4 is read: $\qquad\qquad b_1 + b_2(0.0487773) = -0.103083$
$\qquad\qquad\qquad\qquad\qquad b_1 = 0.00069$

Step 2 is read: $\qquad b_0 + b_1(73.1765) + b_2(23.5824) = 41.3529$
$\qquad\qquad\qquad\qquad\qquad b_0 = 91.47$

SUMMARY

This unit has emphasized the interpretation of the estimated sequential and partial regression coefficients. Geometrically, 41.35 is the mean or average height of the plane in Figure 10.1, -0.10 is the slope of the plane in Figure 10.2 where the fitted model includes the x_0 and $x_1 - \bar{x}_1$ variables, and -2.13 is the change in y per unit change in the x_2 direction for the plane in Figure 10.3 where the fitted model includes the x_0, $x_1 - \hat{x}_1(0)$, and $x_2 - \hat{x}_2(0, 1)$ variables. The sequence of figures follows the sequence of stages in ABDO from which the sequential fitted model was calculated. At each stage a regression coefficient is fitted by regressing y on an entering x variable, corrected or adjusted for any association with previously entered x variables. Consequently, each sequential coefficient is the straight-line fit between y and the remaining part of the entering x variable that has not been computationally usurped by previously entered x variables.

An alternative fitted model can also be calculated from ABDO. Nonsequential in spirit, the estimated coefficients are called partials and are calculated as though each x variable were adjusted for association with all the other x variables.

The fitted model can be expressed either in terms of the sequential coefficients,

$$\hat{y}(0,1,2) = 41.35x_0 - 0.10(x_1 - \bar{x}_1) - 2.13[x_2 - \hat{x}_2(0,1)]$$

or in terms of the partial regression coefficients,

$$\hat{y}(0,1,2) = 91.47x_0 + 0.00069x_1 - 2.13x_2$$

The first equation has the advantage of indicating the exact nature of the x variable associated with each estimated coefficient. The form commonly used in reporting experimental results is the second equation. The partial regression coefficients do not depend on the ordering of the x variables, and the estimates are part of any standard regression program output. Predicted values can be easily calculated in terms of the original x_1 and x_2 values.

The interpretation of the two kinds of estimated regression coefficients is summarized in Table 10.2.

Table 10.2 Interpretation of Estimated Regression Coefficients

	SEQUENTIAL		PARTIAL	
Notation	Regression		Notation	Regression
$\bar{y}$	y on x_0		b_0	y on $x_0 - \hat{x}_0(1,2)$
$b_{1 \cdot 0}$	y on $x_1 - \hat{x}_1(0)$		b_1	y on $x_1 - \hat{x}_1(0,2)$
$b_{2 \cdot 0,1}$	y on $x_2 - \hat{x}_2(0,1)$		b_2	y on $x_2 - \hat{x}_2(0,1)$

Unit 11

Data Reduction by Regression

Objectives

- To summarize experimental data by meaningful combinations of the observations.
- To express regression coefficients as weighted sums of observations.
- To estimate a quadratic polynomial model.

In previous units the ABDO algorithm has been used for calculating regression coefficients and sums of squares of a sequence of models. Although it hasn't been emphasized, ABDO is a methodology for summarizing or reducing the data set from n observations to a relatively few meaningful combinations of the original observations. Indeed, a regression coefficient is a weighted sum of the response data.

This unit is a transition between the previous units that emphasized data description by fitting a sequence of candidate models and future units that will generalize beyond the actual data set. One motivation for further insight into ABDO is its use in Unit 12, where it provides input to a new algorithm for calculating both estimates and their standard errors. However, it is a useful methodology in its own right, and the data reduction aspects will be emphasized now.

LINEAR COMBINATIONS OF THE OBSERVATIONS

Data reduction refers to the process of summarizing the observations into a few meaningful combinations like the sum (or mean) and other regression coefficients. The remaining variability in the n observations then goes into the residual term. Regression coefficients of models fitted in earlier units are weighted sums; that is, each observation in the sum is multiplied by a negative, zero, or positive weight. Formally defined, a weighted sum of the observations expressed as

$$\ell_1 y_1 + \ell_2 y_2 + \cdots + \ell_n y_n$$

where the ℓ_i are weights (constants), is called a *linear combination* of the y_i. A linear combination multiplied or divided by a number is also a linear combination:

$$\frac{3y_1 + 2y_2 - 2y_3}{5} = \frac{3}{5}y_1 + \frac{2}{5}y_2 - \frac{2}{5}y_3$$

If each $\ell_i = 1$, then the linear combination is the sum. Similarly, suppose that each $\ell_i = (x_{ij} - \hat{x}_{ij})$, where $\hat{x}_{ij}$ is the predicted value of x_{ij} using other x variables. For sequential regression coefficients $\ell_i = [x_{ij} - \hat{x}_{ij}(0, 1, \ldots, j-1)]$, and the linear combinations are the last elements in the odd rows of ABDO.

Each regression coefficient may then be expressed as $\Sigma \ell_i y_i / \Sigma \ell_i^2$, or

$$\left(\frac{1}{\Sigma \ell_i^2}\right)(\ell_1 y_1 + \ell_2 y_2 + \cdots + \ell_n y_n)$$

where $\ell_i = (x_{ij} - \hat{x}_{ij})$.

RULES FOR LINEAR COMBINATIONS

In order to establish a groundwork for data reduction methodology and ultimately for generalizing beyond the actual data set, we need to accept several rules for linear combinations of the observations. The rules presented here assume that the observations, the components of the response vector **Y**, denoted as $y_1, y_2, \ldots, y_n$, are uncorrelated with means $\mu_1, \mu_2, \ldots, \mu_n$ and variances $\sigma_1^2, \sigma_2^2, \ldots, \sigma_n^2$. Then the linear combination $\ell_1 y_1 + \ell_2 y_2 + \cdots + \ell_n y_n$ will have a frequency distribution with mean

$$\ell_1 \mu_1 + \ell_2 \mu_2 + \cdots + \ell_n \mu_n = \Sigma \ell_i \mu_i \qquad \text{(Mean Rule)}$$

and variance

$$\ell_1^2 \sigma_1^2 + \ell_2^2 \sigma_2^2 + \cdots + \ell_n^2 \sigma_n^2 = \Sigma \ell_i^2 \sigma_i^2$$
(Unequal Variance Rule)

When the variances are all equal, that is, $\sigma_1^2 = \sigma_2^2 = \cdots = \sigma_n^2 = \sigma^2$, the variance is

$$\left(\ell_1^2 + \ell_2^2 + \cdots + \ell_n^2\right)\sigma^2 = \left(\Sigma \ell_i^2\right)\sigma^2$$
(Equal Variance Rule)

When the variances are all equal, one linear combination, denoted as

$$m_1 y_1 + m_2 y_2 + \cdots + m_n y_n$$

is uncorrelated with another linear combination,

$$\ell_1 y_1 + \ell_2 y_2 + \cdots + \ell_n y_n$$

provided

$$\ell_1 m_1 + \ell_2 m_2 + \cdots + \ell_n m_n = 0 \quad \text{(Orthogonality Rule)}$$

These rules apply regardless of the distribution of the observations. However, if the y_i are normally distributed, then any linear combination of the y_i is also normally distributed (Normal Distribution Rule).

Pairs of linear combinations of the observations that follow the Orthogonality Rule are called *orthogonal linear combinations*. A good data reduction methodology gives a few orthogonal linear combinations containing the explainable part of the variability among the observations. For efficient computation and for an interpretation of the data that is free of confusing redundancies, each linear combination or summary statistic of the data set should be orthogonal to every other linear combination. Of even more importance, the availability of orthogonal linear combinations allows the use of our simple variance rules in establishing the variance properties of linear combinations of estimated regression coefficients in Unit 12. The calculation of standard errors is the foundation of developing inference methodology.

NUMERICAL ILLUSTRATION OF DATA REDUCTION USING ABDO

To illustrate numerically the data reduction properties of ABDO, we let

$$\mathbf{X} = \begin{bmatrix} 1 & 1 & 1 \\ 1 & 2 & 4 \\ 1 & 3 & 9 \\ 1 & 4 & 16 \\ 1 & 5 & 25 \end{bmatrix} \quad \text{and} \quad \mathbf{Y} = \begin{bmatrix} y_1 \\ y_2 \\ y_3 \\ y_4 \\ y_5 \end{bmatrix}$$

Numerical values are not used for **Y** so that the demonstration will apply to any values that might be observed. Starting with $\mathbf{X'X}|\mathbf{X'Y}$, the resulting ABDO calculations are as follows:

$$\begin{array}{ccc|l}
5 & 15 & 55 & y_1 + y_2 + y_3 + y_4 + y_5 \\
 & 55 & 225 & y_1 + 2y_2 + 3y_3 + 4y_4 + 5y_5 \\
 & & 979 & y_1 + 4y_2 + 9y_3 + 16y_4 + 25y_5 \\
\hline
5 & 15 & 55 & y_1 + y_2 + y_3 + y_4 + y_5 \quad = \Sigma x_0 y \\
1 & 3 & 11 & (y_1 + y_2 + y_3 + y_4 + y_5)/5 = \bar{y} \\
\hline
 & 10 & 60 & -2y_1 - y_2 \quad\quad\quad + y_4 + 2y_5 \quad = \Sigma(x_1 - \bar{x}_1)y \\
 & 1 & 6 & (-2y_1 - y_2 \quad\quad\quad + y_4 + 2y_5)/10 = b_{1 \cdot 0} \\
\hline
 & & 14 & 2y_1 - y_2 - 2y_3 \quad - y_4 + 2y_5 \quad = \Sigma[x_2 - \hat{x}_2(0,1)]y \\
 & & 1 & (2y_1 - y_2 - 2y_3 \quad - y_4 + 2y_5)/14 = b_{2 \cdot 01}
\end{array}$$

Reinforced here is that each sequential regression coefficient is simply a linear combination of the observations, $\Sigma \ell_i y_i$, divided by $\Sigma \ell_i^2$. Specifically, the jth coefficient ($j = 0, 1, 2$) can be written as

$$b_j = \frac{1}{\Sigma(x_{ij} - \hat{x}_{ij})^2} \left[(x_{1j} - \hat{x}_{1j}) y_1 + (x_{2j} - \hat{x}_{2j}) y_2 + \cdots + (x_{nj} - \hat{x}_{nj}) y_n \right]$$

where $\hat{x}_{ij}$ is $\hat{x}_{ij}(0, 1, \ldots, j-1)$. In the numerical illustration the negative, zero, and positive weights, the $\ell_i = (x_{ij} - \hat{x}_{ij})$ of the three orthogonal linear combinations, are the coefficients of the observations as shown by the right-hand elements of the odd rows of the ABDO calculations.

			Coefficient (ℓ_i)			
Variable	$i=1$	$i=2$	$i=3$	$i=4$	$i=5$	$\Sigma(x_{ij} - \hat{x}_{ij})^2$
x_0	1	1	1	1	1	5
$x_1 - \bar{x}_1$	-2	-1	0	1	2	10
$x_2 - \hat{x}_2(0, 1)$	2	-1	-2	-1	2	14

The **X** matrix used in the numerical illustration is an example of fitting a quadratic polynomial model. This commonly occurring data analysis problem is now developed in more detail.

QUADRATIC POLYNOMIAL MODEL

When the relationship between a response variable and an input or explanatory variable is known to be curvilinear, a plot of the data indicates that a fitted straight line will not be adequate. In addition, the residual plot from the straight-line model with one x variable will be U-shaped. One approach to solving this problem is to change the scale of the x variable, as shown in Unit 8, and to fit a slope of the reexpressed x variable. If the original scale is to be retained, then the curvilinear relationship needs to be modeled.

Suppose we could divide the total range of the x variable into several parts and calculate a straight line within each part. If the slopes are smoothly changing across the different parts of x, it is said that y is increasing (decreasing) at an increasing (decreasing) rate. The necessary curvature needed to describe the data can be summarized in β_2 of the quadratic polynomial model written as a regression model with two independent variables,

$$y = \beta_0 x_0 + \beta_1 x_1 + \beta_2 x_2 + \varepsilon$$

where $x_2 = x_1^2$. Classic examples include fertilizer and nutrition experiments. For example, within a certain range of applied nitrogen fertilizer, crop yield is expected to increase at a decreasing rate, reach a maximum yield, and then decrease. Polynomials are also used in studies where the levels of x are not controlled, as they are in a designed experiment.

Suppose we had measured the response of y to five levels of x. One example would be the **X** matrix of the previous section, where x_1 includes the five actual or coded levels and $x_2 = x_1^2$.

$$X = \begin{bmatrix} 1 & 1 & 1 \\ 1 & 2 & 4 \\ 1 & 3 & 9 \\ 1 & 4 & 16 \\ 1 & 5 & 25 \end{bmatrix} \quad \text{and} \quad Y = \begin{bmatrix} y_1 \\ y_2 \\ y_3 \\ y_4 \\ y_5 \end{bmatrix}$$

In the previous section these five independent observations were reduced to three independent linear combinations of the observations with ABDO. As an alternative to the ABDO calculations, we could find the coefficients of the linear combinations directly as the first step of the data analysis. This alternative approach would be especially attractive if we had several studies all having the same values for the x variables. To eliminate going through all the ABDO calculations for each data analysis, we construct a matrix **L** in the following manner:

1. Write the first column of **L** as the first column of **X**.
2. Regress the second column of **X** on the first column of **L**. The residual column from this regression becomes the second column of **L**.
3. Regress the third column of **X** on the first two columns of **L**. The residual column from this regression becomes the third column of **L**.

Application of these steps yields a matrix of sequential x variable residuals:

$$L = \begin{bmatrix} 1 & -2 & 2 \\ 1 & -1 & -1 \\ 1 & 0 & -2 \\ 1 & 1 & -1 \\ 1 & 2 & 2 \end{bmatrix}$$

If the sum of cross products between the elements of two columns in a matrix is equal to zero, the columns are said to be orthogonal. Notice that each pair of columns of **L** is orthogonal. To verify that the second and third columns are orthogonal, we have

$$(-2)(2) + (-1)(-1) + (0)(-2) + (1)(-1) + (2)(2) = 0$$

The columns in **L** are recognized as the coefficients of the three orthogonal linear combinations found by ABDO. Calculating **L** is an alternative to ABDO and is clearly doing the same job as ABDO in solving for the $x_{ij} - \hat{x}_{ij}$ residuals or coefficients. These residual columns, sequentially calculated and listed in the matrix **L**, show the actual fate of the original columns in **X** during the stages of the ABDO calculations. For example, after stage 0 of ABDO the x_1 values have been changed to

$x_1 - \hat{x}_1(0) = x_1 - \bar{x}_1$. After stages 0 and 1 the x_2 values have been changed to $x_2 - \hat{x}_2(0, 1)$.

The same three linear combinations, found earlier with ABDO, can also be found by multiplying **L′** by **Y**:

$$\mathbf{L'Y} = \begin{bmatrix} 1 & 1 & 1 & 1 & 1 \\ -2 & -1 & 0 & 1 & 2 \\ 2 & -1 & -2 & -1 & 2 \end{bmatrix} \begin{bmatrix} y_1 \\ y_2 \\ y_3 \\ y_4 \\ y_5 \end{bmatrix} = \begin{bmatrix} y_1 + y_2 + y_3 + y_4 + y_5 \\ -2y_1 - y_2 + 0 + y_4 + 2y_5 \\ 2y_1 - y_2 - 2y_3 - y_4 + 2y_5 \end{bmatrix}$$

Not surprisingly, the first linear combination is the sum. If one had to choose one linear combination of the data as the most revealing single number, one would usually choose the sum (or mean). The second linear combination says to contrast the responses associated with the highest levels with those of the lowest levels; specifically, give twice the weight to the extremes and zero weight to the middle. From our knowledge of straight-line regression in Unit 5, we can view this number as an indication of the nonzero slope of the relationship between y and x. In the third combination the extreme values are contrasted with the middle three values, a measure of the curvature or lack of fit to a straight-line slope.

Having replaced the **X** matrix of our regression format by **L**, we can calculate **L′L** easily since the columns of **L** are orthogonal. The matrix **L′L** has the sum of squares of the coefficients of each column of **L** in the main diagonal (5, 10, and 14 for our example) and zeros elsewhere. The estimated sequential regression coefficients are easily calculated from **L′L|L′Y**, since each element in **L′Y** (each linear combination of the observations) is divided by the corresponding sum of squares of the coefficients from **L′L**. For our example we have

$$\bar{y} = \frac{y_1 + y_2 + y_3 + y_4 + y_5}{1^2 + 1^2 + 1^2 + 1^2 + 1^2} = 0.2 y_1 + 0.2 y_2 + 0.2 y_3 + 0.2 y_4 + 0.2 y_5$$

$$b_1 = \frac{-2y_1 - y_1 + 0 + y_4 + 2y_5}{(-2)^2 + (-1)^2 + 0 + 1^2 + 2^2} = -0.2 y_1 - 0.1 y_2 + 0 y_3 + 0.1 y_4 + 0.2 y_5$$

$$b_2 = \frac{2y_1 - y_2 - 2y_3 - y_4 + 2y_5}{2^2 + (-1)^2 + (-2)^2 + (-1)^2 + 2^2}$$

$$= 0.143 y_1 - 0.071 y_2 - 0.143 y_3 - 0.071 y_4 + 0.143 y_5$$

The values in the **X** matrix can be called polynomial coefficients, while the values in **L** are commonly called orthogonal polynomial coefficients and are tabled for equally spaced levels of the x variable. The table values usually given in statistical methodology books can be used when each level of the x variable is equally replicated, but the methodology of solving for **L** can be used for unequally spaced or unequally replicated studies. Here **L** has been used with polynomial problems. This orthogonalizing computing procedure can be generalized to any number of columns in **X** and for data analysis problems other than polynomials. Henceforth the procedure for calculating **L** will be called ORTHO. In routine multivariable regression calculations we would not use the ORTHO procedure unless we had a natural or conjectured ordering of the x variables, such as we had for the polynomial example. Then we are equally at ease using the original columns in the **X** matrix or the orthogonalized columns of the **L** matrix. Calculation of **L** and subsequent statistics can be tedious but easily done with a hand or desk calculator. The ORTHO procedure will also be helpful in later units on the analysis of sample means.

SOYMILK DATA

Data from an experiment measuring the effect of boiling time (hours) on the protein content of soymilk (grams per 100 milliliters) are shown in Table 11.1 and Figure 11.1. In experimental design terminology each boiling time is a treatment and each treatment was replicated three times in a completely randomized design. The objective of the experiment is to estimate the functional relationship between protein content of soymilk and boiling time, which is known to be curvilinear. Therefore a quadratic polynomial model should reasonably approximate the relationship. Consequently a quadratic polynomial model sequence, $x_0|x_1|x_2$, where x_1 is boiling time (hours) and x_2 is x_1^2, was formulated for the data analysis. As usual, we will start with **X'X**|**X'Y** and work through the ABDO calculations of Table 11.2.

Table 11.1 Protein Content of Soymilk (g/100 ml) Versus Boiling Time (h)

	Boiling Time			
Replication	0	0.2	0.5	1.0
1	2.74	3.14	3.44	3.68
2	2.25	2.68	3.53	3.75
3	2.34	2.83	3.63	3.51
Mean	2.44	2.88	3.53	3.65

Source: E. E. Escueta, "Effect of Boiling Treatment and Gata (Coconut Cream) Addition to Soymilk on the Chemical, Rheological and Sensory Properties of Tofu (Soybean Curd)" (Ph.D. thesis, Cornell University, 1979).

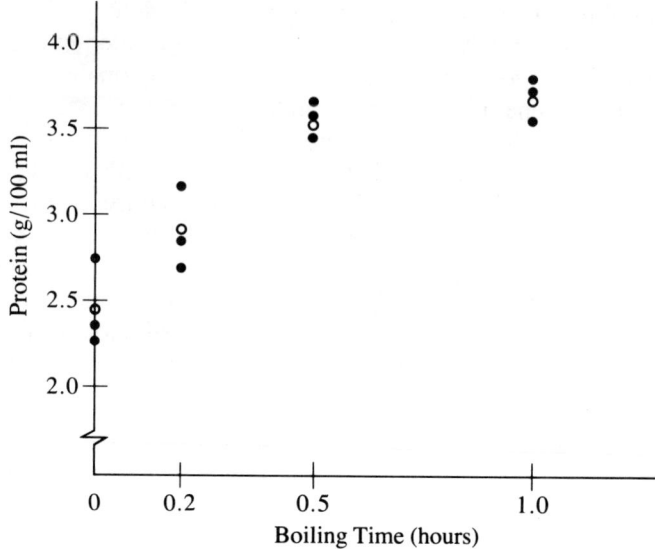

Figure 11.1 Soymilk Data Plot of Protein Content Versus Boiling Time
(0 indicates mean response)

Table 11.2 ABDO Calculations for Fitting the Quadratic Polynomial Model with the Soymilk Data

12	5.1	3.87	37.52
	3.87	3.399	17.97
		3.1923	13.936
12	5.1	3.87	37.52
	0.425	0.3225	3.1267
	1.7025	1.75425	2.024
	1	1.030396	1.18884
		0.136653	−0.249722
		1	−1.82742

The model sum of squares is 120.17656 with three degrees of freedom, leaving a residual sum of squares of 120.493−120.17656=0.31644 with nine degrees of freedom. However, we know that the experimental error of a completely randomized design with four treatments and three replications is associated with eight degrees of freedom. With four levels of the boiling time variable we realize that a cubic polynomial could have been fitted. If indeed we had added a cubic term, with $x_3 = x_1^3$, the resulting prediction equation would be

$$\hat{y}(0,1,2,3) = 2.44333x_0 + 1.9775x_1 + 1.58417x_2 - 2.35833x_3$$

The predicted values for any one of the boiling times will be the same, of course, for each replication. In addition, the predicted value for each boiling time is the same as the treatment mean of the three replications. Thus we have

Boiling time	$\hat{y}(0,1,2,3)$	Treatment Mean
0	2.44333	=(2.74+2.25+2.34)/3
0.2	2.88333	=(3.14+2.68+2.83)/3
0.5	3.53333	=(3.44+3.53+3.63)/3
1	3.64667	=(3.68+3.75+3.51)/3

Consequently, we see that the cubic polynomial curve will connect all four treatment means in Figure 11.1, and each $y - \hat{y}(0,1,2,3)$ residual is the difference between a replicate value and a treatment mean. Then the experimental error sum of squares is

$$\Sigma[y - \hat{y}(0,1,2,3)]^2 = 0.294667$$

Previously we had fitted the quadratic polynomial with a residual sum of squares of $0.31644 = \Sigma[y - \hat{y}(0,1,2)]^2$. The difference between the two model fits is the failure of the quadratic polynomial to fit the treatment means.

With the addition of x_3 to the model sequence, the ANOVA table in Table 11.3 can be constructed reflecting the sequence of candidate models,

Table 11.3 ANOVA Table for a Candidate Model Sequence for the Soymilk Data

Source	Degrees of Freedom	Sum of Squares	Mean Square
$R(0)$	1	117.314	
$R(1\|0)$	1	2.40621	
$R(2\|0,1)$	1	0.45635	
Residual (1)	9	0.31644	0.03516
$R(3\|0,1,2)$	1	0.02177	
Residual (2)	8	0.29467	0.03683

specifically, the quadratic polynomial followed by the general model, the cubic polynomial. As demonstrated by the closeness of the residual mean squares, the cubic polynomial is not needed. Formally, the test for evaluating the cubic is $F=0.02177/0.03683$, with one and eight degrees of freedom. Hence the quadratic polynomial is adequate to describe the relationship between the response variable and boiling time. The sum of squares associated with the difference between the reduced model and the general model is sometimes called *lack of fit*. In our example the general model is the cubic polynomial, so the lack of fit is associated with only one degree of freedom.

In Units 15 through 19 sample means will be analyzed with emphasis on evaluating single-degree-of-freedom comparisons among the means. When the treatments are levels of a quantitative (continuous) factor, the desired comparisons need to reflect the functional relationship between the response variable and the quantitative factor. These comparisons can be calculated with the use of orthogonal polynomial coefficients. As discussed in the previous section of this unit, these coefficients are the columns in the **L** matrix. For the soymilk data the coefficients cannot be found in easily available tables of orthogonal polynomial coefficients since the levels of boiling time are not equally spaced. For a quadratic polynomial the **L** matrix will include x_0, $x_1 - \hat{x}_1(0)$, and $x_2 - \hat{x}_2(0,1)$. The ORTHO procedure of the previous section operates on the columns of **X**, so we can utilize our ABDO calculations from Table 11.2. Then

$$x_1 - \hat{x}_1(0) = x_1 - \bar{x}_1 = x_1 - 0.425$$
$$x_2 - \hat{x}_2(0,1) = x_2 - [0.3225 + 1.0304(x_1 - 0.425)]$$

The resulting **L** matrix is

x_0	$x_1 - \bar{x}_1$	$x_2 - \hat{x}_2(0,1)$
1	−0.425	0.11542
1	−0.425	0.11542
1	−0.425	0.11542
1	−0.225	−0.05066
1	−0.225	−0.05066
1	−0.225	−0.05066
1	0.075	−0.14978
1	0.075	−0.14978
1	0.075	−0.14978
1	0.575	0.08502
1	0.575	0.08502
1	0.575	0.08502

Now each pair of columns in **L** is orthogonal, and the four values in each of the last two columns are called the *orthogonal polynomial coefficients*. If we calculate **L'L|LY** and solve for the regression coefficients, we will find 3.1267, 1.18884, and -1.82742, the same values as given for the estimated sequential coefficients in Table 11.2.

MEAN AND VARIANCE OF LINEAR COMBINATIONS IN REGRESSION

This last section is a transition to the inferential material of Unit 12. The polynomial example has shown the value of summarizing n data points in a relatively few linear combinations, the elements in **L'Y** or the right-hand elements in the odd rows of ABDO. In generalizing beyond the data in Unit 12, we will need to know the mean and variance of each of the summary linear combinations in the intermediate stages of the regression calculations. For the quadratic polynomial example we have

$$\mathbf{L'Y} = \begin{bmatrix} \Sigma x_0 y \\ \Sigma (x_1 - \bar{x}_1) y \\ \Sigma [x_2 - \hat{x}_2(0,1)] y \end{bmatrix} = \begin{bmatrix} 1 & 1 & 1 & 1 & 1 \\ -2 & -1 & 0 & 1 & 2 \\ 2 & -1 & -2 & -1 & 2 \end{bmatrix} \begin{bmatrix} y_1 \\ y_2 \\ y_3 \\ y_4 \\ y_5 \end{bmatrix}$$

Statistically, the y_i are random variables. The three linear combinations are also random variables, since they are combinations of the original five. Being random variables, each has a frequency distribution with a mean and a variance. These linear combinations are orthogonal, as may be shown using the Orthogonality Rule.

The sum of the observations, $\Sigma x_0 y$, is considered first. For the polynomial example the five observation equations, written in the regression model format of the last section of Unit 3, are

$$y_1 = \beta_0 + \beta_1 + \beta_2 + \varepsilon_1$$
$$y_2 = \beta_0 + 2\beta_1 + 4\beta_2 + \varepsilon_2$$
$$y_3 = \beta_0 + 3\beta_1 + 9\beta_2 + \varepsilon_3$$
$$y_4 = \beta_0 + 4\beta_1 + 16\beta_2 + \varepsilon_4$$
$$y_5 = \beta_0 + 5\beta_1 + 25\beta_2 + \varepsilon_5$$

In regression terminology each observation is an estimator of a linear combination of β_0, β_1, and β_2, as defined by the observation equations. Then the sum of the five observations is estimating the sum of the five linear combinations of β_0, β_1, and β_2, or

$$5\beta_0 + 15\beta_1 + 55\beta_2$$

If a large number of studies, each with the same five levels of the x variable, were carried out, the sums of the five observations would have a frequency distribution with

$$\text{mean}(\Sigma x_0 y) = 5\beta_0 + 15\beta_1 + 55\beta_2$$

In a regression problem each observation is assumed to have a mean μ that is equal to $\mathbf{X}\boldsymbol{\beta}$, as shown in Unit 3. Application of the Mean Rule, $\Sigma \ell_i \mu_i$ for the three linear combinations of the observations, gives

$$\begin{bmatrix} \text{mean}(\Sigma x_0 y) \\ \text{mean}(\Sigma (x_1 - \bar{x}_1) y) \\ \text{mean}(\Sigma [x_2 - \hat{x}_2(0,1)] y) \end{bmatrix} = \begin{bmatrix} 1 & 1 & 1 & 1 & 1 \\ -2 & -1 & 0 & 1 & 2 \\ 2 & -1 & -2 & -1 & 2 \end{bmatrix} \begin{bmatrix} \mu_1 \\ \mu_2 \\ \mu_3 \\ \mu_4 \\ \mu_5 \end{bmatrix}$$

$$= \begin{bmatrix} 1 & 1 & 1 & 1 & 1 \\ -2 & -1 & 0 & 1 & 2 \\ 2 & -1 & -2 & -1 & 2 \end{bmatrix} \begin{bmatrix} 1 & 1 & 1 \\ 1 & 2 & 4 \\ 1 & 3 & 9 \\ 1 & 4 & 16 \\ 1 & 5 & 25 \end{bmatrix} \begin{bmatrix} \beta_0 \\ \beta_1 \\ \beta_2 \end{bmatrix}$$

$$= \begin{bmatrix} 5 & 15 & 55 \\ 0 & 10 & 60 \\ 0 & 0 & 14 \end{bmatrix} \begin{bmatrix} \beta_0 \\ \beta_1 \\ \beta_2 \end{bmatrix} = \begin{bmatrix} 5\beta_0 + 15\beta_1 + 55\beta_2 \\ 10\beta_1 + 60\beta_2 \\ 14\beta_3 \end{bmatrix}$$

We can see again that the first linear combination of the observations, the sum, is estimating a certain linear combination of the betas, namely, $5\beta_0 + 15\beta_1 + 55\beta_2$. Also note that the coefficients of the betas in the mean values are the same numbers found in the odd rows of the ABDO stages, that is,

Row 1:   5   15   55 $\Sigma x_0 y$

Row 3:   10   60 $\Sigma (x_1 - \bar{x}_1) y$

Row 5:   14 $\Sigma [x_2 - \hat{x}_2(0,1)] y$

Hence ABDO has not only produced independent linear combinations of the observations but has also preserved expressions for their mean values. Application of the equal variance rule gives

$$\text{variance}(\Sigma x_0 y) = (1^2 + 1^2 + 1^2 + 1^2 + 1^2)\sigma^2 = 5\sigma^2$$

$$\text{variance}(\Sigma (x_1 - \bar{x}_1) y) = [(-2)^2 + (-1)^2 + 0^2 + 1^2 + 2^2]\sigma^2 = 10\sigma^2$$

$$\text{variance}(\Sigma [x_2 - \hat{x}_2(0,1)] y) = [(2)^2 + (-1)^2 + (-2)^2 + (-1)^2 + 2^2]\sigma^2$$

$$= 14\sigma^2$$

Note that the coefficients of σ^2 in the variance expressions are simply the first nonzero numbers in the expressions for the means, that is, the first numbers in the odd-numbered steps of ABDO. Hence ABDO has produced independent linear combinations and has preserved expressions for not only their mean values but also for the variances of the linear combinations.

In summary, the matrix (hereafter call the ABDO matrix)

$$\begin{bmatrix} 5 & 15 & 55 \\ 0 & 10 & 60 \\ 0 & 0 & 14 \end{bmatrix} \begin{vmatrix} y_1+y_2+y_3+y_4+y_5 \\ -2y_1-y_2+y_4+2y_5 \\ 2y_1-y_2-2y_3-y_4+2y_5 \end{vmatrix}$$

extracted from the ABDO output (the odd rows from each stage), along with the residual sum of squares and the number of observations, contains all the information in the original data. Specifically, we have the following information:

1. The right-most column of the ABDO matrix gives orthogonal linear combinations of the original observations, that is, $\Sigma x_0 y$, $\Sigma(x_1-\bar{x}_1)y$, and $\Sigma[x_2-\hat{x}_2(0,1)]y$.
2. The remaining part of the ABDO matrix contains the coefficients of the elements of β for the mean values of these new random variables (the linear combinations of the observations). The diagonal elements of the remaining part are the coefficients of σ^2 in the variance of these linear combinations of the observations.

SUMMARY

The ABDO algorithm reduces a data set to a few summary statistics, the estimated regression coefficients. We know that an estimated regression coefficient, b_j, has the form

$$\sum_{i=1}^{n}(x_{ij}-\hat{x}_{ij})y_i \Big/ \sum_{i=1}^{n}(x_{ij}-\hat{x}_{ij})^2$$

where $\hat{x}_{ij}$ is the predicted value of x_{ij} using other x variables. The numerator can be expressed as a linear combination of the observations, $\Sigma \ell_i y_i$, where $\ell_i = (x_{ij}-\hat{x}_{ij})$. The denominator is $\Sigma \ell_i^2$, or since the denominator is a constant, the estimated coefficients found by ABDO can also be expressed as a linear combination of the observations, $\Sigma \ell_i y_i$, where $\ell_i = (x_{ij}-\hat{x}_{ij})/\Sigma(x_{ij}-\hat{x}_{ij})^2$. By the Variance Rule we know the variance of the estimated coefficients. Application of the Variance Rule gives

$$\text{variance}(b_j) = \sigma^2 \Big/ \sum_{i=1}^{n}(x_{ij}-\hat{x}_{ij})^2$$

The elements in the odd rows of the sequential ABDO calculations contain all the known information on the explainable part of the response variability. The summary ABDO matrix forms the basis for calculation of both individual and linear combinations of regression coefficients. From the orthogonal linear combinations of observations, variances of linear combinations of estimated regression coefficients can also be calculated by the Variance Rule. The estimates of linear combinations of regression coefficients and their standard errors are used in constructing confidence intervals and testing hypotheses, as developed in the next unit.

Unit 12

Generalizing Beyond the Data

Objectives

- To present a framework for inference.
- To express inference parameters as linear combinations of regression coefficients.
- To develop a flow of operations for data analysis and generalization beyond the data.

Up to now our regression methodology has been applied primarily with one objective in mind, a quantitative description of the data. The conceptual concerns of using our fitted model to make inferences to similar situations were briefly introduced in Unit 3. However, we haven't really attempted to generalize results beyond the existing data set—for instance, to predict the response for combinations of x variables not included in our data.

In research the more stimulating use of regression is for inference. In a test of the effectiveness of a drug for treating people with hypertension, the main interest lies not with the individuals in the experiment but with the large population of people who could potentially benefit from this drug. Thus the principal objective of the next two units will be the development of regression methodology for inference. Emphasized is the computation and interpretation of standard errors (standard deviations of statistics estimated from the data), which are used for confidence intervals and tests of hypotheses.

With the firefly data of Unit 10 we fitted a two-variable model. The emphasis was on estimating a sequence of models and interpreting estimated partial regression coefficients. We did not worry about the variability of an estimated partial regression coefficient. Now we will. If we envision the same light intensity and temperature model for other firefly situations, several questions about the firefly data come to mind.

- What magnitude of variability would we expect in the two estimated partial regression coefficients? In the predicted values?
- Do the light intensity and temperature variables together play an important role in explaining the variability in first flash times?
- Is one variable more important than the other?

These will be called inference problems when we are dealing with the real or conceptual mode of repeating the current experiment many times. The need of a measure of variability is then apparent, and the most usual one is the standard error. With the firefly problem the experiment will not be repeated exactly, but we might reasonably assume that the current situation is representative or, conceptually, is from a population of real but not yet realized situations. More explicitly, we envision situations when the range and distribution of values for light intensity and temperature in the experiment are similar to the range and distribution for other situations included in the inferential umbrella—for example, other days with similar conditions. Also assumed is that the influence of other variables that affect variability in the response stays the same. This framework was introduced in Unit 3, using the statistical inference terminology of target population, parameters, random samples, and estimators.

THE INFERENCE MODEL

In order to make inferences beyond our data, we have to make some assumptions. Fortunately the starting point is the same model we have been using all along. The model assumed in the earlier units to be a reasonable representation of the data is also a reasonable model to use in making inferences beyond the data set. Expressed in the matrix notation of Unit 3, the model is

$$Y = X\beta + \varepsilon$$

where the terms are defined as follows:

- **Y** is a vector of responses measured on the response variable.
- **X** is a matrix of the explanatory variables.
- β is a vector of unknown parameters.
- ε is a vector of uncorrelated, normally distributed random variables, each having mean zero and variance σ^2.

That is, we assume that the vector of observations on the response variable is the sum of two parts, but we observe only the sum and not the individual components. The first term is the part of y that may be explained by the explanatory variables. The vector β may be regarded as the coefficients from fitting the model if we had data from the entire population instead of just a sample. The second term is a random component. While the individual elements of this vector are not predictable, the relative frequencies of different values taken over the entire population are assumed to follow the normal probability distribution with a zero mean and a variance of σ^2. With these assumptions the population characteristics are determined by the parameters σ^2 and β. Hence an inference about σ^2 or β is an inference about the population.

Inferential considerations to some are exemplified by the output for existing package regression programs. The first part is an evaluation of the regression model's worth. Some data analysts interpret the R^2 value, while others rely on the F statistic for comparing the two-variable regression model with the mean model. The next section of the output usually is a listing of the estimated partial regression coefficients, along with the associated standard errors and t statistics. All these statistics have some value in a preliminary evaluation of the regression analysis. However, they are not without problems, especially when there are many x variables. The overall statistics can be misleading, and interpretation of the individual beta estimates and standard errors can be dangerous. The routine computer output certainly should not be construed as the total analysis of the data. It is argued here that in-depth data analysis and inference making in most practical situations revolve around linear combinations of the partial regression coefficients. Usually our interest in σ^2 is to the extent that it affects our inference about the coefficients. The emphasis of the remainder of this unit, then, deals with inference about linear combinations of the coefficients.

LINEAR COMBINATIONS OF REGRESSION COEFFICIENTS

A linear combination of the partial regression coefficients, denoted as $c\beta$, where c is a row vector whose elements are numerical values assigned by the data analyst, is

$$c\beta = [c_0 \, c_1 \, \cdots \, c_k] \begin{bmatrix} \beta_0 \\ \beta_1 \\ \vdots \\ \beta_k \end{bmatrix} = c_0 \beta_0 + c_1 \beta_1 + \cdots + c_k \beta_k$$

Theoretically, the possibilities for $c\beta$ are almost without limit. Common forms, as exemplified by the two-variable problem where the elements in β are β_0, β_1, and β_2, include the following:

(1) $\quad c = [1 \ 0 \ 0], \quad c = [0 \ 1 \ 0], \quad$ or $c = [0 \ 0 \ 1]$

These three linear combinations are nothing more than the three individual partial regression coefficients, for example, β_1 with the second c,

$$c\beta = [0 \ 1 \ 0] \begin{bmatrix} \beta_0 \\ \beta_1 \\ \beta_2 \end{bmatrix} = \beta_1$$

Another common form is

(2) $\quad c = [0 \ 1 \ -1]$

A linear combination where the sum of the elements of c equals zero, as in (2), is called a *contrast*. Estimating the difference between β_1 and β_2 is primarily of value when the regression model is being used to analyze data that can also be described by mean models. As shown in Units 15–19, indicator x variables will be constructed to handle this type of data. When the x variables are measured, we could also have interest in contrasts among the coefficients when the variables have equivalent units of measurement.

$$(3) \quad \mathbf{c} = [x_{i0} \quad x_{i1} \quad x_{i2}]$$

The parameter $\mathbf{c}\boldsymbol{\beta}$ for (3) is the population mean response to a set of x values, which usually is one of the rows in the $\mathbf{X}$ matrix, but the x values in $\mathbf{c}$ need not be any of the rows of the $\mathbf{X}$ data matrix. In either case the estimator $\mathbf{cb}$ is called the predicted value.

For any of the situations in (1), (2), or (3) we could be interested in estimation or hypothesis testing. The estimation procedure will be to estimate $\mathbf{c}\boldsymbol{\beta}$, followed by the calculation of the confidence interval on $\mathbf{c}\boldsymbol{\beta}$ by using the standard error of the estimator $\mathbf{cb}$. A null hypothesis concerning $\mathbf{c}\boldsymbol{\beta}$ is written as H_0: $\mathbf{c}\boldsymbol{\beta} = 0$. For the second and third cases of (1), $\mathbf{c} = [0\ 1\ 0]$ or $\mathbf{c} = [0\ 0\ 1]$, the null hypothesis concerns a test of the difference between a one-variable model with x_0 included and a two-variable model with x_0 included. From subject matter considerations the zero value of H_0 could be replaced by another constant. For example, if an improved practice would be adopted only if the changeover costs would be recovered by potential savings, the contrast between the two practices would be built into $\mathbf{c}$, zero in H_0 would be replaced by the necessary difference, and the alternative hypothesis would be one-tailed.

For linear combinations of estimated regression coefficients, it is hoped that we can utilize the rules for linear combinations of observations. In Unit 11 the variance rules assumed that the observations were uncorrelated, a good assumption for estimated sequential coefficients. Unfortunately, the same doesn't hold in general for a linear combination of estimated partials.

A GENERAL APPROACH TO DATA ANALYSIS AND GENERALIZATION

The previous section has established the feasibility of using a linear combination of partial regression coefficients, $\mathbf{c}\boldsymbol{\beta}$, as a unifying approach to data analysis and generalizations beyond the data. Needed now is a methodology for handling a wide variety of inference problems. The general approach of the methodology includes the following:

- Estimation of $\mathbf{c}\boldsymbol{\beta}$. The estimator is denoted as $\mathbf{cb}$, where the elements in $\mathbf{b}$ are the estimated partial regression coefficients from ABDO.
- Calculation of the standard error of $\mathbf{cb}$. The variance of $\mathbf{cb}$ is of the form $v_c \sigma^2$, where v_c is a constant that depends only on the $\mathbf{X}$ matrix and $\mathbf{c}$ (not

on **Y**). The square root of $v_c\sigma^2$ is the standard error of **cb**. From the output of ABDO a new algorithm, ELIM, is developed in the next section to give **cb** and the standard error.
- Construction of confidence intervals and test of hypotheses for generalization beyond the data. A confidence interval is of the form

$$\mathbf{cb} \pm \text{SQRT}(v_c s^2) t_{\alpha/2}$$

where SQRT stands for square root, s^2 is the residual mean square, and $t_{\alpha/2}$ is an upper percentage point of Student's t distribution with the degrees of freedom of s^2. The value of α is chosen by the data analyst so that $1-\alpha$ is the desired confidence coefficient. To test H_0: $\mathbf{c}\boldsymbol{\beta}=0$, we calculate the t test statistic

$$t = \mathbf{cb}/\text{SQRT}(v_c s^2)$$

with degrees of freedom of the residual mean square, and compare the calculated value with $t_{\alpha/2}$.

ELIMINATION (ELIM) ALGORITHM FOR CALCULATING ESTIMATES AND STANDARD ERRORS

Having demonstrated some of the useful features of the ABDO algorithm in Unit 11, we now are in a position to address the major objectives of this unit, the estimation of a linear combination of the regression coefficients, $\mathbf{c}\boldsymbol{\beta}$, and calculation of a standard error of the estimator **cb**. A simple algorithm applied to the ABDO matrix from Unit 11 will give the basic statistics, **cb** and $v_c s^2$, for calculating confidence intervals and evaluating tests of hypotheses for any **c**—for example, a simple regression coefficient, a contrast among the coefficients, or a predicted value as outlined in the previous section. We have seen in Unit 11 that application of ABDO will give the residual sum of squares and certain linear combinations of the y observations, the number being equal to the number of columns in the **X** matrix. In this sense a data reduction has been accomplished. Each of these linear combinations of the observations has a mean that can be expressed in terms of the beta parameters of the regression model. From the output of ABDO we can form a matrix, the ABDO matrix, whose elements will include not only the relevant linear combinations of the observations but also the values needed to calculate their means and variances. Once we have this information, any linear combination of the partial regression coefficients, $\mathbf{c}\boldsymbol{\beta}$, can be estimated along with the standard error.

For the polynomial problem of Unit 11 suppose we are interested in

$$\beta_0 + 2\beta_1 + 4\beta_2$$

This expression is a linear combination of the elements of β. It also happens to be the mean response for the second observation, that is, when $x_1 = 2$. If we define c as [1 2 4], then this linear combination may be expressed as $c\beta$. Our objective may be any of the following:

1. Estimation of $c\beta$ and its standard error.
2. Prediction of a future observation, whose mean response is $c\beta$, and its standard error.
3. Testing the hypothesis that $c\beta$ equals some specified value.

Regardless of the objective we must first calculate $cb = \hat{y}_2$ and v_c. The estimator will be a linear combination of the observations since the estimators of β_0, β_1, and β_2 are linear combinations of the observations. From ABDO in Unit 11 we have already obtained three linear combinations of the observations (now denoted as z_1, z_2, and z_3); they are

$$z_1 = y_1 + y_2 + y_3 + y_4 + y_5$$
$$z_2 = -2y_1 - y_2 + 0y_3 + y_4 + 2y_5$$
$$z_3 = 2y_1 - y_2 - 2y_3 - y_4 + 2y_5$$

We now want to find a combination of the three z's that will estimate $\beta_0 + 2\beta_1 + 4\beta_2$.

In Unit 11 we found the means and variances of the z's, as shown below.

z	Mean $= c\beta$	Variance $= v_c \sigma^2$
z_1	$5\beta_0 + 15\beta_1 + 55\beta_2$	$5\sigma^2$
z_2	$10\beta_1 + 60\beta_2$	$10\sigma^2$
z_3	$14\beta_2$	$14\sigma^2$

Knowing that

z_1	estimates	$5\beta_0 + 15\beta_1 + 55\beta_2$
z_2	estimates	$10\beta_1 + 60\beta_2$
z_3	estimates	$14\beta_2$

then our goal of estimating $c\beta$, where $c = [1\ 2\ 4]$, can be stated as

$$?z_1 + ?z_2 + ?z_3 \quad \text{estimates} \quad \beta_0 + 2\beta_1 + 4\beta_2$$

Proceeding systematically, we obtain the following:

$z_1/5$	estimates	$\beta_0 + 3\beta_1 + 11\beta_2$
$z_1/5 - z_2/10$	estimates	$\beta_0 + 2\beta_1 + 5\beta_2$
$z_1/5 - z_2/10 - z_3/14$	estimates	$\beta_0 + 2\beta_1 + 4\beta_2$

Elimination (ELIM) Algorithm for Calculating Estimates and Standard Errors

As we know from Unit 11, the three linear combinations of the observations, z_1, z_2, and z_3, are orthogonal linear combinations. Furthermore, $z_1/5$ is $\bar{y}$, $z_2/10$ is the sequential $b_{1\cdot 0}$, and $z_3/14$ is the sequential $b_{2\cdot 0,1}$. Therefore, $\hat{y}_2 = \bar{y} - b_{1\cdot 0} - b_{2\cdot 0,1}$, and, using our variance rule, we have

$$\text{variance}(\hat{y}_2) = \text{variance}(\bar{y}) + \text{variance}(b_{1\cdot 0}) + \text{variance}(b_{2\cdot 0,1})$$
$$= (1/5)^2(5\sigma^2) + (1/10)^2(10\sigma^2) + (1/14)^2(14\sigma^2)$$
$$= (13/35)\sigma^2$$

In hindsight we see that we could have calculated **cb** by $b_0 + 2b_1 + 4b_2$, but our simple variance rule couldn't be used to find the variance of a linear combination of estimated partial regression coefficients. Thus our approach has been to calculate **cb** in terms of the estimated sequential coefficients so that the variance could be easily formulated and calculated.

An algorithm called ELIM provides a systematic method of using the ABDO output to calculate the quantities needed for inference about $c\beta$, namely, **cb** and v_c. First, write down the leading $p \times p$ elements of the ABDO matrix as given in Unit 11 and augment it below by the coefficients of the linear combination, **c**, to be estimated. Add a column with elements z_1, z_2, z_3, and 0. For the example from Unit 11 these are the first four rows of Table 12.1. Divide the first element in the fourth row by the lead element in the first row; this result is called the first multiplier. From the fourth row subtract the product of the first multiplier and the first row and write the result in the fifth row. The second multiplier is the ratio of the second element of row five to the second element of row 2. From the fifth row subtract the product of the second multiplier and the second row and write the result in the sixth row. The pattern should now be clear. Applying the operation again will give three zeros in the seventh row followed by the negative of the estimate **cb** as the last element.

Table 12.1 ELIM Calculations for Quadratic Polynomial Example

ABDO Matrix				Multipliers
5	15	55	z_1	1/5
	10	60	z_2	$-1/10$
		14	z_3	$-1/14$
1	2	4	0	
0	-1	-7	$-z_1/5$	
0	0	-1	$-z_1/5 + z_2/10$	
0	0	0	$-z_1/5 + z_2/10 + z_3/14$	

The coefficient of σ^2 in the variance of the estimator, v_c, is the sum of products of squared multipliers and diagonal elements of the ABDO matrix. For this example the variance $v_c\sigma^2$ is

$$[(1/5)^2(5)+(-1/10)^2(10)+(-1/14)^2(14)]\sigma^2$$
$$=[(1/5)+(1/10)+(1/14)]\sigma^2=(13/35)\sigma^2$$

In brief, the ELIM algorithm is a systematic procedure for calculating estimates of linear combinations of the estimated partial regression coefficients and their standard errors. In Unit 11 we have seen that each estimated partial regression coefficient is, in turn, a linear combination of the observations. Consequently, the predicted values $\hat{y}_i$ can be expressed as a linear combination of the observations since each predicted value can be written as **cb**. These linear combinations can be found directly from the ELIM calculations. In our example the estimate of $\hat{y}_2 = \mathbf{cb} = [1\ 2\ 4]\mathbf{b}$ has been shown to be

$$(z_1/5 - z_2/10 - z_3/14)$$

Substituting the linear combinations of the observations for each z, we find

$$\hat{y}_2 = 0.26 y_1 + 0.37 y_2 + 0.34 y_3 + 0.17 y_4 - 0.14 y_5$$

The variance of $\hat{y}_2$ can now be calculated directly with the Variance Rule:

$$[0.26^2 + 0.37^2 + 0.34^2 + 0.17^2 + (-0.14)^2]\sigma^2 = 0.37\sigma^2$$

which is also part of the ELIM output, $(13/35)\sigma^2$. The variance σ^2 is estimated by the residual mean square.

Like ABDO, the ELIM algorithm is also sequential. Estimates of $\hat{y}_2$ for two reduced models are nested within Table 12.1; namely, $\hat{y}_2$ is $z_1/5$ for x_0 alone and is $z_1/5 - z_2/10$ for the model with x_0 and x_1. These estimates are expressed as linear combinations of the observations in Table 12.2.

Table 12.2 Linear Combinations of the Observations

Model	c	$\hat{y}_2$	Variance $(\hat{y}_2) = v_c\sigma^2$
x_0 alone	[1]	$\hat{y}_2 = 0.2 y_1 + 0.2 y_2 + 0.2 y_3 + 0.2 y_4 + 0.2 y_5$	$0.2\sigma^2$
x_0 and x_1	[1 2]	$\hat{y}_2 = 0.4 y_1 + 0.3 y_2 + 0.2 y_3 + 0.1 y_4 + 0 y_5$	$0.3\sigma^2$
x_0, x_1, and x_2	[1 2 4]	$\hat{y}_2 = 0.26 y_1 + 0.37 y_2 + 0.34 y_3 + 0.17 y_4 - 0.14 y_5$	$0.37\sigma^2$

It is interesting to note that the observed y_2 is of increasing importance as the model includes more variables. This aspect will be developed more fully in Unit 20.

INTERESTING LINEAR COMBINATIONS FOR THE FIREFLY DATA

In this unit's introduction we listed several inference questions regarding the firefly data. These will now be considered, using our newly developed methodology for linear combinations. A general starting point is to calculate the estimates and standard errors of the three partial regression coefficients (including the intercept) for the two-variable problem. For completeness the stagewise results from applying the ABDO algorithm to the firefly data in Unit 10 are given here.

		17	1,244	400.9	703
$X'X\|X'Y=$			109,328	30,228.9	49,557
				9,555.65	16,363.1
$Y'Y=$					30,211
Stage 0					
Step 1:		17	1,244	400.9	703
Step 2:		1	73.1765	23.5824	41.3529
Stage 1					
Step 3:			18,296.5	892.453	−1,886.06
Step 4:			1	0.0487773	−0.103083
Stage 2					
Step 5:				57.9532	−123.297
Step 6:				1	−2.12753
Stage 3					
Step 7:					683.143
Step 8:					1

The first step will be to set up the ABDO matrix by selecting certain key values from the ABDO output—namely, the values in steps 1, 3, and 5—and adding **c** and zero as the fourth row. The multipliers and subsequent operations of ELIM are described in the previous section and illustrated below for estimating β_1, that is, estimating $c\beta$, where $c = [0\ 1\ 0]$.

Row					Multipliers
1	17	1,244	400.9	703	$m_1 = 0$
2		18,296.5	892.453	-1886.06	$m_2 = 1/18{,}296.5$
3			57.9532	-123.297	$m_3 = -0.0487773/57.9532$
4	0	1	0	0	
5	0	1	0	0	[row 4 $-(m_1)$row 1]
6	0	0	-0.0487773	0.103083	[row 5 $-(m_2)$row 2]
7	0	0	0	-0.000692	[row 6 $-(m_3)$row 3]

Consequently, **cb** is $b_1 = 0.000692$ and the standard error is the square root of

$$\left[\frac{1}{18{,}296.5} + \frac{(-0.0487773)^2}{57.9532}\right]\left[\frac{683.143}{17-3}\right] = (0.0683)^2$$

Note that the value in the second set of brackets is our estimator of σ^2, the residual sum of squares divided by the degrees of freedom, the number of observations n minus the number of nonredundant x variables in the regression model, $p = k + 1$. A 95% confidence interval on β_1 is

$$0.000692 \pm (2.145)(0.0683)$$

where 2.145 is the appropriate value from the t distribution with 14 degrees of freedom. A more complete interpretation will be given in the next unit, but from this estimated confidence interval we see that it is difficult to be excited about the importance of light intensity when included in the two-variable model.

Similar ELIM calculations can be performed for the intercept β_0 and for β_2, the partial regression coefficient for temperature. The results can be summarized as shown below.

c	cβ	cb	Standard Error	95% Confidence Interval
[1 0 0]	β_0	$b_0 = 91.5$	18.8	$91.5 \pm (2.145)(18.8)$
[0 0 1]	β_2	$b_2 = -2.13$	0.918	$-2.13 \pm (2.145)(0.918)$

Interesting Linear Combinations for the Firefly Data

Suppose we want to estimate the population mean response for $x_0 = 1$, $x_1 = 87$, and $x_2 = 24.4$. This is one of the data points, the twelfth row in the **X** matrix. Then **c** = [1 87 24.4], and we have the following computations:

				Multipliers
17	1,244	400.9	703	1/17
	18,296.5	892.453	−1,886.06	13.8235/18,296.5
		57.9532	−123.297	0.143372/57.9532
1	87	24.4	0	
0	13.8235	0.817647	−41.3529	
0	0	0.143372	−39.9280	
0	0	0	−39.6229	

Consequently, the estimator **cb** is $\hat{y} = 39.6$, and the standard error is the square root of $v_c s^2$,

$$\text{SQRT}\left[\frac{1}{17} + \frac{(13.8235)^2}{18,296.5} + \frac{(0.143372)^2}{57.9532}\right]\left[\frac{683.143}{17-3}\right]$$

$$= \text{SQRT}(3.396) = 1.843$$

These results and others from interesting combinations of the x's are summarized in Table 12.3.

Table 12.3 Summary of Results for Firefly Data

c					95% Confidence Interval	
x_0	x_1	x_2	cb=$\hat{y}$	Standard Error=SQRT($v_c s^2$)	Lower	Upper
1	87	24.4	39.6	1.843	35.6	43.6
1	55	23.3	41.9	2.015	37.6	46.2
1	73.2	23.58	41.4	1.694	37.8	45.0
1	27	26	36.2	5.188	25.1	47.3
1	135	18	53.3	8.678	34.7	71.9
1	140	26.7	34.8	3.847	26.5	43.1

The last two columns are the upper and lower limits of the 95% confidence interval, the predicted value plus and minus the product of the standard error and 2.145 (the appropriate value from the t distribution with 14 degrees of freedom). The standard errors exhibit more variability than expected. The interval of 1.694 to 8.678 is especially surprising, since all the values of x_1 and x_2 in the table are within the intervals of 26 to 140 for x_1 and 17.8 to 26.7 for x_2 of the data set. The lowest standard error is at the mean combination $x_1 = 73.2$ and $x_2 = 23.58$. The first and last c's are actual data points, but the standard error for $c = [1\ 140\ 26.7]$ is more than twice the magnitude of the $[1\ 87\ 24.4]$ combination. The fourth and fifth combinations are particularly bothersome. They are roughly three and five times the standard error at the mean combination.

The reason for these large standard errors becomes clear when x_2 is plotted against x_1. The combinations of x_1 and x_2 in the data set are all in a diagonal band going from the lower left to upper right. In other words, low light intensity is associated with low temperature and high light intensity with high temperature in this set of data. Thus predicting for the fourth and fifth combinations is an extrapolation practice not to be encouraged without caution, even when we assume that the model holds for areas outside our data. The standard errors give quantitative information on the instability of our estimator, especially when predicting outside the area of the data points. Consequently, a standard error should always accompany an estimated value. However this isn't always easy to get from some computer package programs. For example, some programs don't give the standard error of an estimated value when the combination of x's is not part of the data set. If an appropriate option is not available in a regression program, the ELIM algorithm is an efficient and systematic method for calculating the estimate and associated standard error for any linear combination of the partial regression coefficients $c\beta$.

The $c = [1\ 55\ 23.3]$ in the table is an interesting combination. The x_1 and x_2 values are well within the intervals of the data set. The predicted value of 41.9 and standard error of 2.015 seem reasonable. However, interpolation can lead to interpretation problems if the population of inference is not well defined. In the data originally reported by the experimenter were three observations not included in the data analyzed here because the general weather description for those days was recorded as overcast, as compared with hazy, cloudy, and clear for the other days. The observed value was 10 for $c = [1\ 55\ 23.3]$, giving a residual of $10 - 41.9$, a relatively large value considering that the estimated residuals for the analyzed data have a range of -13 to 10. Consequently, any interpretation of the analyzed data has to clearly state the range of conditions over which the model holds. Unit 21 develops methodology for coping with large standard errors and unrealistic estimates.

SUMMARY

In generalizing beyond the experimental data to the target population, the starting point is the data analysis methodology developed in previous units. From a sequence of candidate models we select the most appropriate model on which to base our inference methods. Confidence interval estimation and tests of hypothesis are cast in the form of linear combinations of the regression coefficients, denoted as $c\beta$ and estimated by **cb**. Three general groups of **c** emerge. When **c** includes values of the explanatory x variables, **cb** is a predicted value. Other groups include contrasts among the coefficients, where the **c** elements sum to zero, and more generally any meaningful linear combinations of the regression coefficients where the elements of **c** are zero or one.

Fundamental to making inferences is a variability measure of **cb**. Each element in **b** is a linear combination of the observations and **cb** is, in turn, a linear combination of the estimated coefficients. Since **cb** can be expressed in terms of the sequential regression coefficients, formulation of the standard error of **cb** is a straightforward application of methodology developed in Units 11 and 12. But still needed is a systematic procedure to handle the required calculations. Various approaches are available. A computer statistical package program, with a provision for calculation of both **cb** and the standard error of **cb**, is usually preferred. If such a program is not available, STAN, a microcomputer program for regression models (Unit 24), or ELIM, for hand and desk calculators, can be used.

Unit 13

Comparison of Candidate Models

Objectives

- To extend analysis of variance (ANOVA) methodology for several explanatory variables.
- To extend the sequencing approach for model selection.
- To develop ANOVA tables with sequential or partial sums of squares.

In previous units with the firefly data we have fitted the two-variable model as

$$\hat{y}(0,1,2) = 91.5 + 0.000692 x_1 - 2.13 x_2$$
$$(0.0683) \qquad (0.918)$$

where y is the time of first flash, x_1 is light intensity, and x_2 is temperature. The values in the parentheses under the estimated coefficients are the estimated standard errors as calculated from the ELIM algorithm developed in Unit 12. The ratio of estimated coefficient to estimated standard error for x_2, equal to $2.13/0.918 = 2.32$, is one indication that temperature is an important variable in a fitted two-variable model. On the other hand, the estimated standard error for b_1 is larger than the estimated coefficient, leading us to conjecture that a reduced model with temperature alone would be an adequate fitted model. This unit uses the analysis of variance (ANOVA) techniques developed in Unit 6 to compare candidate models. Specifically, we will compare the two-variable model with reduced models including one, or perhaps neither, explanatory variable.

DECOMPOSITION OF THE FIREFLY DATA

The basis of the ANOVA technique is the decomposition of each observation into a predicted part and a residual part. For a comparison of models a sequence of candidate models is hypothesized. Then the predicted part of the decomposition is also partitioned into parts.

In Unit 6 where a single reduced model was compared with the full model, $\hat{y}$ for the full model was written as a sum of two parts,

$$\hat{y}_r + (\hat{y} - \hat{y}_r)$$

where $\hat{y}_r$ is the predicted observation based on the reduced model. With two variables the basic decomposition is

$$y = \hat{y}(0,1,2) + [y - \hat{y}(0,1,2)]$$

where $\hat{y}(0,1,2)$ is the predicted value using the fitted two-variable model.

After fitting the sequence of models, denoted as $x_0|x_1|x_2$, we can partition $\hat{y}(0,1,2)$ as

$$\hat{y}(0)+[\hat{y}(0,1)-\hat{y}(0)]+[\hat{y}(0,1,2)-\hat{y}(0,1)]$$

In other words, the predicted part is a sum of a predicted value for the most reduced model plus two additional components, one for each increasingly general model. Each term is a measure of the extra amount of variability in y that can be associated with the more general model.

The decomposition of y for the analysis of variance table is the equation

$$y=\hat{y}(0)+[\hat{y}(0,1)-\hat{y}(0)]+[\hat{y}(0,1,2)-\hat{y}(0,1)]+[y-\hat{y}(0,1,2)]$$

where

$$\hat{y}(0)=\bar{y}=41.35$$
$$\hat{y}(0,1)=\bar{y}+b_{1\cdot 0}(x_1-\bar{x}_1)=41.35-0.10(x_1-\bar{x}_1)$$
$$\hat{y}(0,1,2)=\bar{y}+b_{1\cdot 0}(x_1-\bar{x}_1)+b_{2\cdot 0,1}[x_2-\hat{x}_2(0,1)]$$
$$=41.35-0.10(x_1-\bar{x}_1)-2.13[x_2-\hat{x}_2(0,1)]$$
$$=91.47+0.00069x_1-2.13x_2$$

for the firefly data as estimated in Unit 10. Then we have

y		$\hat{y}(0)$		$\hat{y}(0,1)-\hat{y}(0)$		$\hat{y}(0,1,2)-\hat{y}(0,1)$		$y-\hat{y}(0,1,2)$
45.00		41.35		4.86		0.39		−1.60
40.00		41.35		3.94		−4.64		−0.65
58.00		41.35		3.42		8.86		4.37
50.00		41.35		3.32		0.03		5.30
31.00		41.35		2.90		−0.20		−13.06
52.00		41.35		1.87		−1.29		10.06
54.00		41.35		1.87		4.67		6.10
38.00		41.35		1.77		−5.86		0.74
40.00	=	41.35	+	0.33	+	3.68	+	−5.36
28.00		41.35		−0.19		−6.44		−6.72
38.00		41.35		−0.60		−2.41		−0.34
36.00		41.35		−1.42		−0.31		−3.62
36.00		41.35		−2.77		5.51		−8.10
46.00		41.35		−2.77		−1.30		8.71
40.00		41.35		−3.80		−2.81		5.25
31.00		41.35		−5.86		1.82		−6.31
40.00		41.35		−6.89		0.30		5.23

The model $y = \mu x_0 + \varepsilon$ is the most reduced model we could fit and does account for a large part of each observation. After fitting the Mean Model, we find the residuals to be $y - \hat{y}(0) = y - \bar{y}$. In general, our interest in the two-variable model is the extent to which $y - \bar{y}$ can be associated with light intensity and temperature. And each of the two columns after the $\hat{y}(0)$ column appear to account for an important segment of $y - \bar{y}$. For example, $y_3 - \bar{y} = 58 - 41.35 = 16.65$. Thus more than two-thirds of the $y_3 - \bar{y}$ residual is associated with the sequential fitting of x_1 (3.42) and x_2 (8.86).

The columns of the observational decomposition can be associated with the stages of ABDO. The linear combinations of the observations at each sequential stage are orthogonal (Unit 11), and we would expect the observational decomposition columns to be orthogonal. Furthermore, the sequential parts of each observation will be additive, and we would also expect the sum of squares of each column in the decomposition to add to the total sum of squares.

ANALYSIS OF VARIANCE FOR THE FIREFLY DATA

Squaring and summing the observations for each column in the observational decomposition table will give the necessary sums of squares for an ANOVA table, as shown in Table 13.1. Alternatively, the sum of squares can be calculated directly from the ABDO stages.

Table 13.1 Analysis of Variance of the Firefly Data for the Model Sequence $x_0 | x_1 | x_2$

Source	Sum of Squares	Degrees of Freedom	Mean Square
$R(0)$	29,071.12	1	29,071.12
$R(1\|0)$	194.4210	1	194.4210
$R(2\|0, 1)$	262.3184	1	262.3184
Residual	683.1429	14	48.79592

The $R(\ )$ notation in the table should be read as the regression sum of squares for fitting a particular model. Specifically,

$R(0)$ = sum of squares associated with the Mean Model

$R(1|0)$ = additional sum of squares for fitting light intensity

= difference between the one-variable model with light intensity and the Mean Model

$R(2|0, 1)$ = additional sum of squares for fitting temperature

= difference between the two-variable model with light intensity and temperature and the one-variable model with light intensity

Each regression sum of squares is associated with one degree of freedom since one additional variable is included with each progressively more general model in the sequence. The residual degrees of freedom is the difference between the number of observations and the number of variables in the most general or full model.

In the calculation of the regression sum of squares associated with the two-variable model—that is, the combined effect of light intensity and temperature—the two component parts, $R(1|0)$ and $R(2|0, 1)$, are added and denoted as $R(1, 2|0)$. Dividing $R(1, 2|0)$ by the total corrected sum of squares gives R^2, a summary measure of the overall goodness of fit.

COMPARING CANDIDATE MODELS

In Unit 12 we used the t distribution for constructing a confidence interval on a beta coefficient. The same distribution is now used to test the hypothesis that a linear combination of coefficients is equal to zero. For example, the null hypothesis and test statistic for evaluating the importance of temperature are

$$H_0: \beta_2(\text{of the two-variable model}) = 0$$

and

$$t = \frac{cb}{\text{SQRT}(v_c s^2)} = \frac{-2.13}{0.918} = -2.32$$

with 14 degrees of freedom. Both numerator and denominator can be calculated from ELIM by using $c = [0\ 0\ 1]$, as demonstrated in the previous unit.

An equivalent test is available from the ANOVA table. First we need to recognize that the differences between the predicted values for the two-variable model and the one-variable model with light intensity, $\hat{y}(0, 1, 2) - \hat{y}(0, 1)$, will be nothing more than random variability if the beta coefficient for temperature is equal to zero. The mean square (the sum of squares with one degree of freedom) is then an estimator of σ^2. The residual mean square is also assumed to be an estimator of σ^2. From statistical theory it is known that the ratio of two independently estimated mean squares under the null hypothesis of no difference follows the F distribution. And with one degree of freedom in the numerator of F, the square root of F is the t statistic.

From the ANOVA table

$$F = \frac{R(2|0,1) \text{ mean square}}{\text{residual mean square}} = \frac{262.3184}{48.79592} = 5.38$$

with 1 and 14 degrees of freedom. From an F table we find that the probability of observing a value of this magnitude or larger on chance alone is 0.04. Note that SQRT(5.38)=2.32, the t value.

Now the question arises about evaluating light intensity. If x_2 is judged to be important, a test on β_1 of the model $y = \mu x_0 + \beta_1(x_1 - \bar{x}_1) + \varepsilon$ doesn't make much sense when we already have shown that x_2 needs to be included in the model. Most data analysts, however, would want to test $\beta_1 = 0$ with both x_1 and x_2 in the model. This can be done by fitting the sequence of models $x_0|x_2|x_1$. The subsequent decomposition of y is

y		$\hat{y}(0)$		$\hat{y}(0,2) - \hat{y}(0)$		$\hat{y}(0,1,2) - \hat{y}(0,2)$		$y - \hat{y}(0,1,2)$
45.00		41.35		5.27		−0.02		−1.60
40.00		41.35		−0.67		−0.03		−0.65
58.00		41.35		12.27		0.01		4.37
50.00		41.35		3.36		−0.01		5.30
31.00		41.35		2.72		−0.01		−13.06
52.00		41.35		0.60		−0.01		10.06
54.00		41.35		6.54		0.01		6.10
38.00		41.35		−4.07		−0.02		0.74
40.00	=	41.35	+	3.99	+	0.01	+	−5.36
28.00		41.35		−6.61		−0.02		−6.72
38.00		41.35		−3.01		−0.00		−0.34
36.00		41.35		−1.73		0.00		−3.62
36.00		41.35		2.72		0.03		−8.10
46.00		41.35		−4.07		0.01		8.71
40.00		41.35		−6.61		0.01		5.25
31.00		41.35		−4.07		0.03		−6.31
40.00		41.35		−6.61		0.03		5.23

The striking feature of this y decomposition is the very small magnitudes of the additional variability associated with light intensity after fitting the temperature variable. With the sequence $x_0|x_2|x_1$ almost all the variability explained by the two-variable model is actually associated with temperature, whereas with the sequence $x_0|x_1|x_2$ both x_1 and x_2 were contributing to the model sum of squares. Of course, the total of the two sums of squares for the two variables is the same regardless of the sequence of the fitted models. This result is indicated by the residual column being the same for both sequences.

One doesn't need a calculator or an ANOVA table to see that light intensity will not be significant. In fact, the sum of squares of the temperature variable, $R(2|0) = (5.27)^2 + (-0.67)^2 + \cdots + (-6.61)^2 = 456.7344$, is almost the same magnitude as the sum of squares due to both variables from the ANOVA table, 456.7394, and the sum of squares for light intensity after fitting temperature is $R(1|0,2) = 0.005$. Consequently, light intensity after fitting temperature (x_1 of the two-variable model) is not important, and the significance test of interest concerns β_2 in the one-variable model with x_2.

The F statistic for testing the significance of temperature is

$$F = \frac{R(2|0) \text{ mean square}}{\text{residual mean square}} = \frac{456.7344}{(683.1429 + 0.0050)/15} = 10.03$$

In the denominator the sums of squares for the last two columns in the $x_0|x_2|x_1$ decomposition of y have been added together and divided by the combined degrees of freedom of $14+1$, which is equivalent to the residual mean square based on the $y - \hat{y}(0,2)$ residuals. An alternative would be the continued use of the residual mean square from the two-variable model with the associated 14 degrees of freedom. However, we have decided that the correct model for the firefly data is $\beta_0 x_0 + \beta_2 x_2$; consequently, the $y - \hat{y}(0,2)$ residuals are used. An additional practical consideration with most computer programs is the computational feasibility of using the reduced model residual mean square for testing and confidence interval estimation.

It should not be surprising that this F, 10.03, is different from the previous F, 5.38, for testing the same coefficient, β_2. The solution to the paradox is that two different hypotheses are being tested. Using the notation of Unit 10, our first test involved $H_0: \beta_{2 \cdot 0, 1} = 0$ for the two-variable model, whereas the second test involved $H_0: \beta_{2 \cdot 0} = 0$ for a one-variable model with x_2.

The demise of light intensity in the two-variable sequence $x_0|x_2|x_1$ can be more easily understood by considering the x_1 residuals after the first two stages of ABDO. From Unit 10 we know that each estimated regression coefficient, regardless of the number of explanatory variables, is identical to the regression coefficient estimated from a single-variable regression. Specifically, the partial regression coefficient b_1 in the model with x_0, x_1, and x_2 is identical to the coefficient obtained from regressing y on the single variable $x_1 - \hat{x}_1(0,2)$. Figure 13.1 compares the plot (a) of y on x_1, where a negative relationship is apparent, with the plot (b) of y on $x_1 - \hat{x}_1(0,2)$. The y values are the same in both plots, of course, but the x_1 values are different from the $x_1 - \hat{x}_1(0,2)$ values. The large values of y shift from the left side of the plot in (a) to the middle of the plot in (b). Hence no trend is apparent after adjusting x_1 for x_0 and x_2.

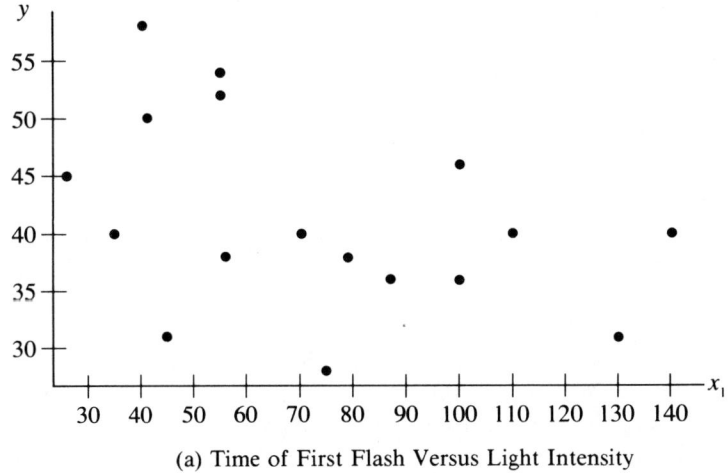

Figure 13.1 Plots of Time of First Flash Versus Light Intensity

ELECTRICITY LOAD DATA

The general objective of the ANOVA technique is the comparison of candidate models. In classic multiple regression with measured x variables, selecting a model sequence for the data analysis can be arbitrary if there is no previous knowledge of relative importance among the x variables. Fortunately the nature of the data or subject matter considerations can lead directly to a sequencing of models, starting with the most reduced model and proceeding to the most general model. A data set from the Kentucky Utilities Company is an example of situations where one or more x variables are measured, but the nature of the data set and the proposed use of the results give a structure to the data analysis that is well handled by a single sequence of candidate models.

The Kentucky Utilities Company measured several weather variables in order to predict electricity demand by knowing the forecasted weather data. A prediction equation would be useful to the company in operating their power plant and in intercompany network decision making. Shown in Figure 13.2 are the data for measured electricity loads and average daily temperatures for the period July 1–July 28, 1971, at Lexington, Kentucky. The relationship was expected to be a straight line in the July temperature range. However, examination of the plot shows that the fitted line actually is close to very few points. Scrutiny of the residuals $y - \hat{y}$ showed a number of relatively large negative and positive residuals. Furthermore, the standard error of the slope estimator is fairly large (5.18) relative to the slope estimate (8.13), giving a 90% confidence interval that includes zero.

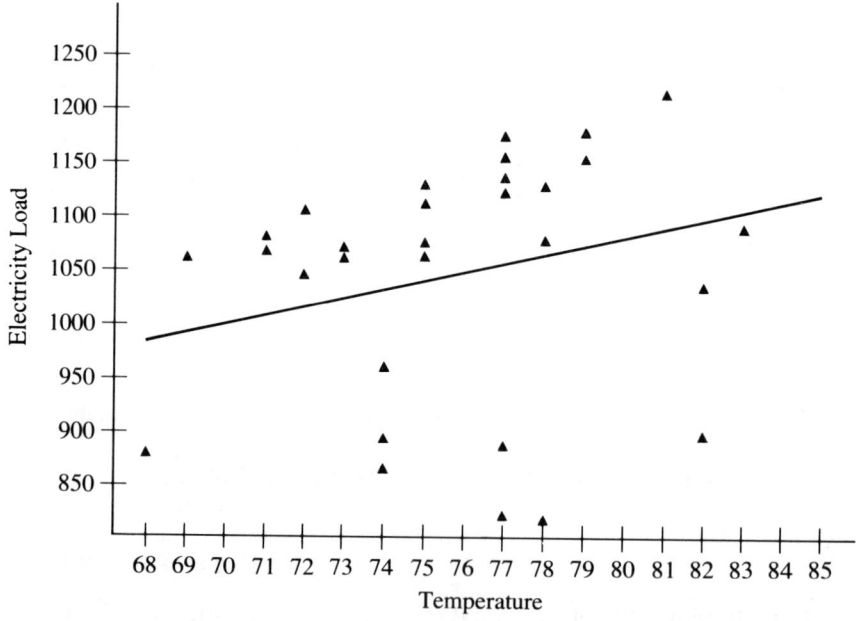

Figure 13.2 Plot of Electricity Load Versus Temperature

The residuals, along with the basic data and predicted loads, are given in Table 13.2 in the order of the day of the month. The outstanding feature of the residuals is that the negative residuals come in groups of two, with

Table 13.2 Electricity Load Data and Residuals

Identification		Temperature	Electricity Load	Predicted Electricity Load	Residuals
1	TH	77	1156	1060	95.72
2	F	75	1073	1044	28.97
3	SA	74	893	1036	−142.9
4	SU	78	818	1068	−250.4
5	M	77	889	1060	−171.3
6	TU	72	1045	1020	25.35
7	W	79	1155	1077	78.47
8	TH	81	1213	1093	120.2
9	F	85	1273	1125	147.7
10	SA	83	1092	1109	−17.03
11	SU	82	899	1101	−201.9
12	M	75	1130	1044	85.97
13	TU	77	1136	1060	75.72
14	W	75	1111	1044	66.97
15	TH	77	1121	1060	60.72
16	F	77	1174	1060	113.7
17	SA	82	1034	1101	−66.91
18	SU	74	866	1036	−169.9
19	M	72	1105	1020	85.35
20	TU	71	1066	1012	54.47
21	W	73	1062	1028	34.22
22	TH	78	1078	1068	9.596
23	F	78	1126	1068	57.60
24	SA	74	961	1036	−74.90
25	SU	77	825	1060	−235.3
26	M	79	1179	1077	102.5
27	TU	73	1068	1028	40.22
28	W	75	1063	1044	18.97
29	TH	69	1061	995	65.72
30	F	71	1081	1012	69.47
31	SA	68	880	987	−107.2

Source: Unpublished data, courtesy of Kentucky Utilities Company, Lexington, Ky.

five days in between. The one exception is the fifth residual. However, Monday, July 5, was a holiday and so the residual pattern tells the story. Furthermore, the Saturday residuals are smaller than the Sunday residuals. Fitting separate straight lines for weekdays, Saturdays, and Sundays is suggested by the residuals. July 5 was included in the Sunday group of data.

In hindsight, the fitting of the one-variable temperature model with all the data should not have been attempted. The problem depicted so vividly by the y residuals could have been anticipated—namely, that the average load demand would vary from weekday to weekend and perhaps between Saturday and Sunday. If differential slopes for weekdays, Saturdays, and Sundays are also possibilities, then a six-variable model should be fitted, allowing a different intercept and slope for each of the three groups. The six variables are constructed by keeping the two x variables for each group in separate columns in the X matrix (36×6) and filling in the remaining positions with zeros. Then we have the model shown in Table 13.3.

With the assumption of straight-line relationships, the six-variable model of three different straight lines is the most general model. It seems reasonable, however, to assume that the slopes will not depend on the day of the week. Unless there is evidence that the slopes do differ, we will want to use all the days to estimate a common slope. A model with four variables, allowing three intercepts and a common slope, would give three lines, each with the same slope. The X matrix for this reduced model has two fewer variables than the most general model and includes x_1, x_3, x_5, and $x_7 = x_2 + x_4 + x_6$. For estimating the one common slope by using all the observations, we combine or collapse the three slope variables into one variable, x_7.

The additional regression sum of squares of the six-variable model compared with the four-variable reduced model is found from a sequence of the four-variable reduced model followed by the six-variable general model. The additional sum of squares divided by the associated degrees of freedom of $6-4=2$, compared with an estimate of σ^2, gives an evaluation of the need for the two additional variables of the general model. If they are judged not to be important, the reduced model with one common slope is said to be adequate.

If the reduced model is to be used in practice, the power company will also want to be sure that the common slope is indeed different from zero. Consequently, the reduced model should be compared with a further reduced model of three lines with a common slope of zero, that is, a model with only the three intercepts. To compare candidate models, we construct a sequence of models so that the additional sum of squares for the more general models can be calculated sequentially. Then the need of the most general model is evaluated first, using the residual mean square from fitting the most general model as an estimator of σ^2. For the electricity load data the needed sequence starts with the three-intercept model, followed by the three-line model with common slope, and then the three-line model with

Table 13.3 Six-Variable Model for Electricity Load Data

Identification		x_1	x_2	x_3	x_4	x_5	x_6
1	TH	1	77	0	0	0	0
2	FR	1	75	0	0	0	0
6	TU	1	72	0	0	0	0
7	W	1	79	0	0	0	0
8	TH	1	81	0	0	0	0
9	F	1	85	0	0	0	0
12	M	1	75	0	0	0	0
13	TU	1	77	0	0	0	0
14	W	1	75	0	0	0	0
15	TH	1	77	0	0	0	0
16	F	1	77	0	0	0	0
19	M	1	72	0	0	0	0
20	TU	1	71	0	0	0	0
21	W	1	73	0	0	0	0
22	TH	1	78	0	0	0	0
23	F	1	78	0	0	0	0
26	M	1	79	0	0	0	0
27	TU	1	73	0	0	0	0
28	W	1	75	0	0	0	0
29	TH	1	69	0	0	0	0
30	F	1	71	0	0	0	0
3	SA	0	0	1	74	0	0
10	SA	0	0	1	83	0	0
17	SA	0	0	1	82	0	0
24	SA	0	0	1	74	0	0
31	SA	0	0	1	68	0	0
4	SU	0	0	0	0	1	78
5	M	0	0	0	0	1	77
11	SU	0	0	0	0	1	82
18	SU	0	0	0	0	1	74
25	SU	0	0	0	0	1	77

different slopes, that is,

$$X=[x_1 x_3 x_5 | x_1 x_3 x_5 x_7 | x_1 x_2 x_3 x_4 x_5 x_6]$$

The subsequent ANOVA table for the electricity load data is shown in Table 13.4.

Table 13.4 ANOVA for Electricity Load Data

Source of Variation	Sum of Squares	Degrees of Freedom	Mean Square
$R(1,3,5)$	34,660,693.99	3	11,553,564.66
$R(7|1,3,5)$	77,555.29	1	77,555.29
$R(2,4,6|1,3,5,7)$	2,667.81	2	1,333.91
Residual	26,647.90	25	1,065.92

At a first glance at the table, we might think that the two degrees of freedom associated with the additional sum of squares for fitting the most general model should be three. However, ABDO or any computer program fitting this sequence of models would show the correct value of two for the difference between the four-variable reduced model and the six-variable general model. Said differently, $x_7 = x_2 + x_4 + x_6$, and once we have included any three of the four variables, the additional sum of squares for the fourth variable is identically equal to zero and is associated with zero degrees of freedom. Details of these considerations, including the fitting of models with variables similar to x_1, x_3, and x_5, are covered in the analysis of sample means starting in Unit 15.

Examination of the ANOVA table indicates that the reduced model with separate intercepts and a common slope of 12.76 (standard error of 1.51) is sufficiently adequate. The resulting prediction equation is

$$\hat{y} = b_1 x_1 + b_3 x_3 + b_5 x_5 + b_7 x_7 = 152.7 x_1 + 0.02903 x_3 - 130.4 x_5 + 12.76 x_7$$

For use in practice, the forecast value of temperature would be x_7, and one of the remaining variables would be one (the other two would be zero), depending on the day of the week to be predicted.

SUMMARY

The ANOVA technique, developed in Unit 6 to evaluate the need for the slope in a one-variable model, was extended in this unit to two or more variables. A sequence of models is conjectured, starting with the most reduced model. More general models follow and are formulated by sequentially adding variables. The rows in the ANOVA table then show the additional sum of squares for each candidate model in the sequence.

For two variables the sequences $x_0|x_1|x_2$ and $x_0|x_2|x_1$ can be conjectured and ANOVA tables calculated. In addition, a nonsequential or partial ANOVA table can be constructed from the third rows of the first two tables. See Table 13.5.

Table 13.5 Sequential and Partial Sums of Squares

Sequential Sums of Squares		
$x_0|x_1|x_2$	$x_0|x_2|x_1$	Partial Sum of Squares
$R(0)$	$R(0)$	$R(0)$
$R(1|0)$	$R(2|0)$	$R(1|0,2)$
$R(2|0,1)$	$R(1|0,2)$	$R(2|0,1)$
Residual (1)	Residual (2)	Residual (3)

The sums of squares for the two sequential ANOVA tables can be calculated from the sequential decomposition of each observation; for instance,

$$y = \hat{y}(0) + [\hat{y}(0,1) - \hat{y}(0)] + [\hat{y}(0,1,2) - \hat{y}(0,1)] + [y - \hat{y}(0,1,2)]$$

Or the sums of squares can be found directly from ABDO calculations. The combined effect of both variables is calculated by adding the second and third rows of the sequential tables:

$$R(1,2|0) = R(1|0) + R(2|0,1) = R(2|0) + R(1|0,2)$$

Unless x_1 is orthogonal to x_2 so that $R(1|0) = R(1|0,2)$ and $R(2|0) = R(2|0,1)$, the combined effect, $R(1,2|0)$, will not be equal to $R(1|0,2) + R(2|0,1)$ from the partial ANOVA table.

In the two sequential tables the residual sum of squares can be calculated by subtracting the additive sequential sums of squares from the total sum of squares. In general, a different residual (3) sum of squares will result from subtracting the nonadditive partial sum of squares of the first three rows. Computer programs equate residual (3) with the residual sums of squares from a sequential table, and, in general, the four partial sums of squares will not equal the total. If the candidate models in the sequence differ by one variable, as in both sequential tables with the two-variable model, the F test statistics calculated from the partial table will be equivalent to the t test statistics given in a computer regression output for testing partial regression coefficients. When the sum of squares decomposition is not additive, as will be the situation in general with the partial table, the numerators of the t or F test statistics will be correlated.

The degrees of freedom associated with each row of the additional regression sums of squares do not have to be equal to one. For example, with six x variables, suppose x_0 gives the most reduced model, x_0, x_4, and x_5 give the reduced model, and x_0, x_1, x_2, x_3, x_4, and x_5 give the general model. Then we have the following:

Sums of Squares	Degrees of Freedom
$R(0)$	1
$R(4,5\|0)$	2
$R(1,2,3\|0,4,5)$	3
Residual	$n-6$

In general, the degrees of freedom associated with the additional regression sum of squares for each row in the table is equal to the additional number of variables in the more general candidate model, as long as none of the additional variables are linear combinations of previous variables. For example, if $x_3 = x_5 - x_1 - x_2$, then x_3 is said to be redundant and $R(1,2,3\|0,4,5)$ would be associated with only two degrees of freedom.

The general procedure for comparing candidate models is to write the most general model consistent with the study. Depending on the data structure and the subject matter objectives, conjecture one or more reduced models. Then write out the x variables in the sequence of candidate models, starting with the most reduced model, augmenting with additional variables for the more general model, and continuing until the additional variables for the most general model are added. Estimation of the regression equation, represented by the sequence of candidate models, is followed by calculation of the sequential ANOVA table. The need for the additional variables in the most general model is first evaluated, and if they are judged not important, the evaluation continues with the reduced models, starting with the least reduced model. The objective is to find the simplest (the most reduced) model adequate for describing the data and for inference. At times, other sequential orderings of the candidate models or partial ANOVA tables can be helpful in the interpretation. An analysis that compared regression lines with electricity load data was carried out in this unit to demonstrate a model sequence approach to data analysis.

Unit 14

Case Study: Land Evaluation Data

Sale prices for unimproved agricultural land in the state of South Dakota depend primarily on climatic and land factors. Westin, Stout, Bannister, and Frazee have compiled data on sale price and climatic and land variables for 2700 individual sales during the period 1967–1969. In this unit we will work with averages for 59 South Dakota counties, excluding the Black Hills area. The major objective of the study is to model the relationship between land sale price and several explanatory variables, including climatic and land variables. Prior study had shown that size of farm or year of sale were not important in explaining variability in sale price. Shown in Table 14.1 are the county averages for climatic variables of precipitation (PPT) and temperature (TEMP), land variables of topography (SLOPE) and soil texture (SOIL), and sale price (VALUE).

Table 14.1 South Dakota Unimproved Agricultural Land Data (1967–1969)

County	PPT	TEMP	SLOPE	SOIL	VALUE
Aurora	20.0	48	1.379	6.276	93
Beadle	19.0	45	1.972	5.366	83
Bennett	17.0	49	3.788	4.485	54
Bon Homme	23.5	49	2.562	5.344	152
Brookings	21.0	44	2.552	5.672	131
Brown	22.0	43	2.000	6.768	123
Brule	19.0	48	2.484	6.032	80
Buffalo	18.0	46	4.190	7.619	55
Butte	14.0	45	3.543	8.543	20
Campbell	16.0	42	3.224	5.735	69
Charles Mix	22.0	49	2.309	6.000	101
Clark	21.0	43	2.873	5.817	81
Clay	25.0	49	1.774	6.774	222
Codington	21.0	42	2.459	5.757	88
Corson	15.5	42	3.510	4.941	49
Davison	22.0	48	2.577	5.077	120

County	PPT	TEMP	SLOPE	SOIL	VALUE
Deuel	21.0	43	2.860	5.820	91
Dewey	16.0	45	3.868	5.763	43
Douglas	22.0	48	2.422	5.600	120
Edmunds	18.0	43	2.725	5.435	64
Faulk	18.0	44	2.604	5.167	70
Grant	22.0	44	2.500	5.750	109
Haakon	15.0	46	3.405	6.703	42
Hamlin	22.0	43	2.438	5.813	98
Hand	18.0	45	2.415	6.000	76
Hanson	22.0	46	2.553	5.043	128
Harding	13.0	42	3.745	8.275	21
Hughes	16.5	46	2.679	6.607	85
Hutchinson	23.0	48	2.260	5.500	171
Hyde	17.0	45	2.484	5.613	57
Jackson	16.0	47	3.714	8.286	37
Jerauld	19.0	46	2.444	5.528	77
Jones	17.0	48	3.759	8.000	53
Kingsbury	22.0	45	2.053	5.790	112
Lake	23.0	45	2.854	5.951	147
Lincoln	25.5	47	2.280	5.760	214
Lyman	17.0	47	3.842	7.868	65
McCook	24.0	46	2.286	5.204	158
McPherson	17.5	42	2.982	5.570	65
Marshall	20.0	42	3.373	6.706	92
Meade	13.5	45	3.731	7.096	32
Mellette	17.5	49	4.643	6.571	51
Miner	20.5	45	2.133	5.867	88
Minnehaha	25.0	45	2.947	5.982	217
Moody	24.0	45	2.545	6.000	175
Pennington	14.0	46	4.208	7.500	41
Perkins	14.0	42	2.980	6.720	31
Potter	16.5	45	2.574	5.894	80
Roberts	21.5	44	2.467	6.533	90
Spink	19.5	45	1.914	7.011	92
Stanley	16.0	46	3.346	8.000	38

County	PPT	TEMP	SLOPE	SOIL	VALUE
Sully	16.0	46	2.181	5.927	86
Tripp	18.5	47	2.926	6.481	92
Turner	25.0	47	1.900	5.675	188
Union	25.5	49	2.741	6.370	261
Walworth	16.5	45	3.278	5.917	68
Washabaugh	16.0	48	5.128	7.128	33
Yankton	24.0	49	2.225	5.212	175
Ziebach	15.0	45	4.000	5.019	37

PPT = annual average precipitation (in.)
TEMP = annual average temperature (°F)
SLOPE = average slope index of sold farms, determined from soil association maps
SOIL = average soil texture index of sold farms, determined from soil association maps
VALUE = average sale price (dollars/acre)

Source: F. C. Westin, M. Stout, D. L. Bannister, and C. J. Frazee. "Land Sale Prices in South Dakota and Their Relationship to Some Soil, Climatic and Productivity Factors," *Soil Science Society of America Proceedings* 37 (1973): 606–611.

EVALUATION OF GENERAL MODEL

Preliminary analysis of the data emphasizes data plotting; that is, we need to "stop, look, and listen" before initiating the formal analysis. The range of each x variable and the distribution of the values within the range should be checked. Also, the nature of the relationship between VALUE and each x variable should be examined for adequacy of a straight-line approximation and regression analysis assumptions. For example, VALUE reflects a curvilinear relationship with PPT, whereas with TEMP the relationship is a straight line, although a cone-shaped pattern is evident, with VALUE more variable at higher temperatures. A summary for the data plots of VALUE versus each explanatory variable is given in Table 14.2.

In the regression of VALUE on the x variables, we can choose the sequence of the variables, and part of the evaluation will be the comparison of sequential and partial sums of squares. It is known that precipitation is the most important variable in explaining variability in sale price. So each x variable is related to VALUE individually, with the major question being the additional explanatory ability of TEMP, SLOPE, and SOIL after PPT. Specifically, do the land variables collectively contribute after the climatic variables, and does TEMP contribute in addition to PPT? The SLOPE and

Table 14.2 Summary of Preliminary Analysis for the Land Evaluation Data

Plot	Relationship	Remarks on the x Variable		
		Range	Mean	Distribution
VALUE vs. PPT	Strong, positive, and curvilinear	13–26	19	Uniform
VALUE vs. TEMP	Weak, positive, and straight line	42–49	46	Uniform
VALUE vs. SLOPE	Moderate, negative, and straight line	1.4–5.1	2.4	Concentrated around 2.4
VALUE vs. SOIL	Weak, negative, and straight line	4.5–8.5	6.2	Uniform

SOIL values have to be determined from county soil association maps and are not as readily available as the weather data. More importantly, the land variables, especially SOIL, can be important factors within counties but are not thought to be as important when working with county averages. Patterns of these land factors may be found to be repeating within counties, so the range across counties does not reflect the definite trends shown by PPT from west to east and TEMP from north to south. Consequently, the land variables, with an ordering of SLOPE and then SOIL, are included in the general model sequence after the climatic variables.

Table 14.3 Sequential and Partial Sums of Squares from Four-Variable General Model

Source	Degrees of Freedom	Sums of Squares	
		Sequential	Partial
PPT	1	145,110	67,097
TEMP	1	1,175	1,165
SLOPE	1	96	141
SOIL	1	124	124
RESIDUAL	54	26,891	
Residual mean square = 498			

The PPT variable is included first, and thus the curvilinear relationship between VALUE and PPT needs to be modeled. However, if we naively run a four-variable model with PPT, TEMP, SLOPE, and SOIL as the general model, the output will include the sequential and partial sums of squares, as shown in Table 14.3, and the residual plot, $y - \hat{y}$ versus $\hat{y}$, as shown in Figure 14.1.

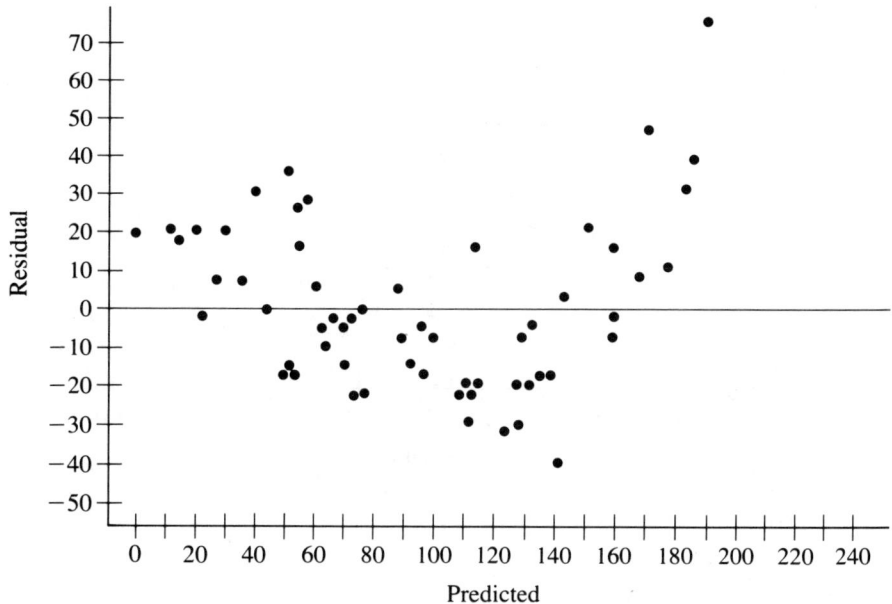

Figure 14.1 Residual Plot for Four-Variable General Model with PPT, TEMP, SLOPE, and SOIL

Even though $R^2 = 0.85$, the residual standard deviation of 22 is larger than we would like to see. The residual plot has a U-shaped tendency, and a high proportion of the large positive residuals are southeastern counties. The comparison of the sequential and partial sums of squares gives us some information on the four explanatory variables; namely, precipitation is the important variable, the contribution of the pair of land variables is even worse than our marginal expectations, and serious correlation problems among the explanatory variables are not obvious.

The PPT partial sum of squares of 67,097 is less than half the sequential sum of squares. Each of the other variables, when taken individually, can explain an important proportion of the VALUE variability. For example, consider the relationship between VALUE and SLOPE, shown in Figure 14.2. Even though the values of SLOPE are concentrated in a relatively narrow range and the relationship is not strong, a straight-line fit will explain an important proportion ($R^2=0.34$). When the SLOPE values are adjusted for PPT, as discussed in Units 10 and 13 (see Figure 13.1), the range of the adjusted SLOPE variable in Figure 14.2 is half the original SLOPE range, the adjusted SLOPE values are concentrated in the middle of the range, and the relationship shows nothing more than random variability.

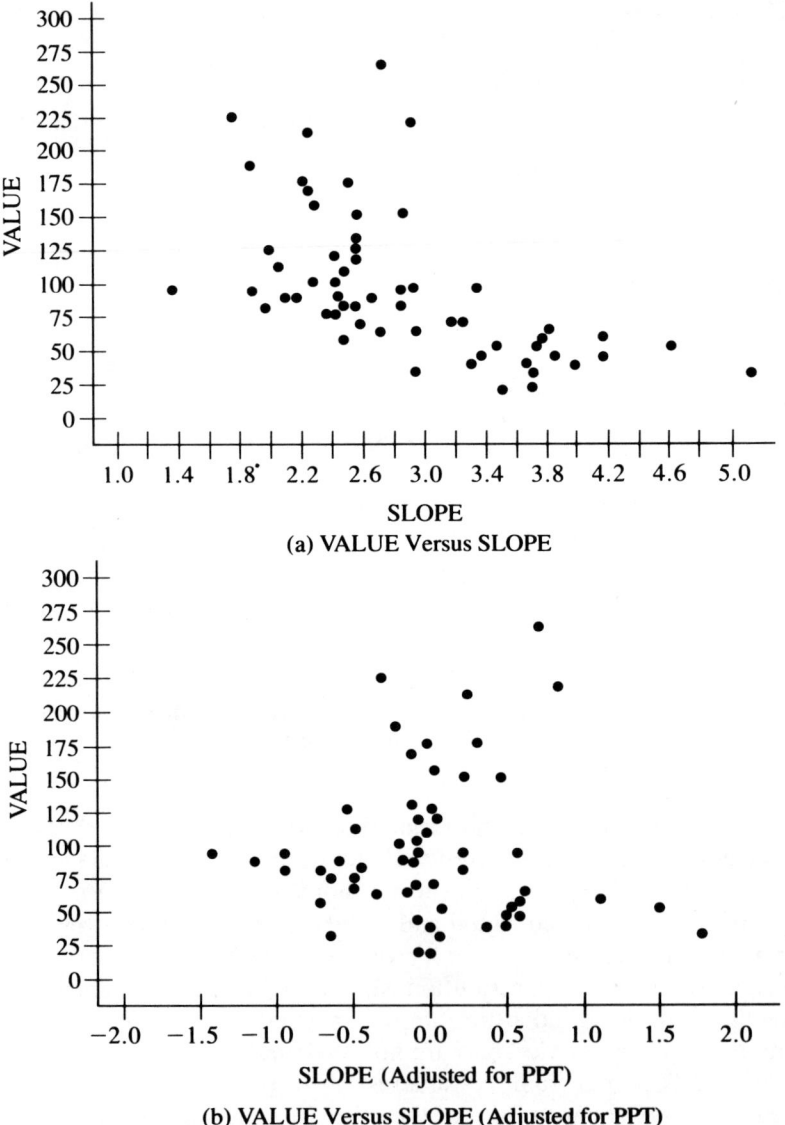

Figure 14.2 Plots of VALUE Versus SLOPE and SLOPE Adjusted for PPT

The overall evaluation is that the general model with the four explanatory variables is not adequate. Improved modeling of the precipitation variable is needed, but the land variables do not warrant further consideration for now. The status of the temperature variable is less certain.

IMPROVED MODELING

One interpretation of the U-shaped residual plot of Figure 14.1 is that a curvilinear relationship exists with one of the x variables. Plotting the residuals against PPT reflects the same U-shaped pattern. Approaches for handling curvilinear relationships include reexpression of PPT (Unit 8) or a direct modeling of the curvilinear relationship between VALUE and PPT. One option is to fit a polynomial—specifically, a quadratic polynomial model (Unit 11). A quadratic polynomial in PPT should help in the underestimation of land price in the southeastern counties. In addition to having higher precipitation, this section of the state also has higher temperature. Consequently, a quadratic polynomial model in PPT and TEMP, including squared terms and the product term of PPT and TEMP, is run. For this model the standard deviation is reduced from 22 to 16. The residual plot is improved in comparison to Figure 14.1, but a U-shaped pattern is still evident. The partial sums of squares are less than the sequentials, especially for PPT (5,094 compared with 145,110). But interpretation of the estimated partial regression coefficients is difficult; for instance, the partial coefficient for PPT is -66. Overall, we have a better fit, reflected by an R^2 of 0.92, but addition of quadratic terms has not given a sufficient improvement in the magnitude or the pattern of the residuals. In general, inclusion of the temperature variable has become dubious.

Another alternative is to model the relationship between VALUE and PPT as a higher-order polynomial. However, a close scrutiny of the plot between VALUE and PPT reveals that the range of PPT can be divided into two segments. The relationship above 22 inches of precipitation could be fitted with a straight line and the slope of the line would be greater than the slope of the straight line fitted for precipitation of 22 or less. The 11 counties with more than 22 inches of rainfall are the southeastern counties, and most of these counties also have warmer temperatures and favorable slopes and soil textures. Therefore it seems reasonable to expect that the relationship between VALUE and PPT would have a different slope for a straight-line model than for the lower segment of the precipitation range.

In a technique called *segmented* or *spline regression*, we can envision fitting two straight lines with different intercepts and slopes. The column format of the **X** matrix, with x_1 and x_2 for the lower segment and x_3 and x_4 for the upper segment, would be similar to the electricity load problem of Unit 13. It also seems reasonable that the two lines should join at PPT=22. If the parameters β_1 and β_2 are the intercept and slope for the lower segment and β_3 and β_4 are the intercept and slope for the upper segment, then

$$\beta_1+\beta_2(22)=\beta_3+\beta_4(22)$$

or

$$\beta_1 x_1+\beta_2 x_2-\beta_3 x_3-\beta_4 x_4=y=0$$

This additional information can be incorporated into the data analysis by adding an observation to the data set with $x_1=1$, $x_2=22$, $x_3=-1$, $x_4=-22$, and VALUE=0. If we indicate that this additional observation should carry much more weight than the other 59 county observations, we ensure that the two estimated straight lines will join at PPT=22. This unequal weighting of the observations (Unit 7) can be handled by including a weight column with values of 1 for each of the 59 county observations and a very large value, say, 10,000, for the added observation. We now have the tabulation shown below.

County	x_1	x_2	x_3	x_4	Weight	Value
Aurora	1	20	0	0	1	93
Beadle	1	19	0	0	1	83
Bennett	1	17	0	0	1	54
Bon Homme	0	0	1	23.5	1	152
⋮	⋮	⋮	⋮	⋮	⋮	⋮
Yankton	0	0	1	24.0	1	175
Ziebach	1	15	0	0	1	37
	1	22	−1	−22	10,000	0

Running this regression, we obtain a residual sum of squares of 11,060 and a standard deviation of 14. The range of the residuals is −24 to 34, and the residual plot does not reflect any discernible pattern. Figure 14.3 shows the two estimated straight lines with slopes of 9.7 (lower) and 34 (upper) and with associated standard errors of 0.73 and 2.1. Underestimated counties are now restricted to counties along the Missouri River and Brookings, Hutchinson, and Union counties in the east, where special circumstances prevail.

Overall, a reduced model deleting TEMP, SLOPE, and SOIL from the general model is found to be satisfactory, but only after additional modeling of the relationship between VALUE and PPT.

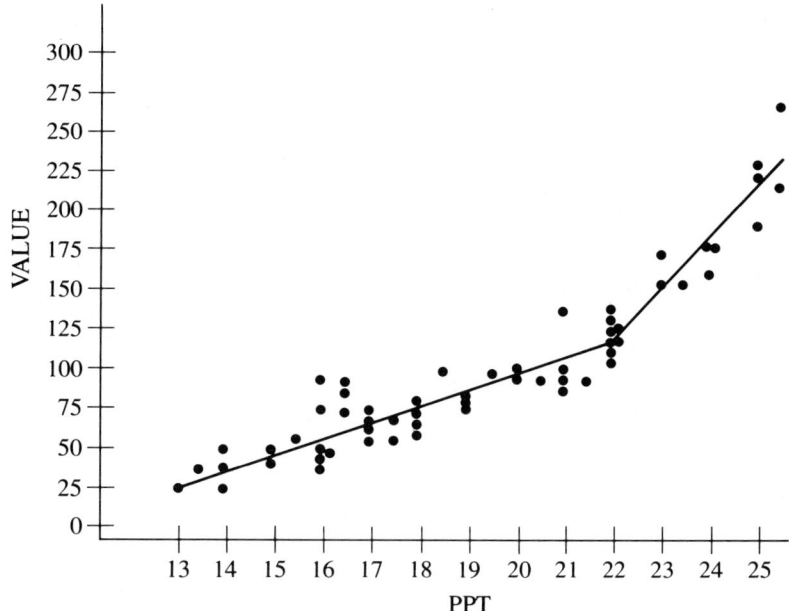

Figure 14.3 Segmented Regression Model; the Slopes Are 9.70 (Lower Segment) and 34 (Upper Segment)

SUMMARY: AN APPROACH TO DATA ANALYSIS FOR SEVERAL x VARIABLES

The following outline is intended to serve as a suggested guide to the analysis of several x variables but should not be considered exhaustive.

1. Perform a preliminary data analysis.
   a. Determine the adequacy of the range and distribution of each x_j.
   b. Determine if a straight-line approximation is reasonable within the range of each x_j. Plots are helpful. Motivated by the subject matter knowledge, a reexpression of x_j, the addition of higher-order terms, or segmented regression can be helpful.
2. Select a general model and assess its adequacy.
   a. Fit the general model and compute $\hat{y}$ and $y - \hat{y}$.
   b. Examine the frequency distribution of the residuals.
   c. Examine the plot of $y - \hat{y}$ versus $\hat{y}$ following these steps: (1) consider the magnitude of the residuals both in general and for single extreme values; (2) examine the pattern of the scatter for any trends or a cone shape.

d. If the residuals are not adequate, then try plotting the residuals versus selected x_j's to attempt to discern the nature of the inadequacy.
e. Determine the overall importance of the x variables. Compare the model mean square with the residual mean square.

3. Formulate one or more candidate model sequences.
   a. If the general model has been assessed to be adequate, arrange a model sequence from the most reduced to the most general based on a natural or conjectured order of importance. Subject matter knowledge is of extreme importance in this step. Within a candidate model sequence you might want to consider several permutations of the ordering.
   b. Evaluate the need for the general model. Obtain the sequential sum of squares (SS) for each candidate model. For the candidate model sequence $x_0 | x_0 x_1 x_2 x_4 | x_0 x_1 x_2 x_4 x_5 x_3$ these are as follows:

$R(0)$	SS for the most reduced model
$\left.\begin{array}{l} R(1\|0) \\ R(2\|0,1) \\ R(4\|0,1,2) \end{array}\right\}$	The sum yields the additional SS for fitting the next reduced model, equal to $R(0,1,2,4) - R(0)$
$\left.\begin{array}{l} R(5\|0,1,2,4) \\ R(3\|0,1,2,4,5) \end{array}\right\}$	The sum is the additional SS due to fitting the general model, equal to $R(0,1,2,3,4,5) - R(0,1,2,4)$

4. If the general model is not deemed necessary, then sequentially evaluate the need for the reduced models by comparing the additional mean square with the residual mean square until one of the candidate models is judged important.
   a. Fit the selected candidate model and examine the residuals.
   b. For each x_j, (1) compare the slope of the one-variable model ($b_{j \cdot 0}$) with the associated partial coefficient from the candidate model; calculation of $x_j - \hat{x}_j$ may prove useful; (2) examine the standard errors of each.

5. Proceed with inference.
   a. Interpret the estimated coefficients, standard errors, and confidence intervals.
   b. Discuss $\hat{y}$, associated standard errors, and confidence intervals.

Unit 15

Analysis of Sample Means

Objectives

- To use regression methodology for estimating sample means and standard errors.
- To construct indicator variables and model sequences for evaluating equality of means.
- To express contrasts among sample means as estimated regression coefficients.

With the two-variable development of Units 10–13 and the computational techniques of Unit 7, we are now in a position to handle data sets with several x variables. Before tackling the problems of many measured explanatory variables (Unit 20), we will first apply our regression methodology to data with a y variable but no measured x variables. Consider a comparative drug trial where three drugs have been allocated to three groups of subjects. No quantitative (measured) x variable exists; instead we have a qualitative (categorical) variable identifying the drug received by each subject. Possibilities include the values 1, 2, 3, the labels A, B, C, or the actual drug names. Even if numeric values are used, ordinality is not assumed. The heart of analyzing the response data is the estimation of means and inference on linear combinations of means.

At first glance, our regression model seems to have no place in the analysis of sample means since no x variables have been measured. There is the example where the treatments are levels of a quantitative variable and we are fitting a functional relationship—for example, a quadratic polynomial—between the sample means of the response variable and the preselected levels of the treatment variable. If the treatments were allocated to the experimental units in a completely randomized design, then the analysis is basically a regression analysis. However, we will use our regression techniques as an umbrella methodology for analyzing a wide variety of research studies including designed experiments and surveys. The algebraic formulations presented in some statistical methodology books for calculating sums of squares for experimental and survey designs can mislead the data analyst into thinking that the analysis of each design is different. We will use one general approach for all designs, and our regression methodology will be the unifier.

From the preceding discussion it is becoming clear that the data analyst will need to substitute selected numerical values for the usual measured x variables. Thus this unit presents the basic procedure used in constructing x variables, known as *indicator variables*. Use of these indicator variables will enable us to estimate means, meaningful contrasts among the means, and standard errors, along with the appropriate sums of squares for an analysis of variance table. The flow of the calculations will be similar to what has been done before, but, in the terminology of sample means, we will

- Calculate the sample means and the residual mean square.
- Evaluate the adequacy of a few linear combinations of the means, expressed as contrasts, to account for the variability among the means.

HISTORY

Our first presentation of an analysis of variance (ANOVA) table was in Unit 6. There we saw that each observation is decomposed into predicted and residual parts, and the predicted part can be further subdivided. The sums of squares for all the parts add to the total sum of squares for the observations. The ANOVA table is a convenient format of accountability for the decomposition of the total sum of squares since it lists the sums of squares, the degrees of freedom, and the mean squares (variances) for the various parts.

The early history of ANOVA tables for designed experiments centered on the calculation of the ANOVA table directly, with only an implicit awareness of a formal model. But recognition of a model with a mean(s) plus a residual term is important. However, we will generally avoid specifying complex classification models involving parameters other than the means—and then worrying about the complications of the model having more parameters than can be estimated. Consequently, our focus will be on inferences concerning means based on estimated means models.

A GENERAL MEANS MODEL

In Unit 1 we described a simple model by expressing an observation as the sum of two parts, a subgroup mean and a residual. Two common situations lead to this simple model:

1. An experimental study with a completely randomized design. Each of the treatments is allocated at random to one or more experimental units and a response variable is measured on each unit.
2. An observational study with a natural division of the units into subgroups and a response variable measured on each unit. Classification or stratification of the units into subgroups is based on presence or absence of one or more characteristics of the units. For instance, in an example of male-female and juvenile-adult status of an animal population, four subgroups are identifiable. Stratification is not limited to dichotomous classifications.

In each of these situations the focus is on the treatment or subgroup means, which can be modeled as

$$y_{j\ell} = \mu_j + \varepsilon_{j\ell} \quad j = 1, 2, \ldots, t; \quad \ell = 1, 2, \ldots, r_j$$

where $y_{j\ell}$ is the observed response of the jth treatment (subgroup) on the ℓth unit, μ_j is the population mean of the jth treatment (subgroup), and $\varepsilon_{j\ell}$ is the random error component. The r_j notation indicates that the number of replications (observations) for each treatment (subgroup) can vary.

RESIDUAL AND EXPERIMENTAL ERRORS

The ε term in regression models has been called the residual term and is calculated as the difference between the observed and the predicted values. The residual represents that part of the observation that has to be ascribed to unknown sources, and those sources are assumed to be random variability. In a controlled experiment factors known to affect the variability in the observations, except for the treatment, of course, are controlled at a constant level. In addition, randomization of the treatments to the experimental units eliminates a bias due to an unknown factor from being associated with any single treatment. Consequently, the variability in the observations among the r_j replications for each treatment can be ascribed to random variability, and the corresponding sum of squares is based on the difference between the observation and the treatment mean, $y_{j\ell} - \bar{y}_j$. Divided by $(r_j - 1)$, the resulting mean square is an estimator of σ_j^2. It is usually assumed that the distribution of the ε's for each treatment is the same, that is, $\sigma_1^2 = \sigma_2^2 = \cdots \sigma_t^2 = \sigma^2$, and the estimators of the σ_j^2 are pooled together to form a pooled estimator of σ^2. In regression terminology the pooled estimator is called the *residual mean square*, and for a completely randomized design the same quantity is called the *experimental error*. They are one and the same provided the regression model is correctly specified. However, as we proceed to more complex situations, two things can happen:

1. The measure of experimental error described for the completely randomized design will not be available for other experimental designs. Additional assumptions or increased experimentation will be needed if we are to have an estimator of σ^2. For example, in the usual randomized complete block design, the units (subjects or plots) are grouped into homogeneous blocks, and each treatment is allocated at random to one unit within each block. Each unit is now associated with a block-treatment combination and no replication of these combinations exists. Consequently, no measure of experimental error is available without replicating at least one of the treatments within the blocks.

2. The residual sum of squares from a regression analysis may include more than the experimental error. Whenever possible, we will want to divide the residual sum of squares into experimental error and a *remainder* sum of squares. This remainder is sometimes called *lack of fit* to indicate that the comparison of the lack-of-fit and experimental error mean squares is a measure of the adequacy of the fitted model. If both the lack-of-fit and

experimental error mean squares are judged to be estimating σ^2, the fitted model is adequately explaining the nonrandom variability among the observations.

CONSTRUCTION OF INDICATOR VARIABLES

Indicator variables really are not new to us. The x_0 variable in our regression model is an indicator variable. Each observation helps in estimating the intercept (or the overall mean if x_0 is the only x variable in the model), and a value of one is assigned to every row in the x_0 column. Remember that each row is uniquely associated with one observation. Therefore, if we set up an x variable for each treatment in an experiment, it will not be a foreign idea if we assign a value of one to every observation (every row) to which that treatment has been allocated, and assign a value of zero if another treatment was allocated. In this way we can construct an **X** matrix with t columns, one for each treatment. (Note that throughout our discussion the terms *treatment* and *subgroup* are used interchangeably.)

Before writing the **X** matrix directly, let's back up a step and write out the five observational equations, based on the general means model

$$y_{j\ell} = \mu_j + \varepsilon_{j\ell}$$

for a small sample with two treatments and with three observations on the first treatment and two on the second:

$$y_{11} = \mu_1 + \varepsilon_{11}$$
$$y_{12} = \mu_1 + \varepsilon_{12}$$
$$y_{13} = \mu_1 + \varepsilon_{13}$$
$$y_{21} = \mu_2 + \varepsilon_{21}$$
$$y_{22} = \mu_2 + \varepsilon_{22}$$

A tabular method for constructing the columns of the **X** matrix for any means model starts by forming a table with the means as the column headings and the observations as the rows. Enter as the tabular values the coefficients of the means in each observational equation. For our example we write a one if μ_1 or μ_2 is associated with the observation and a zero if not, giving us

	μ_1	μ_2
y_{11}	1	0
y_{12}	1	0
y_{13}	1	0
y_{21}	0	1
y_{22}	0	1

Construction of Indicator Variables

Now **X** is the body of the table; that is, the two columns of the table are labeled x_1 and x_2. Also, in matrix form we have

$$\mathbf{Y} = \begin{bmatrix} y_{11} \\ y_{12} \\ y_{13} \\ y_{21} \\ y_{22} \end{bmatrix} \qquad \boldsymbol{\beta} = \begin{bmatrix} \mu_1 \\ \mu_2 \end{bmatrix} \qquad \boldsymbol{\varepsilon} = \begin{bmatrix} \varepsilon_{11} \\ \varepsilon_{12} \\ \varepsilon_{13} \\ \varepsilon_{21} \\ \varepsilon_{22} \end{bmatrix}$$

We can see that the matrix equation

$$\mathbf{Y} = \mathbf{X}\boldsymbol{\beta} + \boldsymbol{\varepsilon} = \begin{bmatrix} 1 & 0 \\ 1 & 0 \\ 1 & 0 \\ 0 & 1 \\ 0 & 1 \end{bmatrix} \begin{bmatrix} \mu_1 \\ \mu_2 \end{bmatrix} + \begin{bmatrix} \varepsilon_{11} \\ \varepsilon_{12} \\ \varepsilon_{13} \\ \varepsilon_{21} \\ \varepsilon_{22} \end{bmatrix}$$

reproduces the observational equations. The row values in **X** relate each observation to a mean, and the two columns are called indicator variables.

As we have already noted, this **X** matrix could have been written directly, by defining the two x variables as

$x_1 = 1$ with presence of treatment 1
$\quad = 0$ with absence of treatment 1
$x_2 = 1$ with presence of treatment 2
$\quad = 0$ with absence of treatment 2

It is not difficult to see that the example can be generalized to t treatments and r_j replications. The **X** matrix will have t columns, one column for each treatment. The operating rule for the assignment of the ones and zeros within each column is straightforward.

The general means model can be written in our more familiar regression model format, but without x_0, as

$$y_i = \beta_1 x_{i1} + \beta_2 x_{i2} + \varepsilon_i \qquad i = 1, 2, \ldots, 5$$

Note that the single index for y runs from 1 to 5, inclusive. There is no loss of information from using a single subscript since the x's now indicate the treatment associated with each observation. The order of the observations is not important, but usually the observations are listed by treatments, by the measurement sequence, or by other practical considerations.

ESTIMATION OF MEANS AND STANDARD ERRORS

Using the **X** matrix from the previous section for the example of three observations on treatment 1 and two observations on treatment 2, we can apply the ABDO algorithm. Starting with the calculation of **X'X|X'Y**, the calculations for this example will be carried through without numerical values for the observations. In this way we can see that each sequential regression coefficient is indeed a familiar linear combination of the observations.

	Model:	$y_i = \beta_1 x_{i1} + \beta_2 x_{i2} + \varepsilon_i$		
	X'X|X'Y:	$\Sigma x_1^2 = 3$	$\Sigma x_1 x_2 = 0$	$\Sigma x_1 y = y_1 + y_2 + y_3$
			$\Sigma x_2^2 = 2$	$\Sigma x_2 y = y_4 + y_5$
Step 1:		3	0	$y_1 + y_2 + y_3$
Step 2:		1	0	$(y_1 + y_2 + y_3)/3 = b_1$
Step 3:			2	$y_4 + y_5$
Step 4:			1	$(y_4 + y_5)/2 = b_2$

Since the **X'X** matrix is a diagonal matrix, the partials can be read directly from steps 2 and 4; that is, b_1 is the mean of the treatment 1 observations, denoted as $\bar{y}_1$, and b_2 is the mean of the treatment 2 observations, $\bar{y}_2$. Both b_1 and b_2 can be expressed as linear combinations of the observations. From the rules given in Unit 11 each linear combination has a mean of $\ell_1 \mu_1 + \ell_2 \mu_1 + \ell_3 \mu_1 + \ell_4 \mu_2 + \ell_5 \mu_2$ and a variance of $\sigma^2 \Sigma \ell_i^2$. For our example the ℓ_i coefficients for the two linear combinations, from the last column of steps 2 and 4, are

$$[1/3 \quad 1/3 \quad 1/3 \quad 0 \quad 0] \quad \text{and} \quad [0 \quad 0 \quad 0 \quad 1/2 \quad 1/2]$$

Therefore,

b_1 has a mean of $\beta_1 = \mu_1$ and a variance of $\sigma^2/3$

b_2 has a mean of $\beta_2 = \mu_2$ and a variance of $\sigma^2/2$

where σ^2 is estimated by the residual mean square.

So far, we have seen that the indicator variables in the **X** matrix can be constructed so that the estimated regression coefficients are exactly the linear combinations $\bar{y}_1$ and $\bar{y}_2$ and the variances are σ^2/r_1 and σ^2/r_2, respectively. Usually the inference problem would be concerned with the difference between the two means, $\mu_1 - \mu_2$. If we are prepared to accept that the estimator of $\mu_1 - \mu_2$ is $\bar{y}_1 - \bar{y}_2$, then the variance of $\bar{y}_1 - \bar{y}_2$ could be found from the Variance Rule of a linear combination:

$$\text{Var}(\bar{y}_1 - \bar{y}_2) = (1)^2 \text{Var}(\bar{y}_1) + (-1)^2 \text{Var}(\bar{y}_2) = \sigma^2(1/r_1 + 1/r_2)$$

To see a full development of these results, the methodology of Unit 12 is employed. The contrast $(\mu_1 - \mu_2)$ can be expressed as $\mathbf{c}\boldsymbol{\beta}$, where $\mathbf{c} = [1\ -1]$. Application of the ELIM algorithm, starting with the ABDO matrix, gives the following array:

			Multipliers
3	0	$y_1 + y_2 + y_3$	$m_1 = 1/3$
	2	$y_4 + y_5$	$m_2 = -1/2$
1	-1	0	
0	-1	$-\bar{y}_1$	[row 3 $-(m_1)$row 1]
0	0	$-\bar{y}_1 + \bar{y}_2$	[row 4 $-(m_2)$row 2]

Consequently, $\mathbf{cb}$ is $-(-\bar{y}_1 + \bar{y}_2) = \bar{y}_1 - \bar{y}_2$, and the standard error is the square root of

$$\sigma^2\left[(1/3)^2(3) + (-1/2)^2(2)\right] = \sigma^2(1/3 + 1/2)$$

Actually, we didn't have to go through ABDO in order to find the elements of the ABDO matrix. The general means model leads to a diagonal $\mathbf{X'X}$. Consequently, the elements of the ABDO matrix are the elements of $\mathbf{X'X|X'Y}$. Proceeding directly to ELIM would give the estimate and standard error for the following parameters:

Parameter	c
μ_1	(1, 0)
μ_2	(0, 1)
$\mu_1 - \mu_2$	(1, -1)

SEQUENCING CANDIDATE MODELS

In the previous section ELIM was used to find the estimate of $\mu_1 - \mu_2$ and the estimated standard error of $\bar{y}_1 - \bar{y}_2$. Formally testing the null hypothesis of the equality of means for two treatments is testing the null hypothesis of $\mu_1 - \mu_2 = 0$. The t test statistic, equal to $\bar{y}_1 - \bar{y}_2$ divided by its estimated standard error, is commonly used.

Another approach to the inference problem of $\mu_1 - \mu_2$ is the sequencing and evaluation of candidate models. If the model

$$y_{j\ell} = \mu_j + \varepsilon_{j\ell} \quad \text{or} \quad y_i = \beta_1 x_{i1} + \beta_2 x_{i2} + \varepsilon_i$$

is the general means model, the question arises about the adequacy of a reduced means model. One natural candidate is the Mean Model of Unit 4,

$$y_{j\ell} = \mu + \varepsilon_{j\ell} \quad \text{or} \quad y_i = \mu x_{i0} + \varepsilon_i \quad i = 1, 2, \ldots, 5$$

The reduced model, which utilizes the x_0 column of ones, conjectures that one overall mean ($\mu_1 = \mu_2 = \mu$) will be fitted by the data as well as the more general model with means fitted for each of the two treatments. Consequently, the sequencing of the two models proceeds in the same spirit as in previous units. For example, in Unit 6 the overall mean model was compared with the Intercept and Slope Model; in Unit 10 with the two-variable firefly problem the sequence $x_0 | x_0 x_1 | x_0 x_1 x_2$ was investigated; and in Unit 13 with the electricity load data the adequacy of a common slope for straight lines was evaluated. Now the model sequence of interest is

$$x_0 | x_1 x_2 = \begin{bmatrix} 1 & 1 & 0 \\ 1 & 1 & 0 \\ 1 & 1 & 0 \\ 1 & 0 & 1 \\ 1 & 0 & 1 \end{bmatrix}$$

which is to be read as fitting the zero-variable model first, followed by what appears to be a two-variable model. In means model terminology the question is stated as follows: "Is the general means model $y_{j\ell} = \mu_j + \varepsilon_{j\ell}$ needed to describe the data, or is the equal means model $y_{j\ell} = \mu + \varepsilon_{j\ell}$ sufficient?"

From what we have learned in previous units, we will want to use one of two directions, ORTHO or ABDO, for handling the means model sequence:

$$x_0 | x_1 x_2$$

As we might expect, both computational approaches will give us the same final results. ORTHO was developed in Unit 11 and is a procedure for replacing columns of an **X** matrix by orthogonal columns.

In our means model sequence of three columns, three steps are needed:

Step 1: The first column of the orthogonalized sequence is the first column of $x_0 | x_1 x_2$.

Step 2: Regress x_1 on x_0. The residuals, $x_1 - \hat{x}_1(0)$, will be the new x_1.

Step 3: Regress x_2 on x_0 and the new x_1. The residuals will be the new x_2.

The details for calculating the sequential residuals include the calculations shown in Table 15.1.

Table 15.1 Calculating Sequential Residuals

$x_0\|x_1 x_2$	New x_0	New $x_1 = x_1 - \hat{x}_1(0)$	New $x_2 = x_2 - \hat{x}_2(0,1)$
1 1 0	1	$1-(3/5)=2/5$	$0-[(2/5)-(1)(2/5)]=0$
1 1 0	1	$1-(3/5)=2/5$	$0-[(2/5)-(1)(2/5)]=0$
1 1 0	1	$1-(3/5)=2/5$	$0-[(2/5)-(1)(2/5)]=0$
1 0 1	1	$0-(3/5)=-3/5$	$1-[(2/5)-(1)(-3/5)]=0$
1 0 1	1	$0-(3/5)=-3/5$	$1-[(2/5)-(1)(-3/5)]=0$

Combining the three new columns into one matrix, denoted as **L** in Unit 11,

$$\begin{bmatrix} 1 & 2/5 & 0 \\ 1 & 2/5 & 0 \\ 1 & 2/5 & 0 \\ 1 & -3/5 & 0 \\ 1 & -3/5 & 0 \end{bmatrix}$$

we have a surprising result, namely, that the third column elements are all zero. This shows that all the information in the means model sequence is contained in just two orthogonal columns and the third column of zeros can be deleted. For simplicity we will continue to call the two new columns x_0 and x_1. Consequently, our new means model sequence will be

$$x_0|x_1$$

and we can evaluate the need for the general means model relative to the equal means model. In the terminology of Unit 6 the reduced model is the equal means model and the full model is the general means model. The additional sum of squares for fitting the full model can be found from an ANOVA table with the following sums of squares:

$R(0) =$ sum of squares associated with the equal means model

$R(1|0) =$ additional sum of squares associated with x_1

  = difference between the equal means model and the general means model

residual = difference between $Y'Y$, the total uncorrected sum of squares, and $[R(0) + R(1|0)]$ with $(r_1 - 1) + (r_2 - 1)$ degrees of freedom

A formal test of the need of the general means model, or the adequacy of the equal means model, is accomplished by comparing the additional sum of squares for fitting the general means model to the residual by the F test statistic,

$$F = R(1|0) \text{ mean square/residual mean square}$$

with 1 and $(r_1-1)+(r_2-1)$ degrees of freedom.

From an estimation point of view, what are b_0 and b_1 estimating in terms of the means? Since the $X'X$ matrix is diagonal, b_0 and b_1 can be read directly from $X'X|X'Y$,

$$X'X|X'Y = \begin{bmatrix} 5 & 0 & | & y_1+y_2+y_3+y_4+y_5 \\ & 6/5 & | & (2y_1+2y_2+2y_3-3y_4-3y_5)/5 \end{bmatrix}$$

Consequently,

$$b_0 = \bar{y} \quad \text{and} \quad b_1 = \bar{y}_1 - \bar{y}_2$$

where $\bar{y}$ is the overall mean. Indeed, for our example with two treatments the F statistic used in testing the difference in the two models is equivalent to the t statistic used in testing the difference between μ_1 and μ_2 or in testing that β_1 is equal to zero. The t statistic is

$$t = b_1/(\text{estimated standard error of } b_1)$$

with $(r_1-1)+(r_2-1)$ degrees of freedom.

Returning to the original $x_0|x_1 x_2$ model sequence and using ABDO directly, we initiate the data analysis by calculating the sum of squares and cross products of $X'X|X'Y$ directly from the sequence of candidate models by adding the y variable at the end of the sequence:

$$X'X|X'Y = [x_0|x_1 x_2]'[x_0|x_1 x_2| y] = \begin{bmatrix} 5 & 3 & 2 & | & \Sigma y_{j\ell} = 5\bar{y} \\ & 3 & 0 & | & \Sigma y_{1\ell} = 3\bar{y}_1 \\ & & 2 & | & \Sigma y_{2\ell} = 2\bar{y}_2 \end{bmatrix}$$

The three ABDO stages are as follows:

Step 1:	5	3	2	$5\bar{y}$
Step 2:	1	3/5	2/5	$\bar{y}$
Step 3:		6/5	−6/5	$3\bar{y}_1 - 3\bar{y}$
Step 4:		1	−1	$5\bar{y}_1/2 - 5\bar{y}/2$
Step 5:			0	0
Step 6:			0	0

The implicit equation in step 6 is $0b_2 = 0$, and any value of b_2 is a solution. For simplicity, take $b_2 = 0$. Noting that $5\bar{y} = 3\bar{y}_1 + 2\bar{y}_2$, we have

$$b_1 - b_2 = 5\bar{y}_1/2 - 5\bar{y}/2 \quad \text{which gives} \quad b_1 = \bar{y}_1 - \bar{y}_2$$

from step 4. Reading step 2, $b_0 + 3b_1/5 + 2b_2/5 = \bar{y}$, together with $b_2 = 0$ and $\bar{y} = (3\bar{y}_1 + 2\bar{y}_2)/5$, we obtain

$$b_0 = \bar{y}_2$$

As we would expect, the ABDO calculations give the same results for b_1. The ANOVA calculations would be the same with the convention that the sum of squares and degrees of freedom associated with any pair of rows of all zeros would be set equal to zero.

If y is plotted against x_1, considering the $x_0 | x_1 x_2$ model sequence, all the y values are either at $x_1 = 0$ or at $x_1 = 1$. Thus the intercept b_0 is $\bar{y}_2$, and b_1, the change in y for one-unit change in x_1, is $\bar{y}_1 - \bar{y}_2$. The difference in the means is equivalent to the difference between $\hat{y}(0, 1)$ at $x_1 = 0$ and $\hat{y}(0, 1)$ at $x_1 = 1$, using the estimated model $\hat{y}(0, 1) = b_0 x_0 + b_1 x_1$. The graph is shown in Figure 15.1. A line is not drawn between the two estimated means since no meaning can be attached to predicted values between $x_1 = 0$ and $x_1 = 1$.

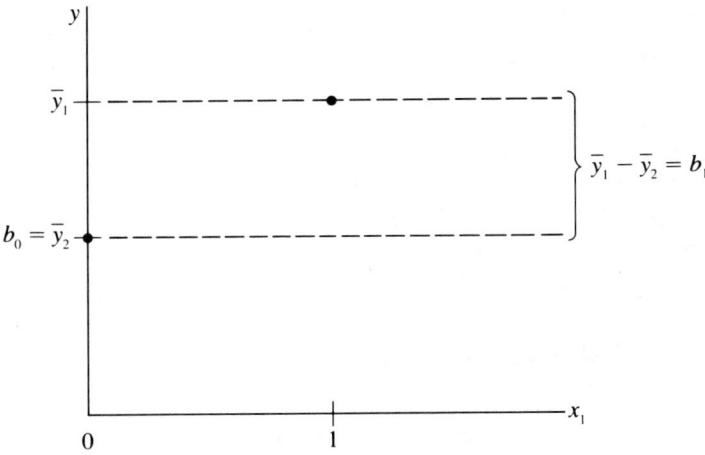

Figure 15.1 Plot of y Versus x_1

In retrospect, we fitted the general means model with x_1 and x_2 in the previous section. The model sum of squares for this case is the sum $R(1) + R(2)$. The upshot of fitting the sequence of models $x_0 | x_1$ after using ORTHO (or the sequence $x_0 | x_1 x_2$ with ABDO) in this section is the same model sum of squares as found in the previous section, with two degrees of freedom. In the model sequence the sum of squares is broken down into two parts, $R(0)$ and $R(1|0)$, so that the equality of the two means can be evaluated.

Several points in the interpretation of the estimated regression coefficients should be emphasized. For example, b_1 in the general means model does not have the same interpretation as b_1 in the sequenced model. Also, b_0 in the sequence with ORTHO is the overall mean, $\bar{y}$, while b_0 in the $x_0|x_1x_2$ sequence with ABDO is $\bar{y}_2$. The estimated regression coefficients in terms of the means can be summarized as shown in Table 15.2.

Table 15.2 Estimated Regression Coefficients in Terms of Means

	General Means Model	Model Sequence with ORTHO	Model Sequence with ABDO
b_0	—	$\bar{y}$	$\bar{y}_2$
b_1	$\bar{y}_1$	$\bar{y}_1 - \bar{y}_2$	$\bar{y}_1 - \bar{y}_2$
b_2	$\bar{y}_2$	—	—

If the data analyst wishes to determine $\bar{y}_1$ and $\bar{y}_2$ directly from the analysis, then the general means model is ideal. The estimated standard errors will be part of the regression program output. If the program has an appropriate option, the contrast $\mu_1 - \mu_2$ and the standard error of $\bar{y}_1 - \bar{y}_2$ can also be estimated, or these can be handled through ELIM. The model sequence approach evaluates the need of the general means model and, for two treatments, also gives the estimate of $\mu_1 - \mu_2$ and the associated estimated standard error as routine output.

EXTENSION TO THREE TREATMENTS

The construction of indicator x variables for the general means model with three treatments proceeds exactly the same as before, with the addition of a third variable x_3, where

$$x_3 = 1 \text{ with presence of treatment 3}$$
$$= 0 \text{ with absence of treatment 3}$$

The resulting regression model now has x_1, x_2, and x_3, and the estimated regression coefficients are $b_1 = \bar{y}_1$, $b_2 = \bar{y}_2$, and $b_3 = \bar{y}_3$. For a test of the sequence of models $x_0|x_1 x_2 x_3$, the input for ABDO is

$$\mathbf{X'X|X'Y} = \begin{bmatrix} n & r_1 & r_2 & r_3 & \Sigma y_{j\ell} \\ r_1 & 0 & 0 & & \Sigma y_{1\ell} \\ r_2 & & 0 & & \Sigma y_{2\ell} \\ r_3 & & & & \Sigma y_{3\ell} \end{bmatrix}$$

where r_1, r_2, and r_3 are the number of observations for the three treatments, n is the total number of observations, and the right-hand column includes the sum of all the observations followed by the sum of the observations for each of the treatments. The first row of the matrix is a result of the overall mean model appearing first in the sequence. The remainder of the matrix is the $\mathbf{X'X|X'Y}$ of the general means model including the sums of squares of x_1, x_2, and x_3 and the sums of cross products with y.

With x_0, $\mathbf{X'X}$ is not diagonal, and the estimated partial regression coefficients, using ABDO and the sequence $x_0|x_1 x_2 x_3$ change; they are

$$b_0 = \bar{y}_3 \qquad b_1 = \bar{y}_1 - \bar{y}_3 \qquad b_2 = \bar{y}_2 - \bar{y}_3 \qquad b_3 = 0.0$$

An important change from the two-treatment case is the testing for the need of the general means model (testing the adequacy of the equal means model). The first regression sum of squares from ABDO will be $R(0)$, the sum of squares associated with the equal means model. However, the additional sum of squares for the general means model will now be the sum of $R(1|0)$, $R(2|0,1)$, and $R(3|0,1,2)=0$, with $(1+1+0)$ degrees of freedom. In general, the number of means with the general means model minus one is the number of degrees of freedom associated with the mean square in the numerator of the F test statistic. For three treatments the null hypothesis is

$$H_0: \mu_1 = \mu_2 = \mu_3 = \mu$$

and the test statistic is

$$F = \frac{[R(1|0) + R(2|0,1)]/2}{\text{residual mean square}}$$

with two and $(r_1-1)+(r_2-1)+(r_3-1)$ degrees of freedom.

When the F test statistic is sufficiently large to indicate the need of the general means model, or in some situations even when F is relatively small, examination of linear combinations of the treatment means will give a more complete interpretation of the data. For example, if the three treatments included a standard treatment and two new treatments, then two meaningful comparisons would be (1) the standard (treatment 1) compared with the average of the new treatments and (2) one of the new treatments compared with the other new treatment. In terms of the parameters of the general means model, the two comparisons are

$$\mu_1 - \left(\frac{\mu_2 + \mu_3}{2}\right) \quad \text{and} \quad \mu_2 - \mu_3$$

The estimates and standard errors could be calculated by using the ABDO matrix from the general means model followed by ELIM for the two $c\beta$, where $c = [1 \ -1/2 \ -1/2]$ and $c = [0 \ 1 \ -1]$. Alternatives are developed in the next unit.

SUMMARY

Our regression models are adaptable for analysis of sample means by constructing indicator variables for the x variables. In the spirit of Unit 13, the first step is evaluating the adequacy of a reduced model. With the two means models of this unit, an equal means model is compared with a general means model by sequencing the two models. Interpretation of the estimated regression coefficients in terms of the sample means depends on the model sequence. Usually the general means model is needed, or the study has been designed to analyze selected contrasts among the means. Then the data analysis concentrates on estimating the contrasts and associated standard errors.

Unit 16

Case Study: Potato Leafhopper Survival Data

In studying the survival and behavioral responses of the potato leafhopper, Dahlman (1963) modified the basic synthetic feeding medium by additions of various sugars. The data in Table 16.1 (Table 15 from Dahlman) are the times (in days) when 50% of the insects in each cage were still living. The basic synthetic medium was 2% agar (control) and the treatments were additions of fructose, glucose, or sucrose to the control agar.

Table 16.1 Potato Leafhopper Survival Data

Treatment	Observation	
Control	2.3	1.7
Sucrose	3.6	4.0
Glucose	3.0	2.8
Fructose	2.1	2.3

Source: Douglas L. Dahlman, "Survival and Behavioral Responses of the Potato Leafhopper, *Empoasca Fabae* (Harris), on Synthetic Media (M.S. thesis, Iowa State University, 1963).

GENERAL MEANS MODEL

With 4 treatments and 8 observations, the general means model is

$$y_{j\ell} = \mu_j + \varepsilon_{j\ell} \quad j=1,2,3,4; \quad \ell=1,2$$

and the x variable matrix is

$$X = \begin{bmatrix} 1 & 0 & 0 & 0 \\ 1 & 0 & 0 & 0 \\ 0 & 1 & 0 & 0 \\ 0 & 1 & 0 & 0 \\ 0 & 0 & 1 & 0 \\ 0 & 0 & 1 & 0 \\ 0 & 0 & 0 & 1 \\ 0 & 0 & 0 & 1 \end{bmatrix}$$

Going through ABDO, we obtain

$$X'X|X'Y = \begin{bmatrix} 2 & 0 & 0 & 0 & | & 2.3+1.7 \\ & 2 & 0 & 0 & | & 3.6+4.0 \\ & & 2 & 0 & | & 3.0+2.8 \\ & & & 2 & | & 2.1+2.3 \end{bmatrix}$$

Step 1: 2   0   0   0   4.0
Step 2: 1   0   0   0   2.0

Step 3:   2   0   0   7.6
Step 4:   1   0   0   3.8

Step 5:   2   0   5.8
Step 6:   1   0   2.9

Step 7:   2   4.4
Step 8:   1   2.2

The means can be read directly from the right-hand elements of the even-numbered steps: $b_1 = \bar{y}_1 = 2.0$, $b_2 = \bar{y}_2 = 3.8$, $b_3 = \bar{y}_3 = 2.9$, and $b_4 = \bar{y}_4 = 2.2$. The ANOVA table is shown in Table 16.2.

Table 16.2 ANOVA for General Means Model

Source of Variation	Degrees of Freedom	Sum of Squares
Model	4	$(4.0)(2.0)+(7.6)(3.8)+(5.8)(2.9)+(4.4)(2.2)=63.38$
Residual	4	$\Sigma(y_{j\ell}-\hat{y}_{j\ell})^2=0.30$

In this example the residual sum of squares, the difference between the total and model sums of squares, is the experimental error sum of squares and could have been calculated directly by the pooled within-treatment sum of squares,

$$\sum_{j=1}^{4}\left[\sum_{\ell=1}^{2}(y_{j\ell}-\bar{y}_j)^2\right]$$

As shown in Unit 15, the standard errors for the means or linear combinations of the means can be calculated by using ELIM or by using the

General Means Model

Variance Rule for a linear combination. For example, the variance of the difference between two means is $2\sigma^2/r$ if the number of observations in each mean, r, is the same. For the leafhoppers both the standard deviation and the standard error of a difference between two means are estimated as the square root of $0.30/4$. With a regression program from a statistical computing package, the output would give the means and standard errors of the means.

Before considering the details of selecting an adequate model, we will pursue the estimation of contrasts and standard errors, using the general means model. Actually, if the experiment were designed with certain contrasts among the treatment means in mind, then one data analysis approach would proceed directly to the desired contrasts without evaluating the need of the general means model, as is done in the next section. With the leafhoppers a meaningful set of questions is as follows:

1. What is the difference between no sugar (control) and some sugar?
2. What is the difference between 6 carbon sugars (fructose and glucose) and 12-carbon sugar (sucrose)?
3. What is the difference between the two 6-carbon sugars?

Stated in terms of the population means, we have the following questions:

Contrast 1: $\mu_1 - (\mu_2 + \mu_3 + \mu_4)/3 = ?$

Contrast 2: $\mu_2 - (\mu_3 + \mu_4)/2 = ?$

Contrast 3: $\mu_3 - \mu_4 = ?$

If a regression program that will handle contrasts is available, the data analyst only has to be concerned with the original general means model. If not, one approach is to start with the ABDO matrix (the odd rows of ABDO), which can be written directly since $X'X$ is diagonal, and use ELIM. The contrast $\mu_1 - (\mu_2 + \mu_3 + \mu_4)/3$ can be expressed as $c\beta$, where $c = [1\ -1/3\ -1/3\ -1/3]$. Then we have the following computations:

					Multiplier
2	0	0	0	4.0	$m_1 = 1/2$
	2	0	0	7.6	$m_2 = -1/6$
		2	0	5.8	$m_3 = -1/6$
			2	4.4	$m_4 = -1/6$
1	$-1/3$	$-1/3$	$-1/3$	0	
0	$-1/3$	$-1/3$	$-1/3$	-2.00	[row 5 $-(m_1)$row 1]
0	0	$-1/3$	$-1/3$	-0.73	[row 6 $-(m_2)$row 2]
0	0	0	$-1/3$	0.23	[row 7 $-(m_3)$row 3]
0	0	0	0	0.97	[row 8 $-(m_4)$row 4]

Consequently, **cb** is $-0.97 = \bar{y}_1 - (\bar{y}_2 + \bar{y}_3 + \bar{y}_4)/3$, and the standard error is the square root of

$$[(1/2)^2(2) + (-1/6)^2(2) + (-1/6)^2(2) + (-1/6)^2(2)]\sigma^2 = (2/3)\sigma^2$$

The estimate of σ^2 was calculated to be $0.30/4$. The standard error of our contrast is 0.224, and a 95% confidence interval on the difference between no sugar and some sugar is $-0.97 \pm (2.776)(0.224)$.

Similar calculations with the other two contrasts will give the results shown in Table 16.3.

Table 16.3 Results for Contrasts 2 and 3

cb	Estimate	Standard Error	95% Confidence Interval
$\bar{y}_2 - (\bar{y}_3 + \bar{y}_4)/2$	1.25	0.237	(0.59, 1.91)
$\bar{y}_3 - \bar{y}_4$	0.70	0.274	(-0.06, 1.46)

SELECTION OF AN ADEQUATE MODEL

Is the general means model needed, or do the data indicate that the variability among the means is not much different from the experimental error variability? In other words, we want to test the equality of means hypothesis,

$$\mu_1 = \mu_2 = \mu_3 = \mu_4 = \mu$$

which is equivalent to testing the adequacy of the equal means model,

$$y_{j\ell} = \mu + \varepsilon_{j\ell}$$

We then set up our sequence of models, first the equal means model, followed by the general means model,

$$x_0 | x_1 x_2 x_3 x_4$$

where x_0 is the column of ones. Going through ABDO, we have the

Selection of an Adequate Model

following results:

$$X'X|X'Y = \begin{bmatrix} 8 & 2 & 2 & 2 & 2 & \Sigma y_{j\ell}=21.8 \\ & 2 & 0 & 0 & 0 & \Sigma y_{1\ell}=4.0 \\ & & 2 & 0 & 0 & \Sigma y_{2\ell}=7.6 \\ & & & 2 & 0 & \Sigma y_{3\ell}=5.8 \\ & & & & 2 & \Sigma y_{4\ell}=4.4 \end{bmatrix}$$

Step 1: 8	2	2	2	2	21.8
Step 2: 1	1/4	1/4	1/4	1/4	2.725
Step 3:	3/2	−1/2	−1/2	−1/2	4.0−5.45
Step 4:	1	−1/3	−1/3	−1/3	−0.97
Step 5:		4/3	−2/3	−2/3	7.6−5.45−0.48
Step 6:		1	−1/2	−1/2	1.25
Step 7:			1	−1	5.8−5.93+0.83
Step 8:			1	−1	0.70
Step 9:				0	0
Step 10:				0	0

In the two-treatment development of Unit 15, we saw that the sequence $x_0|x_1 x_2$ gave us all zeros for the last stage. Now we see the same result with four treatments. ABDO is indicating that only three columns, after x_0, are now giving us information on the additional sum of squares for fitting the general means model after fitting the equal means model. The resulting ANOVA table is given in Table 16.4.

Table 16.4 ANOVA for Testing Equality of Means

Source of Variation	Degrees of Freedom	Sum of Squares
Equal means model	1	$R(0)=(21.8)(2.725)=59.405$
Additional sum of squares for fitting general means model	3	$R(1|0)+R(2|0,1)+R(3|0,1,2)$ $=(-1.45)(-0.97)+(1.67)(1.25)+(0.7)(0.7)$ $=3.975$
Residual	4	0.30

Formally testing the need of the general means model involves the F statistic

$$F = (3.975/3)/(0.30/4) = 17.67$$

In conventional experimental design terminology, three degrees of freedom are associated with the between- (among-) treatments variability and four degrees of freedom with the within-treatments variability or experimental error.

Reemphasized is the equality of the sums of squares for the model sequence (59.405 + 3.975) and the general means model fitting alone (63.38). In other words, the model sequence approach is breaking down the model sum of squares associated with fitting the four means into an additive sequence of sums of squares. As with any set of data where the overall mean is not close to zero, the sum of squares associated with the equal means model in the leafhopper example is a large part of the 63.38. From the sequential approach of ABDO it should be clear that $(21.8)(2.725) = 59.405$, the first sum of squares, is associated with the equal means model $y_{j\ell} = \mu + \varepsilon_{j\ell}$.

ESTIMATION AND INTERPRETATION OF ESTIMATED COEFFICIENTS

Generalizing from results in Unit 15 for two and three treatments to t treatments, we see that the interpretation of the estimated partial regression coefficients using the sequence $x_0 | x_1 x_2 \cdots x_t$ follows a consistent pattern, namely,

- b_0 is $\bar{y}_t$, the mean of the treatment listed last.
- $b_1, b_2, \ldots, b_{t-1}$ are $\bar{y}_1 - \bar{y}_t, \bar{y}_2 - \bar{y}_t, \ldots, \bar{y}_{t-1} - \bar{y}_t$, respectively.
- $b_t = 0$.

Computationally, we start with the last regression coefficient (partial = sequential) and then go backward through the even steps of ABDO. For the leafhopper data we proceed as follows:

From step 10: $\quad\quad\quad\quad\quad\quad\quad\quad\quad\quad 0b_4 = 0 \quad\quad (b_4 = 0)$

From step 8: $\quad\quad\quad\quad\quad\quad\quad\quad\quad\quad b_3 - b_4 = 0.7$

$$b_3 = \bar{y}_3 - \bar{y}_4 = 0.7$$

From step 6: $\quad\quad\quad\quad\quad\quad\quad\quad b_2 - b_3/2 - b_4/2 = 1.25$

$$b_2 = \bar{y}_2 - \bar{y}_4 = 1.6$$

From step 4: $\quad\quad\quad\quad\quad\quad b_1 - b_2/3 - b_3/3 - b_4/3 = -0.97$

$$b_1 = \bar{y}_1 - \bar{y}_4 = -0.2$$

From step 2: $b_0+b_1/4+b_2/4+b_3/4+b_4/4=2.725$

$$b_0=\bar{y}_4=2.2$$

These are the partial regression coefficients from the model sequence $x_0|x_1x_2x_3x_4$, and they have a meaningful interpretation if the objective of the study is to compare each treatment with one of the treatments, like the control treatment of the leafhopper data. The control treatment was not last in the list of the data. If we had made the control the fourth column in $\mathbf{X}$, the partials would have been the comparisons with the control.

Fortunately, the same meaningful set of contrasts estimated from the general means model in the first section of this unit can also be estimated with the $x_0|x_1x_2x_3x_4$ model sequence. Simply include a zero for the first element in the $\mathbf{c}$ vector of coefficients, followed by the same coefficients previously used. For the difference between no sugar and some sugar, we have

$$\mathbf{c}=\begin{bmatrix}0 & 1 & -1/3 & -1/3 & -1/3\end{bmatrix}$$
$$\mathbf{cb}=[(0)(\bar{y}_4)+(\bar{y}_1-\bar{y}_4)-(\bar{y}_2-\bar{y}_4)/3-(\bar{y}_3-\bar{y}_4)/3-(1/3)(0)]$$
$$=-0.2-1.6/3-0.7/3=-0.97$$

which is the same as the estimate previously calculated.

The prediction equation,

$$\hat{y}_{j\ell}=2.2x_0-0.2x_1+1.6x_2+0.7x_3+0x_4$$

will give the predicted values, which are the estimated treatment means. In the $\mathbf{cb}$ format these values are as shown in Table 16.5.

Table 16.5 Predicted Values for Four-Treatment Model

Treatment	$\mathbf{c}=[x_{i0}\,x_{i1}\,x_{i2}\,x_{i3}\,x_{i4}]$					Estimated Mean $=\mathbf{cb}$
Control	1	1	0	0	0	$\hat{y}_{1\ell}=2.2-0.2=2.0=\bar{y}_1$
Sucrose	1	0	1	0	0	$\hat{y}_{2\ell}=2.2+1.6=3.8=\bar{y}_2$
Glucose	1	0	0	1	0	$\hat{y}_{3\ell}=2.2+0.7=2.9=\bar{y}_3$
Fructose	1	0	0	0	1	$\hat{y}_{4\ell}=2.2+0=2.2=\bar{y}_4$

From previous units we are aware that the prediction equation can also be written with the sequential regression coefficients, the values in the right-hand column for the even steps in ABDO. For the leafhoppers

$$\hat{y}_{j\ell}=2.725x_0-0.97[x_1-\hat{x}_1(0)]+1.25[x_2-\hat{x}_2(0,1)]+0.70[x_3-\hat{x}_3(0,1,2)]$$

will give the same predicted values as the predicted equation with the partial regression coefficients. Now the question naturally arises as to the interpretation of the sequential b's in terms of the treatment means. The sequential b_0 is clearly $\bar{y}$, the estimator of μ in the equal means model. From the development of ABDO in previous units, each x variable is adjusted for its own mean during stage 0. Consequently, b_1 is the regression on $x_1 - \hat{x}_1(0) = x_1 - \bar{x}_1 = x_1 - (2/8)$. When we write out the rows of the X matrix, the first two columns of the model sequence go from

x_0	x_1		x_0	$x_1 - \hat{x}_1(0)$
1	1		1	3/4
1	1		1	3/4
1	0		1	−1/4
1	0	to	1	−1/4
1	0		1	−1/4
1	0		1	−1/4
1	0		1	−1/4
1	0		1	−1/4

The new columns can be recognized as a result of using ORTHO from the previous unit. Since the two new columns are orthogonal, the interpretation of the sequential b_1 can be read directly from the new x_1 column, namely, b_1 is the comparison or contrast between the first two observations, associated with the control and the remaining observations. If ORTHO is applied to the model sequence $x_0 | x_1 x_2 x_3 x_4$, we can see what is happening to the x variables in ABDO and we can see that each of the new x values represents a contrast among the means. Since each column is orthogonal to each of the other columns, we say that we have an *orthogonal set of comparisons*. The new residual x variables calculated from ORTHO are as follows:

x_0	$x_1 - \hat{x}_1(0)$	$x_2 - \hat{x}_2(0,1)$	$x_3 - \hat{x}_3(0,1,2)$
1	3/4	0	0
1	3/4	0	0
1	−1/4	2/3	0
1	−1/4	2/3	0
1	−1/4	−1/3	1/2
1	−1/4	−1/3	1/2
1	−1/4	−1/3	−1/2
1	−1/4	−1/3	−1/2

The pattern is clear and a generalization can be stated. When an X matrix is constructed from the model sequence of the equal means model followed by the general means model, the first sequential b is $\bar{y}$ and the others are read as follows:

- $b_{1 \cdot 0}$ is the difference between the first treatment mean $\bar{y}_1$ and the average of all the other treatment means. For example, $b_1 = -0.97$ is the difference between $\bar{y}_1$ and $(\bar{y}_2 + \bar{y}_3 + \bar{y}_4)/3$.
- $b_{2 \cdot 0, 1}$ is the difference between the second treatment mean $\bar{y}_2$ and the average of the remaining treatment means.
$\vdots$
- $b_{t-1 \cdot 0, 1, \ldots, t-2}$ is the difference between the next to last treatment mean and the last treatment mean, where t is the number of treatments.

Actually, the interpretation of the sequential b's can be determined without solving for the residual x variables through ORTHO. Looking at the even steps of ABDO,

Step 4:   1 $-1/3$ $-1/3$ $-1/3$
Step 6:   1 $-1/2$ $-1/2$
Step 8:   1 -1

we can write the contrasts, to be called ORTHO contrasts, in terms of the means as follows:

$$\beta_{1 \cdot 0} = \mu_1 - (\mu_2 + \mu_3 + \mu_4)/3$$
$$\beta_{2 \cdot 0, 1} = \mu_2 - (\mu_3 + \mu_4)/2$$
$$\beta_{3 \cdot 0, 1, 2} = \mu_3 - \mu_4$$

These are the same contrasts as formulated earlier in the unit to answer the meaningful questions of the experiment.

When the set of contrasts has the general form of comparing the first treatment with the remaining, the second treatment with the remaining (except the first), the third treatment with the remaining (except the first and second), ..., and the next to last treatment with the last, the estimated sequential coefficients are the desired estimates, and formal testing can be carried out with the sequential sum of squares. In hindsight, we see that we fortuitously had the treatments in the order of control, sucrose, glucose, and fructose. If we hadn't, the x columns could have been rearranged to accommodate the desired comparisons.

CONSTRUCTING CONTRAST VARIABLES

By now we can begin to see that one data analysis strategy will be to replace the last four columns in the model sequence $x_0 | x_1 x_2 x_3 x_4$ by three orthogonal columns. This alternative is especially advantageous if the analysis is to be done by hand or if the only computer program available is a general regression program without the capability of handling contrasts. It will turn out that any three orthogonal columns will account for the additional sum of squares for fitting the general means model after the equal means model. With each column in the **X** matrix orthogonal to each other column, the partial b's will be the same as the sequential b's, since the **X'X** matrix will be diagonal. Any three orthogonal columns will give the correct additional sum of squares for the general means model, but we will want to construct three meaningful comparisons among the treatment means. In fact, each comparison or contrast should have a one-to-one correspondence with three subject matter issues concerning the treatments.

In the first section of this unit three questions were formulated and the following three contrasts stated (the treatments are in the order of control, sucrose, glucose, and fructose):

Contrast 1: $\mu_1 - (\mu_2 + \mu_3 + \mu_4)/3 = ?$

Contrast 2: $\mu_2 - (\mu_3 + \mu_4)/2 = ?$

Contrast 3: $\mu_3 - \mu_4 = ?$

Each contrast can be written as a linear combination of the means, $c\mu$, so that the **c** coefficients sum to zero. For example, $c_1 = [1\ -1/3\ -1/3\ -1/3]$, $c_2 = [0\ 1\ -1/2\ -1/2]$, and $c_3 = [0\ 0\ 1\ -1]$. By associating the elements in the **c**'s with the observations for treatments 1, 2, 3, and 4, respectively, we can construct three contrast columns for our orthogonal **X** matrix (x_0 of the equal means model will be the first column).

x_0	x_1	x_2	x_3
1	1	0	0
1	1	0	0
1	$-1/3$	1	0
1	$-1/3$	1	0
1	$-1/3$	$-1/2$	1
1	$-1/3$	$-1/2$	1
1	$-1/3$	$-1/2$	-1
1	$-1/3$	$-1/2$	-1

If we follow through with this constructed **X** matrix, using the contrast coefficients, the correct sums of squares will be found but the estimated coefficients will be proportional to the actual contrasts among the estimated treatment means. For example, b_3 will be equal to $(\bar{y}_3 - \bar{y}_4)/2$. In order to make the b's and the estimated standard errors equal to the desired values, we must divide each element of the contrast columns by the sum of squares of the contrast coefficients, that is, the sum of squares of the elements of the **c** vectors. Therefore, the x_1 column is divided by $3(-1/3)^2 + 1^2 = 4/3$, the x_2 column by $2(-1/2)^2 + 1^2 = 3/2$, and the x_3 column by $-1^2 + 1^2 = 2$. Then we have

x_0	x_1	x_2	x_3
1	3/4	0	0
1	3/4	0	0
1	−1/4	2/3	0
1	−1/4	2/3	0
1	−1/4	−1/3	1/2
1	−1/4	−1/3	1/2
1	−1/4	−1/3	−1/2
1	−1/4	−1/3	−1/2

which are the same residual x variables found in the previous section. The $X'X$ matrix is diagonal, and the same estimates can be calculated.

We now have a methodology for starting with any set of orthogonal contrasts and constructing an **X** matrix with orthogonal columns. If we were aware of a meaningful set of orthogonal contrasts at the start of the analysis, we could immediately set up an orthogonal **X** matrix if an option for handling contrasts is not part of the regression program. Generalizing from the leafhopper data to t means, we see that the number of contrast variables for representing any set of orthogonal contrasts is $t-1$. The contrast x variables are constructed by dividing the contrast coefficients by their sum of squares. The analysis can be done on a hand or desk calculator, or a regression program will give the estimated contrasts and standard errors as standard output.

SUMMARY

In experimental design situations treatments are selected with a view toward a planned set of contrasts among the treatment means. In this unit, by using regression methodology, we fit the general means model, calculated means and residual mean squares, and, with the availability of an appropriate option, estimated contrasts and associated standard errors.

An alternative approach is to fit the model sequence $x_0|x_1 x_2 \cdots x_t$. The need for the general means model is based on the additional sum of squares for $x_1, x_2, \ldots, x_t$. Contrasts among the means and associated standard errors can be estimated by ELIM or by a regression program with a capability for handling contrasts. Interpretation of the estimated sequential and partial regression coefficients from the model sequence is summarized in Table 16.6.

Table 16.6 Interpretation of Sequential and Partial Coefficients

Variable	Sequential	Partial
x_0	$\bar{y}$	$\bar{y}_t$
x_1	$\bar{y}_1 - (\bar{y}_2 + \bar{y}_3 + \bar{y}_4 + \cdots + \bar{y}_t)/(t-1)$	$\bar{y}_1 - \bar{y}_t$
x_2	$\bar{y}_2 - (\bar{y}_3 + \bar{y}_4 + \cdots + \bar{y}_t)/(t-2)$	$\bar{y}_2 - \bar{y}_t$
$\vdots$	$\vdots$	$\vdots$
x_{t-1}	$\bar{y}_{t-1} - \bar{y}_t$	$\bar{y}_{t-1} - \bar{y}_t$
x_t	0	0

When the sequential set of contrasts is the desired set, and with program output of sequentials and standard errors, the columns of **X** may be arranged to give the appropriate analysis. If the output gives at least the sequential sums of squares, the contrasts can be formally tested and the standard errors estimated by using the Variance Rule. Whereas the sequentials give an orthogonal set of comparisons and additive sums of squares, the partials do not. Partials are restricted to a set of nonorthogonal contrasts, comparing each treatment with one treatment—for example, the control.

Without a contrast option a third alternative is to adapt the $x_1, x_2, \ldots, x_t$ indicator variables of the model sequence as contrast variables. With an orthogonal set of contrasts the sequentials are the same as the partials and the sums of squares are additive. In this case the standard output of estimated partials, standard errors, and t tests will give the data analysis information needed for the interpretation.

Unit 17

Analysis of Cross-classified Sample Means

Objectives

- To analyze factorial experiments with regression methodology.
- To compare analyses of randomized block designs and factorial experiments.
- To develop a model sequence strategy for analyzing cross-classification data.

The regression approach to the analysis of sample means in Unit 15 includes these steps:

- Specification of the most general means model.
- Construction of an indicator **X** matrix.
- Calculation of means and residual mean squares.
- Selection of an adequate model by model sequencing.
- Estimation of meaningful contrasts and standard errors.

The basic concepts and calculations were set forth in Units 15 and 16, with particular emphasis on constructing indicator *x* variables, so that the regression output is a meaningful and straightforward interpretation of the data. This unit will apply the methodology developed in the previous two units to other regularly occurring situations. Specifically, our model sequence approach will be extended to handle data from factorial experiments and randomized block designs.

LYMPHOCYTE DATA

In this section we will look at an experiment conducted by the research team of Nishizawa, Kiskimoto, Kikutani, and Yamamura from the Osaka University Medical School, who reported on cytoplasmic factors that enhance nuclear nonhistone protein (NHP) in rabbit lymphocytes stimulated by antiimmunoglobulin antibodies (anti-Ig). Table II of their paper in the *Journal of Experimental Medicine* shows the effect of exogenously administered adenosine triphosphate (ATP) on phosphorylation activity. The response variable is measured by ^{32}P incorporation in NHP of nuclei incubated with extracts from anti-Ig-stimulated, or nonstimulated cells. The treatments and data from duplicate experiments are given in Table 17.1.

Again the general means model is

$$y_{j\ell} = \mu_j + \varepsilon_{j\ell} \quad j=1,2,3,4; \quad \ell=1,2$$

Note that the four treatments and two replications are the same as for the leafhopper data in the previous unit. Therefore, the **X** matrix for the general

Table 17.1 Lymphocyte Treatments and Data

Treatments	pmol of $^{32}P/\mu g$ of DNA*
Nonstimulated cells, no ATP added	34.0, 30.0
Anti-Ig-stimulated cells, no ATP added	62.5, 67.5
Nonstimulated cells, ATP added	25.5, 24.5
Anti-Ig-stimulated cells, ATP added	46.0, 50.0

Source: Yoshio Nishizawa, Tadamitsu Kishimoto, Hitoshi Kikutani, and Yuichi Yamamura. "Induction and Properties of Cytoplasmic Factor(s) Which Enhance Nuclear Nonhistone Protein Phosphorylation in Lymphocytes Stimulated by Anti-Ig," *Journal of Experimental Medicine* 146 (1977): 653–664.
*pmol = picomole; μg = microgram.

means model would be exactly the same. The resulting means are

$$\bar{y}_1 = 32 \quad \bar{y}_2 = 65 \quad \bar{y}_3 = 25 \quad \bar{y}_4 = 48$$

The ANOVA table is given in Table 17.2.

Table 17.2 ANOVA for Lymphocyte Data

Source of Variation	Degrees of Freedom	Sum of Squares
Model	4	$(64)(32)+(50)(25)+(130)(65)+(96)(48)=16,356$
Residual (experimental error)	4	$\Sigma(y_{j\ell} - \hat{y}_{j\ell})^2 = 29$

In considering meaningful contrasts among the treatment means, we notice that each treatment is comprised of two factors, the nature of the cells and the nature of the ATP addition. Each factor has two states or levels, nonstimulated or anti-Ig-stimulated for the cell factor, and addition or no addition for the ATP factor. Each treatment is a combination of one state of each factor. All possible combinations are present, so the total number of treatments is the product of the number of states or levels for all the factors in the experiment. Here we say that the treatments make up a 2×2 factorial, where the first value is the number of states for the ATP factor followed by the number of states for the cell factor.

With two factors the data are commonly expressed in a cross-classification format, the rows being one factor and the columns the second factor, as shown in Table 17.3. Realizing that each row and column combination is one of the four treatments, we can form a one-to-one correspondence between the originally stated four treatments and the row and column treatment designations. For our example μ_1 is μ_{11}, μ_2 is μ_{12}, μ_3 is μ_{21}, and μ_4 is μ_{22}, where the first subscript is the row number and the second is the column number. However, we will continue to use the single subscript in this section.

Table 17.3 Cross-Classification Format for Lymphocyte Data

	STATE OF CELL	
State of ATP	Nonstimulated	Anti-Ig-stimulated
No addition	34, 30	62.5, 67.5
Addition	25.5, 24.5	46, 50

With the structure of the 2×2 table and the subject matter, certain contrasts among the four treatments are more obvious. Previous experimentation had shown increased incorporation of ^{32}P into NHP of the nuclei that had been incubated with cell-free extract from anti-Ig-stimulated cells. To confirm the previous work, we will want to compare the observations in the first column with those in the second column. From subject matter knowledge the ATP treatments were included to check the conjecture that the increased incorporation might be reduced by the addition of ATP. Therefore, we want to compare the observations in the first row with those in the second row. In terms of the μ_j's of the general means model, these two contrasts are stated as follows:

Contrast 1: $(\mu_1 + \mu_3)/2 - (\mu_2 + \mu_4)/2$
Contrast 2: $(\mu_1 + \mu_2)/2 - (\mu_3 + \mu_4)/2$

In our analysis we will want to partition the model sum of squares into two parts, the sum of squares associated with the equal means model and the additional sum of squares with three degrees of freedom for fitting the general means model. Then we will break down the three degrees of freedom into three contrasts among the means. We have now specified two of the three contrasts. Since we have compared row means in one contrast and column means in another, a third contrast could compare the average of the observations in one diagonal of the 2×2 layout with the average of the observations in the other diagonal. In terms of the four means of the general means model, this third contrast is

$$(\mu_1 + \mu_4)/2 - (\mu_2 + \mu_3)/2$$

This contrast is offering the possibility of a difference between the "none and both" means and the "one at a time" means. Perhaps the contrast is more identifiable in an algebraically equivalent form,

Contrast 3: $\quad (\mu_4 - \mu_2)/2 - (\mu_3 - \mu_1)/2$

In subject matter terms, does the addition of ATP give the same change for both stimulated and nonstimulated cells? If so, the $(\mu_4 - \mu_2)/2$ difference will equal the $(\mu_3 - \mu_1)/2$ difference, and the terminology of no interaction between the two factors is used. Another equivalent version of the interaction contrast is to compare the difference between the two columns for each row:

$$(\mu_4 - \mu_3)/2 - (\mu_2 - \mu_1)/2$$

As summarized in Unit 16, contrasts can be estimated by one of two methods. The first method is to replace the $x_0|x_1 x_2 x_3 x_4$ sequence by x_0 and three additional variables in the **X** matrix, one for each contrast. The values for the new contrast variables come directly from the coefficients of the μ_j's in the three contrasts. The sum of squares of the coefficients for each contrast is one, so that direct use of the coefficients as contrast variables will give the desired estimates. With the rows of the **X** matrix in consecutive order of the four treatments, we have

$$\mathbf{X} = \begin{bmatrix} 1 & 1/2 & 1/2 & 1/2 \\ 1 & 1/2 & 1/2 & 1/2 \\ 1 & -1/2 & 1/2 & -1/2 \\ 1 & -1/2 & 1/2 & -1/2 \\ 1 & 1/2 & -1/2 & -1/2 \\ 1 & 1/2 & -1/2 & -1/2 \\ 1 & -1/2 & -1/2 & 1/2 \\ 1 & -1/2 & -1/2 & 1/2 \end{bmatrix} \quad \mathbf{Y} = \begin{bmatrix} 34.0 \\ 30.0 \\ 62.5 \\ 67.5 \\ 25.5 \\ 24.5 \\ 46.0 \\ 50.0 \end{bmatrix}$$

Note that each variable of the **X** matrix is orthogonal to every other variable; consequently, **X'X** is diagonal, the ABDO calculations are easy, and the sequential b coefficients are the same as the partial b's. Carrying out the details, we obtain

$$X'X|X'Y = \begin{bmatrix} 8 & 0 & 0 & 0 & | & 340 \\ & 2 & 0 & 0 & | & -56 \\ & & 2 & 0 & | & 24 \\ & & & 2 & | & -10 \end{bmatrix}$$

The ABDO calculations are as follows:

Step 1:	8	0	0	0	340
Step 2:	1	0	0	0	42.5
Step 3:		2	0	0	−56
Step 4:		1	0	0	−28
Step 5:			2	0	24
Step 6:			1	0	12
Step 7:				2	−10
Step 8:				1	−5

With this formulation of the **X** matrix, the output from a computer regression program will give exactly what we want: the estimated regression coefficients and their estimated standard errors. For our data we obtain

$$b_0 = \bar{y} = 42.5$$
$$b_1 = \text{column contrast} = (\bar{y}_1 + \bar{y}_3)/2 - (\bar{y}_2 + \bar{y}_4)/2 = -28$$
$$b_2 = \text{row contrast} = (\bar{y}_1 + \bar{y}_2)/2 - (\bar{y}_3 + \bar{y}_4)/2 = 12$$
$$b_3 = \text{interaction contrast} = (\bar{y}_4 - \bar{y}_2)/2 - (\bar{y}_3 - \bar{y}_1)/2 = -5$$

The resulting ANOVA table is given in Table 17.4.

Table 17.4 ANOVA with Breakdown of Model Sum of Squares

Source of Variation	Degrees of Freedom	Sum of Squares
Overall mean	1	$R(0) = (340)(42.5) = 14450$
Nonstimulation versus anti-Ig stimulation	1	$R(1) = R(1|0) = (-56)(-28) = 1568$
No ATP versus ATP addition	1	$R(2) = R(2|0, 1) = (24)(12) = 288$
Interaction	1	$R(3) = R(3|0, 1, 2) = (-10)(-5) = 50$
Residual	4	29

The need of the interaction contrast variable is tested by

$$F = 50/(29/4) = 6.9$$

with 1 and 4 degrees of freedom. The probability of obtaining a value this large or larger on chance alone is between 0.10 and 0.05. If the interaction is judged not to be important, the variability among the means can be adequately described by the row and column contrasts, that is, $\mathbf{X} = [x_0 | x_1 x_2]$. The model testing then proceeds to consider the need of x_1 and (or) x_2. Both the no-ATP versus the ATP addition contrast and the nonstimulation versus the anti-Ig stimulation contrast would be judged important by the magnitude of the two F's,

$$F = 1568/7.25 \quad \text{and} \quad F = 288/7.25$$

The model formulation stage of the data analysis would then conclude that the equal means model is not adequate. The variability among the four treatment means can be explained by the ATP factor and the stimulation factor without the need of an interaction.

If the F of 6.9 for interaction is judged to be important by the data analyst, then the interpretation of the contrasts represented by the x_1 and x_2 contrast variables are not as clear-cut. Strictly speaking, the difference between the row averages does not have a straightforward interpretation since the interaction indicates that the ATP addition varies, depending on the state of the stimulation factor. Looking at the data, we must conclude that the difference $\bar{y}_4 - \bar{y}_2 = 48 - 65 = -17$ is estimating something different than the difference $\bar{y}_3 - \bar{y}_1 = 25 - 32 = -7$. Finding an important interaction is significant in itself, and a reasonable follow-up is the estimation of the individual differences $\mu_4 - \mu_2$ and $\mu_3 - \mu_1$ (or $\mu_4 - \mu_3$ and $\mu_2 - \mu_1$). In our data set both the -17 and the -7 are in the same direction; that is, ATP addition enhanced phosphorylation. Some data analysts would go out on a limb to interpret the average, $[-17 + (-7)]/2 = -12$, even when the interaction, if judged to be important, is indicating that the two individual differences are estimating different quantities.

An alternative methodology for estimating contrasts works well when an option in the regression program is available to obtain output for estimated contrasts and associated standard errors from either the general means model or a model sequence of the equal means model followed by the general means model. This is equivalent to starting with the general means model, $\mathbf{X} = [x_1 x_2 x_3 x_4]$, calculating the ABDO matrix, and proceeding to ELIM (Unit 12). For example, for estimating the interaction contrast $(\mu_4-\mu_2)/2-(\mu_3-\mu_1)/2$, $\mathbf{c\beta}=(\mu_1-\mu_2-\mu_3+\mu_4)/2$ or $\mathbf{c}=[1/2\ -1/2\ -1/2\ 1/2]$. Then we have the following computations:

					Multiplier
2	0	0	0	64	$m_1=1/4$
	2	0	0	50	$m_2=-1/4$
		2	0	130	$m_3=-1/4$
			2	96	$m_4=1/4$
1/2	−1/2	−1/2	1/2	0	
0	−1/2	−1/2	1/2	−16	
0	0	−1/2	1/2	−3.5	
0	0	0	1/2	29	
0	0	0	0	5	

Consequently, $\mathbf{cb}=-5=(\bar{y}_4-\bar{y}_2)/2-(\bar{y}_3-\bar{y}_1)/2$. The standard error of the contrast is the square root of

$$[(1/4)^2(2)+(-1/4)^2(2)+(-1/4)^2(2)+(1/4)^2(2)]\sigma^2=(1/2)\sigma^2$$

which is estimated by the square root of $(29/4)(1/2)$. The 95% confidence interval is

$$-5\pm(2.776)(1.9) \quad \text{or} \quad -10.27 \text{ to } 0.27$$

FAT DIGESTIBILITY DATA

Now we will look at the results of another study. During three periods of time, fat digestibility was measured on young dairy calves fed milk replacer as their sole source of nutrients. The objectives of the experiment were to compare two sources of fat supplement to a skim milk basal diet and to estimate the effect of including lecithin in the diets. It was conjectured that lecithin would act as an emulsifier and improve fat utilization. Four treatments are considered here:

1. tallow
2. cocoanut fat
3. tallow plus lecithin
4. cocoanut fat plus lecithin

Owing to limited physical conditions, the treatments were allocated at random to four calves, and digestibility coefficients were measured over a period of time. The trial was then repeated with other calves during two other periods. This experimental design is commonly called a *randomized complete block design*, where the blocks are the periods and each fat supplement treatment is randomly allocated to calves within each block. The digestibility coefficient data are given in Table 17.5.

Table 17.5 Fat Digestibility Data

Diet	Period 1	Period 2	Period 3	Total
1	64.6	52.4	53.8	170.8
2	66.0	60.1	64.4	190.5
3	85.0	68.9	77.5	231.4
4	96.0	90.4	98.2	284.6
Total	311.6	271.8	293.9	

Source: Daniel T. Hopkins, "The Digestibility of Fats by Dairy Calves," unpublished MNS Report on Special Problem (Cornell University, 1958).

Handling the data analysis without consideration of periods would give treatment means $\bar{y}_1 = 56.93$, $\bar{y}_2 = 63.50$, $\bar{y}_3 = 77.13$, and $\bar{y}_4 = 94.87$, and a mean square of 33.74 with 8 degrees of freedom. However, some of the variability among the 12 observations can be explained by the period-to-period variability, as expressed by the differences among the three period totals. We have no interest in forming period contrasts and making inferences. Our only interest is to recognize period-to-period variability as part of the residual sum of squares. The differences among the three periods can be removed from the residual by constructing one indicator variable for each period, x_5, x_6, and x_7, and fitting the model sequence

equal means|treatment indicators|period indicators

or

$$x_0 | x_1 x_2 x_3 x_4 | x_5 x_6 x_7$$

where x_1, x_2, x_3, and x_4 are indicator variables for the fat supplement treatments.

Starting with $\mathbf{X'X|X'Y}$, the ABDO calculations and the ANOVA table are as given in Table 17.6. Even though the treatment indicator variables are

Fat Digestibility Data

not orthogonal to the period indicator variables, the estimated sequentials, partials, and sums of squares for treatments do not depend on the order of the treatment and period indicator variables. The zeros in steps 3 and 4 of stage 1, steps 5 and 6 of stage 2, and steps 7 and 8 of stage 3 indicate that the estimated partials for treatments are not influenced by the period indicators. These zeros result from the section of constant values (ones in our example) in the $X'X$ matrix. In experimental design terminology each treatment is allocated to a constant number (usually one) of experimental units within each block. This balance is a property of randomized complete block designs. Since order doesn't affect the data analysis, we will write the model sequence for randomized blocks as

equal means|treatment and block indicators

Table 17.6 Calculations for Fat Digestibility Data

			MODEL SEQUENCE						
	Equal Means	Treatment Indicators				Period Indicators			
	x_0	x_1	x_2	x_3	x_4	x_5	x_6	x_7	y
	1	1	0	0	0	1	0	0	64.6
	1	1	0	0	0	0	1	0	52.4
	1	1	0	0	0	0	0	1	53.8
	1	0	1	0	0	1	0	0	66.0
	1	0	1	0	0	0	1	0	60.1
	1	0	1	0	0	0	0	1	64.4
	1	0	0	1	0	1	0	0	85.0
	1	0	0	1	0	0	1	0	68.9
	1	0	0	1	0	0	0	1	77.5
	1	0	0	0	1	1	0	0	96.0
	1	0	0	0	1	0	1	0	90.4
	1	0	0	0	1	0	0	1	98.2
	12	3	3	3	3	4	4	4	877.3
		3	0	0	0	1	1	1	170.8
			3	0	0	1	1	1	190.5
				3	0	1	1	1	231.4
$X'X\|X'Y$:					3	1	1	1	284.6
						4	0	0	311.6
							4	0	271.8
								4	293.9

Table 17.6 (Continued)

MODEL SEQUENCE

	Equal Means		Treatment Indicators			Period Indicators			
	x_0	x_1	x_2	x_3	x_4	x_5	x_6	x_7	y
Step 1:	12	3	3	3	3	4	4	4	877.3
Step 2:	1	1/4	1/4	1/4	1/4	1/3	1/3	1/3	73.108
Step 3:		9/4	−3/4	−3/4	−3/4	0	0	0	−48.525
Step 4:		1	−1/3	−1/3	−1/3	0	0	0	−21.567
Step 5:			2	−1	−1	0	0	0	−45.
Step 6:			1	−1/2	−1/2	0	0	0	−22.5
Step 7:				3/2	−3/2	0	0	0	−26.6
Step 8:				1	−1	0	0	0	−17.733
						0	0	0	0	0
						0	0	0	0	0
Step 11:						8/3	−4/3	−4/3	19.167
Step 12:						1	−1/2	−1/2	7.1875
Step 13:							2	−2	−11.050
Step 14:							1	−1	−5.5249
								0	0
								0	0

ANOVA for Fat Digestibility Data

Source of Variation	Degrees of Freedom	Sum of Squares
Total	12	66,938.59
Equal means model $= R(0)$	1	64,137.94
Treatments $= R(1,2,3,4\|0)$ $= R(1,2,3,4\|0,5,6,7)$	3	2,530.73
Periods $= R(5,6,7\|0,1,2,3,4)$ $= R(5,6,7\|0)$	2	198.81
Residual	6	71.11

By including both the period and treatment indicator variables, we have reduced the residual mean square from 33.74 to $71.11/6 = 11.85$. We have also cut the degrees of freedom from 8 to 6, but this loss has been more than compensated for by the reduction in the residual mean square to be used in calculating standard errors of contrasts among the means. The contrasts given by most of the sequentials or the partials are not of particular subject matter interest. If an option for estimating contrasts is available in a regression program, then the contrast coefficients would be specified, and the estimates and standard errors would become part of the output.

For the fat digestibility data questions of interest include the following:

- What is the difference between lecithin and no lecithin?
- What is the difference between the two fats without lecithin?
- What is the difference between the two fats with lecithin?

Admittedly, other meaningful sets of questions could be formulated. Stated in terms of the population treatment means, the contrasts are as follows:

$$\text{Contrast 1:} \quad (\mu_1 + \mu_2)/2 - (\mu_3 + \mu_4)/2$$
$$\text{Contrast 2:} \quad \mu_1 - \mu_2$$
$$\text{Contrast 3:} \quad \mu_3 - \mu_4$$

Each of the contrasts needs to be formulated as $c\beta$, remembering that our model sequence includes eight x variables: x_0, four indicators for the treatments, and three indicators for the periods. Consequently, the coefficients of the c vectors (the input) and the estimated contrasts and standard errors (the output) are as shown in Table 17.7.

Table 17.7 Estimated Contrasts from a Regression Program

Contrast Coefficients	Estimate	Standard Error
[0 1/2 1/2 −1/2 −1/2 0 0 0]	−25.78	1.99
[0 1 −1 0 0 0 0 0]	−6.57	2.81
[0 0 0 1 −1 0 0 0]	−17.73	2.81

If a regression program for handling contrasts is not available, we could use ELIM or construct contrast variables and use the model sequence

$$x_0 | x_1 x_2 x_3 | x_5 x_6 x_7$$

where x_1, x_2, and x_3 are the contrast variables replacing the indicator variables for treatments. Now the **X** matrix will be

x_0	x_1	x_2	x_3	x_5	x_6	x_7
1	1/2	1/2	0	1	0	0
1	1/2	1/2	0	0	1	0
1	1/2	1/2	0	0	0	1
1	1/2	−1/2	0	1	0	0
1	1/2	−1/2	0	0	1	0
1	1/2	−1/2	0	0	0	1
1	−1/2	0	1/2	1	0	0
1	−1/2	0	1/2	0	1	0
1	−1/2	0	1/2	0	0	1
1	−1/2	0	−1/2	1	0	0
1	−1/2	0	−1/2	0	1	0
1	−1/2	0	−1/2	0	0	1

This set of contrasts is an orthogonal set, and, in addition, the contrast variables are orthogonal to the period indicator variables. The ANOVA sums of squares will be the same as before. And now the sum of squares for treatments can be broken down into three sums of squares, each associated with a contrast as expressed by the contrast variables and each associated with one degree of freedom. The F test results will be equivalent to those from the t test, where t is an estimated contrast divided by the estimated standard error of the contrast.

COMPARISON OF THE LYMPHOCYTE AND FAT DIGESTIBILITY DATA ANALYSES

Both the data sets analyzed in this unit can be presented in a cross-classification form. For example, we can think of the treatments in terms of rows and the periods in terms of columns for the fat digestibility data. In this format a general means model for a cross- or two-way classification could be written for the fat digestibility data as

$$y_{j\ell} = \mu_{j\ell} + \varepsilon_{j\ell} \qquad j=1,2,3,4; \qquad \ell=1,2,3$$

where the $\mu_{j\ell}$'s are the means of each row and column combination. Using our indicator variable approach for the digestibility data, we would set up 12 variables, one for each of the 4×3 treatment-by-period combinations. With one observation for each of the 12 treatment and period combinations, each regression estimate would be equal to the observation for that treatment and period combination, and $\hat{y}_{j\ell} = \bar{y}_{j\ell}$. Consequently, the two subscripts on the $\mu_{j\ell}$ are also the only two subscripts on the epsilon, and we don't have any measure of the residual.

To compensate for the lack of a residual, our approach was to model treatment contrasts and to include period indicators instead of modeling all 12 means. In this way we have used only 6 degrees of freedom for the model, leaving 6 degrees of freedom for the residual. In general two-way classification terminology, the model sequence is

$$\text{equal means|row and column indicators}$$

It is now of interest to note that we could have modeled the lymphocyte data as

$$y_{j\ell m} = \mu_{j\ell} + \varepsilon_{j\ell m} \qquad j = \ell = m = 1, 2$$

Now we see the explicit two-way structure of the factorial arrangement of the four treatments, where the rows ($j = 1, 2$) are the two states of ATP and the columns ($\ell = 1, 2$) are the two states of cell stimulation, as briefly introduced in the first section. For each row and column combination we recorded two observations ($m = 1, 2$). Earlier we analyzed these data with μ_j ($j = 1, 2, 3, 4$) as the general means model and proceeded directly to specific contrasts. An alternative general means model, formally recognizing the cross-classification structure, is the $\mu_{j\ell}$ model ($j = \ell = 1, 2$). For our example the equivalence is shown by

	State of Cell	
State of ATP	Column 1	Column 2
Row 1	$\mu_1 = \mu_{11}$	$\mu_2 = \mu_{12}$
Row 2	$\mu_3 = \mu_{21}$	$\mu_4 = \mu_{22}$

The four indicator variables for the $\mu_{j\ell}$ model are the same as those for the μ_j model. We can now reanalyze the data by focusing on a model sequence approach, with the $\mu_{j\ell}$ model as the general means model.

At an early stage of the reanalysis, we will want to evaluate the necessity of the general means model. If it is not needed, we can envision that (1) only the ATP factor is important, (2) only the stimulation factor is important, (3) neither is important, or (4) both are important. To accommodate these possibilities, we will construct two row indicator variables for ATP and two column indicator variables for stimulation. If the most reduced model is the equal means model, an appropriate model sequence for evaluating the general means model is

equal means|row and column indicators|general means

or

$$x_0 | x_1 x_2 x_3 x_4 | x_5 x_6 x_7 x_8$$

where x_1 is the row 1 indicator variable, x_2 is the row 2 indicator variable, x_3 is the column 1 indicator variable, x_4 is the column 2 indicator variable, x_5 is the (1,1) treatment combination indicator variable, x_6 is the (1,2) treatment combination indicator variable, x_7 is the (2,1) treatment combination indicator variable, and x_8 is the (2,2) treatment combination indicator variable.

In general, and specifically in Unit 18, the ordering of the row and column indicator variables within the model sequence can be important, but

Table 17.8 Calculations for Lymphocyte Data

	Equal Means	Row Indicators		Column Indicators		General Means				
	x_0	x_1	x_2	x_3	x_4	x_5	x_6	x_7	x_8	y
	1	1	0	1	0	1	0	0	0	34.0
	1	1	0	1	0	1	0	0	0	30.0
	1	1	0	0	1	0	1	0	0	62.5
	1	1	0	0	1	0	1	0	0	67.5
	1	0	1	1	0	0	0	1	0	25.5
	1	0	1	1	0	0	0	1	0	24.5
	1	0	1	0	1	0	0	0	1	46.0
	1	0	1	0	1	0	0	0	1	50.0
	8	4	4	4	4	2	2	2	2	340
		4	0	2	2	2	2	0	0	194
			4	2	2	0	0	2	2	146
				4	0	2	0	2	0	114
X'X|X'Y:					4	0	2	0	2	226
						2	0	0	0	64
							2	0	0	130
								2	0	50
									2	96

Table 17.8 (Continued)

	Equal Means	Row Indicators		Column Indicators		MODEL SEQUENCE General Means				
	x_0	x_1	x_2	x_3	x_4	x_5	x_6	x_7	x_8	y
Step 1:	8	4	4	4	4	2	2	2	2	340
Step 2:	1	0.5	0.5	0.5	0.5	0.25	0.25	0.25	0.25	42.5
Step 3:		2	−2	0	0	1	1	−1	−1	24
Step 4:		1	−1	0	0	0.5	0.5	−0.5	−0.5	12
				0	0	0	0	0	0	0
				0	0	0	0	0	0	0
Step 7:				2	−2	1	−1	1	−1	−56
Step 8:				1	−1	0.5	−0.5	0.5	−0.5	−28
						0	0	0	0	0
						0	0	0	0	0
Step 11:						0.5	−0.5	−0.5	0.5	−5
Step 12:						1	−1	−1	1	−10
							0	0	0	0
							0	0	0	0
								0	0	0
								0	0	0
									0	0
									0	0

the ordering is not important in this unit, since we have an equal number of observations for each treatment combination. The terminology of row and column indicators will denote that either order of these two indicator variables within the model sequence can be used. Table 17.8 presents the indicator variables for the model sequence and the subsequent ABDO calculations.

In the first section of this unit we found the following sums of squares associated with three orthogonal contrast variables:

x_1 = stimulation contrast $R(1) = (-56)(-28) = 1568$

x_2 = ATP contrast $R(2) = (24)(12) = 288$

x_3 = interaction contrast $R(3) = (-10)(-5) = 50$

The equivalence between these sums of squares and the sequential sums of squares of Table 17.8 is immediate for rows, where $(\text{ATP}) = R(1|0) = (24)(12) = 288$ (steps 3 and 4), and for columns where $(\text{stimulation}) = R(3|0,1) = (-56)(-28) = 1568$ (steps 7 and 8). With an equal number of observations for each treatment combination, a block of zeros results in steps 3 and 4, ensuring the same sums of squares of 1568 and 288 if rows and columns had been interchanged.

The more interesting point is the recovery of the interaction contrast sum of squares in steps 11 and 12, as $(-5)(-10) = 50$. The additional sum of squares for the general means model in the $x_0, x_1, \ldots, x_8$ sequence is completely contained in x_5, and the associated sum of squares is the sum of squares for interaction between the row factor (ATP) and the column factor (stimulation). Indeed, one definition of an interaction is the sum of squares associated with the general means model fitted after row and column indicator variables.

In Unit 16 interpretations of the sequential coefficients were ascertained directly from the ABDO output. Specifically, the numerical values in the even steps are the coefficients of the ORTHO contrasts among the means of the assumed model. In Table 17.8 variables x_5 through x_8 are the indicator variables for the $\mu_{j\ell}$ means model. The values from the last five columns of the even-numbered steps of stage 1 (step 4), stage 3 (step 8), and stage 5 (step 12),

0.5	0.5	-0.5	-0.5	12
0.5	-0.5	0.5	-0.5	-28
1	-1	-1	1	-10

are the coefficients and estimates for the row, column, and interaction contrasts,

$$0.5\mu_{11} + 0.5\mu_{12} - 0.5\mu_{21} - 0.5\mu_{22}$$

$$0.5\mu_{11} - 0.5\mu_{12} + 0.5\mu_{21} - 0.5\mu_{22}$$

$$\mu_{11} - \mu_{12} - \mu_{21} + \mu_{22}$$

The first two contrasts are exactly the row and column contrasts called contrasts 1 and 2 in the earlier analysis of the lymphocyte data. At that time, the interaction contrast was defined as the difference between two averages, each involving two means. The indicator variables for the general means model do not account for this factor of 2. But if $b_5 = -10$ (from step 12) is divided by 2, we obtain the same estimate as before, -5. As already noted, the interaction sum of squares is not affected.

If the interaction is judged to be not important, the general means model is not needed and can be deleted from the model sequence. The reduced sequence is

$$\text{equal means}|\text{row and column indicators}$$

or

$$x_0|x_1 x_2 x_3 x_4$$

Fortunately the same ABDO calculations of the full model sequence can be used by ignoring the general means model part of the table. Steps 4 and 8 of Table 17.8,

Step 4: 1 -1 0 0 0.5 0.5 -0.5 -0.5 12
Step 8:   1 -1 0.5 -0.5 0.5 -0.5 -28

would then be

$$\begin{array}{cccccc} 1 & -1 & 0 & 0 & & 12 \\ & & 1 & -1 & & -28 \end{array}$$

The first contrast in step 4 is the row (ATP) contrast with an estimate of 12, and the second contrast is the column (stimulation) contrast with an estimate of -28. The need for each is based on the independent assessment of each contrast, using the residual mean square estimated from the reduced model sequence.

Formal testing involves the additional sum of squares from steps 3 and 4 and from steps 7 and 8. The numerators of the F test statistics are uncorrelated. Equivalently, t tests can be used to evaluate the two contrasts. Strictly speaking, if either the row or column indicators are judged to be not important, a further reduction in the model sequence should be considered —for example, equal means|row indicators—so that a new residual mean square can be estimated. In practice, the residual mean square estimated from the equal means|row and column indicators sequence can suffice for all remaining evaluations. In fact, some data analysts would use the estimated residual mean square from the full model sequence for all evaluations. For a regression program output with sequential sums of squares, this evaluation can be accomplished by the one-model sequence,

$$\text{equal means}|\text{row and column indicators}|\text{general means}$$

When the interaction is judged to be important—that is, the general means model is needed—inferences now involve contrasts among the $\mu_{j\ell}$

means within a row or within a column. In general, the same contrasts among the row means for the reduced model sequence will also be the desired contrasts when the interaction is important, but now the contrasts will be estimated within each of the columns. For the 2×2 factorial we would want to estimate a pair of contrasts, $\mu_{11}-\mu_{21}$ and $\mu_{12}-\mu_{22}$, or, equivalently, the differences within each row, $\mu_{11}-\mu_{12}$ and $\mu_{21}-\mu_{22}$. The choice between the two pairs is based on the subject matter interest in the row and column factors.

Actually, the estimated partial regression coefficients for x_1 and x_3 in the model sequence approach in Table 17.8 are of some interest. From steps 4 and 8,

$$b_1 = 12 - 0.5b_5 = 12 - 0.5(-10) = 17 = \bar{y}_{12} - \bar{y}_{22}$$

and

$$b_3 = -28 - 0.5b_5 = -28 - 0.5(-10) = -23 = \bar{y}_{21} - \bar{y}_{22}$$

Each of these differences is interesting, but the pair do not tell the interaction story. Furthermore, with larger factorials the partials from the row and column indicator variables in the full model sequence will continue to be differences between individual means and the mean of the treatment combination of the last row and last column. As suggested by previous interpretations of partials, these simple differences do not give a complete data interpretation unless the specific interest is limited to comparisons with a single treatment combination—for example, the control.

Let's return to the fat digestibility data. It is now clear that our randomized block analysis is part of a full factorial analysis. However, only the row and column indicators were included in the model sequence. If the general means model had been added, our residual sum of squares of the randomized block analysis would have been calculated directly as the interaction sum of squares. Furthermore, to obtain a residual (or experimental error term), we assumed the interaction between blocks and treatments to be nothing more than random error and calculated the residual sum of squares by subtraction from a sequence with only row and column indicators. The no-interaction assumption is a standard assumption in the analysis of data from a randomized block design where each treatment has been allocated at random to one experimental unit within each block.

Even though the computations for a randomized block design and a factorial follow the same model sequence approach, the randomization procedures and inferential structure are different. Blocks are formed by stratifying experimental units and are introduced in the analysis to reduce the residual sum of squares. Inferences concerning blocks usually are not warranted. With factorials the treatment combinations of both factors are allocated at random to the experimental units, and we are interested in inferences on both factors.

SUMMARY

A data analysis strategy for the model sequence

equal means|row and column indicators|general means

is shown in the following diagram:

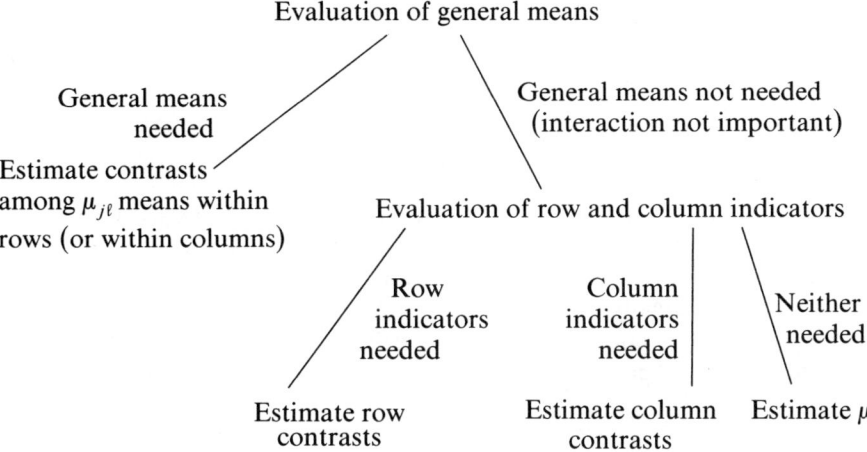

Interpretation of estimated sequential and partial regression coefficients follows the patterns developed in Units 15–17. For example, a 2×3 factorial with equal numbers of observations for each $\bar{y}_{j\ell}$ mean has the format below:

	Columns		
Rows	1	2	3
1	$\bar{y}_{11}$	$\bar{y}_{12}$	$\bar{y}_{13}$
2	$\bar{y}_{21}$	$\bar{y}_{22}$	$\bar{y}_{23}$

With the model sequence

equal means|row and column indicators|general means

the sequential coefficient for rows compares the row 1 means with the row 2 means, while the partial is $\bar{y}_{13} - \bar{y}_{23}$. The first sequential column coefficient compares the first column means with the average of the second and third column means, while the second sequential column coefficient compares the second column means with the third column means (the two partials are

$\bar{y}_{21} - \bar{y}_{23}$ and $\bar{y}_{22} - \bar{y}_{23}$). The first interaction coefficients can be displayed as follows:

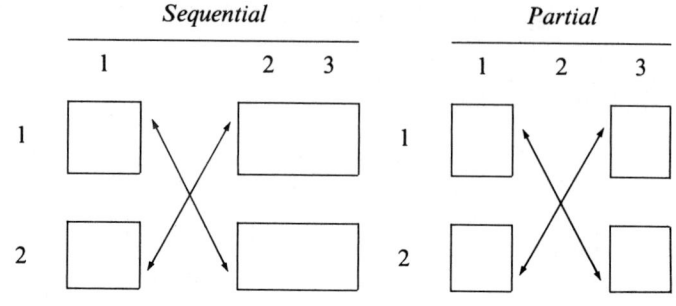

The display indicates that the partial coefficient is an interaction contrast among the four means from the two rows and the first and third columns. The sequential coefficient, however, is an interaction contrast utilizing all six means; specifically, $\bar{y}_{11}$, $\bar{y}_{22}/2$ and $\bar{y}_{23}/2$ are compared with $\bar{y}_{21}$, $\bar{y}_{12}/2$ and $\bar{y}_{13}/2$. The second interaction coefficient is the same for both the sequential and the partial, namely,

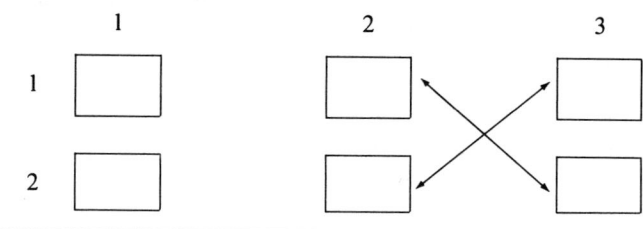

Unit 18

Analysis of Sample Means with Unequal Numbers of Observations

Objectives

- To extend regression methodology and strategy to sample means with unequal numbers of observations.
- To compare procedures for estimating contrasts among the sample means.
- To interpret sequential and partial regression coefficients.

In the previous units on sample means the number of observations for each sample mean has been the same. We will now consider the same kinds of analyses with unequal numbers of observations for each sample mean, sometimes called unbalanced data. The general approach to the analysis will be the same as before; in fact, the output will have the same appearance. However, the computations are more complicated, and some interesting things can happen to affect our interpretations. The seriousness of these effects depends, of course, on the type and magnitude of the imbalance. In experimental studies experiments are designed to have equal sample numbers for each treatment. But sometimes unequal numbers result when cost or variance considerations are important. However, the most common reason for unequal numbers is the loss of experimental units during the course of the experiment. These losses are usually not serious, but a general understanding of the computations will enable us, as data analysts, to evaluate the seriousness. The problem becomes more complex in sampling studies where sample sizes are proportional to population sizes or in observational studies where one sample is selected and the units are categorized into various subgroups. The degree of imbalance can now be rather extreme, especially when the study is planned without forethought about the categories or when there are budget limitations. As in the previous unit, the kinds of problems encountered here are developed through the analysis of two data sets.

PROTEIN NUTRITION DATA

From a data analysis query in the journal *Biometrics*, a nutrition trial evaluated sources of protein supplement on chick growth. Fourteen birds of the same breed, age, and sex were selected for each treatment. Shown here in Table 18.1 are the final weights for the standard horsebean source of protein and two alternative sources. During the experiment some chicks were lost due to unknown causes, resulting in unequal numbers of observations per treatment.

The general means model is

$$y_{j\ell} = \mu_j + \varepsilon_{j\ell} \qquad j=1,2,3;\ \ell=1,2,\cdots,n_j;\ n_1=10,\ n_2=12,\ n_3=14$$

Table 18.1 Final Weights (Grams) of Chicks at Six Weeks in Nutrition Study

	Horsebean	Linseed Oil Meal	Soybean Oil Meal
	179	309	243
	160	229	230
	136	181	248
	227	141	327
	217	260	329
	168	203	250
	108	148	193
	124	169	271
	143	213	316
	140	257	267
		244	199
		271	177
			158
			248
Sample size	10	12	14
Sample mean	160.2	218.75	246.88

Source: "Query," Biometrics 4 (1948): 213–214.

The **X** matrix will be 36×3, with ones in the first ten rows of column 1, in rows 11–22 of column 2, and in rows 23–36 of column 3, and with zeros for all the other elements. Going through ABDO will give the means and the ANOVA table, which is shown in Table 18.2.

Table 18.2 ANOVA for Nutrition Data

Source of Variation	Degrees of Freedom	Sum of Squares
Model	3	$(1{,}602)(160.2)+(2{,}625)(218.75)+(3{,}456)(246.86)$
Residual (experimental error)	33	80,650

A natural set of two contrasts among the treatment means might consist of one to compare the standard horsebean supplement with the average of the two alternative sources and one to compare the two alternative sources. That is,

$$\text{Contrast 1:} \quad \mu_1 - (\mu_2 + \mu_3)/2 \quad \mathbf{c}_1 = [1 \quad -1/2 \quad -1/2]$$

$$\text{Contrast 2:} \quad \mu_2 - \mu_3 \quad \mathbf{c}_2 = [0 \quad 1 \quad -1]$$

Starting with the ABDO matrix, the ELIM calculations for the first contrast are as follows:

```
                                              Multiplier
 10        0         0       1602          m₁ = 1/10
          12         0       2625          m₂ = -1/24
                    14       3456          m₃ = -1/28
----------------------------------------
  1     -1/2      -1/2          0
----------------------------------------
  0     -1/2      -1/2      -160.2
           0      -1/2      -50.825
                     0        72.60
```

$m_1 = 1/10$, $m_2 = -1/24$, $m_3 = -1/28$

The results for the two contrasts are summarized in Table 18.3.

Table 18.3 Results for Two Contrasts

cb	Estimate	Variance	95% Confidence Interval
$\bar{y}_1 - (\bar{y}_2 + \bar{y}_3)/2$	−72.60	(0.1386)(80,650/33)	(−110.09, −35.11)
$\bar{y}_2 - \bar{y}_3$	−28.11	(1/12 + 1/14)(2444)	(−67.72, 11.50)

The t statistic values for testing the two null hypotheses are -3.94 for contrast 1 and -1.445 for contrast 2, both with 33 degrees of freedom. We then have strong evidence that a difference exists between the standard horsebean supplement and the average of the two alternatives, but the evidence for a difference between linseed and soybean oil meals is much weaker.

In Units 15 and 16 the ORTHO orthogonalizing procedure applied to the equal means|general means model sequence produced a set of orthogonal contrasts. Specifically, each new x variable is a contrast between a treatment mean and the average of the remaining treatments. ORTHO applied to the protein nutrition model sequence $x_0 | x_1 \ x_2 \ x_3$, gives the following orthogonalized **X** matrix. The letters in the identification column of row numbers denote horsebean (H), linseed oil meal (L), and soybean oil meal (S).

Row	x_0	x_1	x_2	x_3
H1	1	13/18	0	0
H2	1	13/18	0	0
⋮	⋮	⋮	⋮	⋮
H10	1	13/18	0	0
L11	1	−5/18	7/13	0
L12	1	−5/18	7/13	0
⋮	⋮	⋮	⋮	⋮
L22	1	−5/18	7/13	0
S23	1	−5/18	−6/13	0
S24	1	−5/18	−6/13	0
⋮	⋮	⋮	⋮	⋮
S36	1	−5/18	−6/13	0

This orthogonalized **X** matrix will give the same estimated sequential coefficients and sequential sums of squares as given by applying ABDO to the equal means|general means model sequence with the indicator variables.

In previous units with an equal number of observations per treatment, the ORTHO contrasts among the means could be easily identified from the orthogonalized x variables. Now more care has to be exercised; namely, the ORTHO contrasts are as follows:

Contrast 1: $(9/65)[(13/18)(y_1 + y_2 + \cdots + y_{10})$
$- (5/18)(y_{11} + y_{12} + \cdots + y_{36})] = \bar{y}_1 - [(6/13)\bar{y}_2 + (7/13)\bar{y}_3]$

Contrast 2: $(13/84)[(7/13)(y_{11} + y_{12} + \cdots + y_{22})$
$- (6/13)(y_{23} + y_{24} + \cdots + y_{36})] = \bar{y}_2 - \bar{y}_3$

If we consider only the coefficients of the means, the contrasts do not appear to be orthogonal, but with unequal numbers of observations the orthogonal **X** matrix from ORTHO has to be checked. Then we are assured that the contrasts are orthogonal and that the individual sums of squares will add to the model sum of squares. With the protein nutritional data the contrasts of $\bar{y}_1$ with the average of $\bar{y}_2$ and $\bar{y}_3$, and $\bar{y}_2$ with $\bar{y}_3$, are the desired subject matter contrasts. ABDO then gives the following results:

36	10	12	14	7683
	10	0	0	1602
		12	0	2625
			14	3456
36	10	12	14	7683
1	5/18	6/18	7/18	213.42
	130/18	−60/18	−70/18	−532.17
	1	−6/13	−7/13	−73.69
		84/13	−84/13	−181.65
		1	−1	−28.11
			0	0
			0	0

The value of the sequential x_1 coefficient, -73.69 in step 4, is the estimated ORTHO contrast

$$\bar{y}_1 - [(6/13)\bar{y}_2 + (7/13)\bar{y}_3]$$

that is, it is the difference between horsebean and a weighted average of the two alternative protein sources. To be specific, the 6/13 and 7/13 weights are the relative sample size proportions (12/26 and 14/26). The value -73.69 is, of course, different from the natural contrast, $\bar{y}_1 - (\bar{y}_2 + \bar{y}_3)/2 = 72.60$.

With equal numbers of observations for each treatment, step 4 of ABDO would have been 1, $-1/2$, and $-1/2$, and the same estimates would have been calculated by either approach. With unequal numbers, which is correct? There is no easy answer.

One advantage of ORTHO contrasts is that the sequential sums of squares do add to the general means model sum of squares with three degrees of freedom. The resulting ANOVA table is given in Table 18.4.

The two F statistic values are

$$\frac{(-532.17)(-73.69)}{80{,}650/33} = 16.05 \quad \text{and} \quad \frac{(-181.65)(-28.11)}{80{,}650/33} = 2.09$$

for the comparisons of horsebean with the two alternatives and of linseed with soybean, respectively. The sequential sums of squares are additive, resulting in uncorrelated numerators of the two F statistics, a desirable result that does not hold for natural contrasts. Another statistical argument for the ORTHO contrast is that the variance of the estimated contrast is minimized with the unequal weights.

Table 18.4 ANOVA for ORTHO Approach

Source of Variation	Degrees of Freedom	Sum of Squares
Equal means model	1	$R(0) = (7683)(213.42)$
Additional sum of squares for comparing horsebean with the two alternatives	1	$R(1\|0) = (-532.17)(-73.69)$
Additional sum of squares for comparing linseed with soybean	1	$R(2\|0,1) = (-181.65)(-28.11)$
Residual (experimental error)	33	80,650

On the other hand, suppose the contrast of interest is designed to be a comparison between horsebean and the simple average of the two alternatives. Then the conceptual population sizes are the same, and the varying sample sizes are a result of accidental reasons. Consequently, the 1/2 and 1/2 weights seem more reasonable than the 6/13 and 7/13 weights, which are dependent on the data structure. The ORTHO contrast differs from the natural contrast to the extent of the imbalance of the sample sizes.

One routine approach to analyzing means models with unequal numbers is to use both ORTHO and natural contrasts. If they are in agreement, then a general interpretation of the data is straightforward. However, if the sequential sums of squares approach gives a different evaluation of the data, then this is evidence that the correlation among the numerators of the test statistics is sufficient cause for the data interpretations to be accompanied by guarded statements.

The output from regression programs would give these partial regression coefficients:

$b_3 = 0$ (by convention)
$b_2 = -28$ (the difference between the linseed and soybean means)
$b_1 + (-6/13)(-28.11) = -73.69$
$b_1 = -87$ (the difference between the horsebean and soybean means)
$b_0 + (5/18)(-86.56) + (6/18)(-28.11) = 213.42$
$b_0 = 247$ (the soybean mean)

The t tests in regression outputs are testing the null hypotheses that the betas are equal to zero. Considering b_1 and b_2, we are estimating another set of two contrasts among the means, specifically, the difference between horsebean and soybean and the difference between linseed and soybean. However, the two contrasts are not orthogonal, and the numerators of the two t test statistics will be correlated; that is, the sums of squares for the two contrasts can be sufficiently entangled so that each contrast cannot be independently interpreted.

SWAMP pH DATA

In this section we look at another set of data. As part of an ecological study in 1979, Laura Foster Huenneke identified three types of central New York swamp communities. The three types could be found near a stream and away from a stream. For each of the 2×3 factorial combinations of stream proximity and swamp community, land areas were randomly selected, with the number of selected areas proportional to the total land area. The data in Table 18.5 are pH readings measured on soil samples from the selected areas.

Table 18.5 pH Readings for Swamp Soil Samples

Stream Location	Swamp Community		
	North	Mesic	Shrub
Near	6.6, 7.2, 7.2, 7.0, 6.8, 6.4, 7.0, 6.4 $\sum y_{11m} = 54.6$ $n_{11} = 8$ $\bar{y}_{11} = 6.825$	6.8, 7.0, 6.2, 6.2, 6.4, 6.2 $\sum y_{12m} = 38.8$ $n_{12} = 6$ $\bar{y}_{12} = 6.467$	6.4, 5.2, 6.2, 6.4, 6.4 $\sum y_{13m} = 30.6$ $n_{13} = 5$ $\bar{y}_{13} = 6.12$
Away	6.8, 7.0 $\sum y_{21m} = 13.8$ $n_{21} = 2$ $\bar{y}_{21} = 6.9$	6.2, 5.6, 7.2, 7.2, 7.2, 6.2, 7.2, 7.2 $\sum y_{22m} = 54.0$ $n_{22} = 8$ $\bar{y}_{22} = 6.75$	7.2, 6.8, 7.0, 6.8, 7.0, 6.6 $\sum y_{23m} = 41.4$ $n_{23} = 6$ $\bar{y}_{23} = 6.9$

Source: Laura Foster Huenneke, unpublished data (1979).

As in Unit 17, the general means model is

$$y_{j\ell m} = \mu_{j\ell} + \varepsilon_{j\ell m} \qquad j=1,2;\ \ell=1,2,3;\ m=1,2,\cdots,n_{j\ell}$$

The unequal numbers of observations for each mean will not change our general data analysis approach. Order of the row and column indicators within the model sequence will now be important, and, subsequently, the choice of contrasts is more challenging if both row and column indicators are needed. Indicator variables for equal mean (x_0), stream location (x_1 and x_2), swamp community (x_3, x_4, and x_5), and the general means model (x_6 through x_{11}) are displayed in Table 18.6 followed by **X'X|X'Y**. General means indicators will also be known as interaction indicators, and, in practice, all the indicator variables can be automatically generated by a regression program. The ABDO calculations in Table 18.7 for the model

Table 18.6 Indicator Variables and X'X|X'Y

x_0	x_1	x_2	x_3	x_4	x_5	x_6	x_7	x_8	x_9	x_{10}	x_{11}
Eight lines of											
1	1	0	1	0	0	1	0	0	0	0	0
Six lines of											
1	1	0	0	1	0	0	1	0	0	0	0
Five lines of											
1	1	0	0	0	1	0	0	1	0	0	0
Two lines of											
1	0	1	1	0	0	0	0	0	1	0	0
Eight lines of											
1	0	1	0	1	0	0	0	0	0	1	0
Six lines of											
1	0	1	0	0	1	0	0	0	0	0	1

X'X|X'Y:

35	19	16	10	14	11	8	6	5	2	8	6	233.2
	19	0	8	6	5	8	6	5	0	0	0	124.0
		16	2	8	6	0	0	0	2	8	6	109.2
			10	0	0	8	0	0	2	0	0	68.4
				14	0	0	6	0	0	8	0	92.8
					11	0	0	5	0	0	6	72.0
						8	0	0	0	0	0	54.6
							6	0	0	0	0	38.8
								5	0	0	0	30.6
									2	0	0	13.8
										8	0	54.0
											6	41.4

Table 18.7 ABDO Calculations

Steps	35	19	16	10	14	11	8	6	5	2	8	6	
1													233.2
2	1	0.543	0.457	0.286	0.4	0.314	0.229	0.171	0.143	0.057	0.229	0.171	6.663
3		8.69	−8.69	2.57	−1.6	−0.971	3.66	2.74	2.28	−1.09	−4.34	−3.26	−2.594
4		1	−1	0.296	−0.184	−0.112	0.421	0.316	0.263	−0.125	−0.5	−0.375	−0.299
5			0	0	0	0	0	0	0	0	0	0	0
6			0	0	0	0	0	0	0	0	0	0	0
7				6.38	−3.53	−2.85	4.63	−2.53	−2.10	1.75	−1	−0.750	2.539
8				1	−0.553	−0.447	0.726	−0.396	−0.330	0.274	−0.157	−0.118	0.398
9					6.16	6.16	0.033	2.71	−2.74	−0.033	3.45	3.42	0.445
10					1	−1	0.005	0.440	−0.445	−0.005	0.560	−0.555	0.072
11						0	0	0	0	0	0	0	0
12						0	0	0	0	0	0	0	0
13							1.27	−0.707	−0.563	−1.27	0.707	0.563	0.544
14							1	−0.557	−0.443	−1	0.557	0.443	0.428
15								1.52	−1.52	0	−1.52	1.52	0.754
16								1	−1	0	−1	1	0.497
17									0	0	0	0	0
18									0	0	0	0	0
19										0	0	0	0
20										0	0	0	0
21											0	0	0
22											0	0	0
23												0	0
24												0	0

sequence

equal means|row indicators|column indicators|general means

indicate that one row contrast with one degree of freedom, two column contrasts with two degrees of freedom, and two interaction contrasts with two degrees of freedom can be estimated from the two row indicators, three column indicators, and six interaction indicators.

In Unit 17 contrasts estimated by the sequential coefficients would be readily identified by examining the even ABDO steps in the columns associated with (under) the general means indicator variables. From Table 18.7 we read the following:

Row contrast

$$\text{Step 4: } 0.421\bar{y}_{11} + 0.316\bar{y}_{12} + 0.263\bar{y}_{13} - 0.125\bar{y}_{21} - 0.500\bar{y}_{22} - 0.375\bar{y}_{23} = -0$$

Column contrast 1

$$\text{Step 8: } 0.726\bar{y}_{11} - 0.396\bar{y}_{12} - 0.330\bar{y}_{13} + 0.274\bar{y}_{21} - 0.157\bar{y}_{22} - 0.118\bar{y}_{23} = 0.3$$

Column contrast 2

$$\text{Step 10: } 0.005\bar{y}_{11} + 0.440\bar{y}_{12} - 0.445\bar{y}_{13} - 0.005\bar{y}_{21} + 0.560\bar{y}_{22} - 0.555\bar{y}_{23} = 0.0$$

Interaction contrast 1

$$\text{Step 14: } \bar{y}_{11} - 0.557\bar{y}_{12} - 0.443\bar{y}_{13} - \quad \bar{y}_{21} + 0.557\bar{y}_{22} + 0.443\bar{y}_{23} = 0.4$$

Interaction contrast 2

$$\text{Step 16: } \bar{y}_{12} - \quad \bar{y}_{13} \quad -\bar{y}_{22} + \quad \bar{y}_{23} = 0.4$$

Interpretation of the estimated sequential coefficients in this manner was not emphasized in Unit 17 since we were concerned with only a 2×2 factorial, and with the equal number of observations; order of the row and column indicators was not important. Now the general structure of the contrasts can be recognized from our previous experience with ORTHO contrasts, but, except for the row and last interaction contrasts, the origin of the contrast coefficients is not obvious. The row contrast compares the row 1 means with row 2 means. The positive coefficients add to $+1$ and the negative coefficients add to -1. Furthermore, each coefficient is the proportion of the number of observations for each mean within the row to the total number of observations for the row; for instance, for $\bar{y}_{11}$, $0.421 = 8/19$. Then the estimated sequential, -0.299, is nothing more than the difference between the average of the row 1 observations and the row 2 average and does not utilize any information concerning the columns. The last interaction contrast in the factorial analysis is the interaction contrast among the

Table 18.8 Even Steps in ABDO When Columns Precede Rows in Model Sequence

Steps	Columns		Rows		General Means				Sequentials
4	−0.56	−0.44	0.36	−0.36	0.8	−0.24	−0.2	0.2	−0.32	−0.24	0.248
6	1	−1	−0.026	0.026	0	0.429	−0.455	0	0.571	−0.545	0.083
10			1	−1	0.206	0.442	0.352	−0.206	−0.442	−0.352	−0.415
14					1	−0.557	−0.443	−1	0.557	0.443	0.428
16						1	−1	0	−1	1	0.497

means of the last two rows (the only two in our example) and the last two columns and carries a straightforward interpretation.

The other intermediate contrasts are weighted contrasts of the six means. The contrast coefficients are data-dependent in the sense that the coefficients will change with different numbers of observations for the means. As stressed in the first part of this unit, these coefficients are relevant weights if the sample size proportions approximate the proportions in the six populations. For the swamp data this assumption is reasonable.

As already mentioned several times, the sequential estimates of the row and column contrasts depend on the ordering of the row and column indicators. Table 18.8 displays the pertinent even steps for the model sequence

$$\text{equal means}|\text{column indicators}|\text{row indicators}|\text{general means}$$

The interaction contrasts, following the row and column indicators in each sequence, stay the same, as we would expect, but the column and row contrasts have changed. Now the column contrast coefficients reflect column sample size proportions. The first column contrast compares the north swamp with the other two communities, with

$$\mathbf{c} = [\,0.8 \quad -0.24 \quad -0.2 \quad 0.2 \quad -0.32 \quad -0.24\,]$$

which are derived from

$$8/10 = 0.8 \qquad 6/25 = 0.24 \qquad 5/25 = 0.2$$
$$2/10 = 0.2 \qquad 8/25 = 0.32 \qquad 6/25 = 0.24$$

The second column contrast coefficients,

$$\mathbf{c} = [\,0 \quad 0.429 \quad -0.455 \quad 0 \quad 0.571 \quad -0.545\,]$$

result from 6/14, 5/11, 8/14, and 6/11. Consequently, the sequential estimates are straightforward column differences; for instance, 0.248 is the difference between the average of the north swamp observations and the average of the remaining observations. However, the row contrast, with an estimate of -0.415, is fitted after the column indicators and incorporates column contrast information.

The extent of the row contrast adjustment by the column contrasts depends on the degree of the sample size unequalness. If all the sample sizes are equal, the row contrast is estimated independently of the column contrasts, as in Unit 17. There is also no adjustment in the special case of proportional sample sizes, that is, when the proportion of row 1 sample size to row 2 sample size is the same for all the columns.

The output from a regression program routinely includes the estimated partials and standard errors. With the full model sequence, including the general means model, the partials will not depend on the ordering of the row and column indicators. From our previous experience with partials we might anticipate the following expressions for the partials in terms of the sample means:

$$b_0 = \bar{y}_{23}$$
$$\text{row partial} = \bar{y}_{13} - \bar{y}_{23}$$
$$\text{first column partial} = \bar{y}_{21} - \bar{y}_{23}$$
$$\text{second column partial} = \bar{y}_{22} - \bar{y}_{23}$$
$$\text{first interaction partial} = \bar{y}_{11} - \bar{y}_{13} - \bar{y}_{21} + \bar{y}_{23}$$
$$\text{second interaction partial} = \bar{y}_{12} - \bar{y}_{13} - \bar{y}_{22} + \bar{y}_{23}$$

These partials are not dependent on the data structure, specifically, the sample sizes. That is, they are natural contrasts with straightforward interpretations. However, as we have already determined, the partials have two major disadvantages; namely, the sum of the individual partial sums of squares do not add to the model sum of squares and, in general, the contrasts represented by the partials do not lead to a comprehensive interpretation of the data.

Two ANOVA tables for the sequential sums of squares can be calculated from the last column of the two ABDO calculations. ANOVA(1) is for the model sequence

equal means|row indicators|column indicators|general means

and ANOVA(2) is for the model sequence with the row and column indicators reversed. The sums of squares for the swamp data with M for the x_0 variable, L for stream location variables, C for swamp community variables, and I for interaction variables are as follows:

ANOVA(1) Sums of Squares	ANOVA(2) Sums of Squares
$R(M) = 1553.79$	$R(M) = 1553.79$
$R(L\|M) = 0.7749$	$R(C\|M) = 0.4819$
$R(C\|M, L) = 1.0428$	$R(L\|M, C) = 1.3358$
$R(I\|M, L, C) = 0.6077$	$R(I\|M, L, C) = 0.6077$
residual = 5.6363	residual = 5.6363
total = 1561.85	total = 1561.85

The model sequence evaluation starts with determining the need of the general means model. The necessary sums of squares can be found from either ANOVA table. The F statistic,

$$F = \frac{0.6077/2}{5.6363/29} = 1.56$$

with 2 and 29 degrees of freedom, has a significance level of approximately 0.25, a weight of evidence usually sufficient to judge that the interaction is not important. Evaluation of the reduced model sequence would proceed with both ANOVA tables, using the pooled residual mean square of $(5.6363+0.6077)/(29+2)=0.201$, with 31 degrees of freedom, or the residual mean square from the full model sequence, $5.6363/29=0.194$, with 29 degrees of freedom, depending on the data analyst's pooling methodology.

Assessing the need of stream location (row) or swamp community (column) in the reduced model sequence could result in neither, either, or both factors needed. The approach in Unit 17 with equal numbers of observations was straightforward since the order of row and column indicators did not change the results. Consequently, if both rows and columns were needed, contrasts among rows and contrasts among columns could be estimated independently. With unequal numbers of observations for the means, we will have to exercise more care. Both rows and columns have to be evaluated in the absence and presence of the other factor and a judgment made. For the swamp data the evidence for keeping stream location (row indicators) in the model sequence is stronger than the need for swamp community (column indicators). If the column indicators were judged to be not important, then the model sequence could be further reduced and contrasts among the rows estimated, as was done with the protein nutrition data.

We will proceed as if both row and column indicators were sufficiently important that the accepted model sequence would include both row and column indicators. Attention now shifts to estimating contrasts among rows after fitting columns and estimating contrasts among columns after fitting rows. Consequently, we need the estimates from steps 8 and 10 of Table 18.7 and from step 10 of Table 18.8. These are ORTHO contrasts. The alternative is the use of natural contrasts with the reduced model sequence.

GENERAL DATA ANALYSIS APPROACH

A general approach to the analysis of data from a cross-classification with unequal numbers of observations in the cells is outlined by the following flow of operations.

1. Set up the model sequence by constructing the row, column, and interaction indicator variables. Evaluate the need of the interaction variables in the model by comparing the mean square for the interaction indicator variables, fitted at the end of the sequence, with the residual mean square.

If the regression program output of sequentials or partials for the interaction contrasts are appropriate to the subject matter, the nature of the interaction can be identified. If not, other weighted or natural interaction contrasts can be evaluated through ELIM or a contrast option of the program. In general, when the interaction is judged to be important, the **X** matrix of the indicator variables for the general means model should be the basis for further analysis. Contrasts among the row means within a column (or contrasts among column means within a row, if that is more appropriate to the subject matter) are constructed by using ORTHO or natural **c** coefficients, as discussed in the first part of this unit. For the swamp data, comparing the first column community with the average of the other two communities and then comparing the other two within both stream locations could be handled through the ORTHO set of four **c** vectors:

μ_{11}	μ_{12}	μ_{13}	μ_{21}	μ_{22}	μ_{23}
1	$-6/11$	$-5/11$	0	0	0
0	1	-1	0	0	0
0	0	0	1	$-8/14$	$-6/14$
0	0	0	0	1	-1

Or the natural set of coefficients could be used, where $-1/2$ replaces the fractional values based on the relative sample size proportions within each stream location.

2. If the decision is made that the interaction indicator variables are not needed in the model, a reduced model sequence is sufficient. We can envision two sequences,

$$\text{equal means}|\text{row indicators}|\text{column indicators}$$

or the model sequence with the row and column indicators reversed. We will refer to these sequences as the rows|columns sequence or the columns|rows sequence. The sequential sums of squares should be calculated for each sequence and displayed in ANOVA tables.

Judging the importance (I) or nonimportance (N) of the rows and the columns in the rows|columns sequence can result in one of four conclusions; namely, both are important ($I-I$), neither is important ($N-N$), or one is important, either rows ($I-N$) or columns after rows ($N-I$). Similar outcomes can be envisioned with the columns|rows sequence. Evaluating both sequences can result in one of 14 realistic candidate models. Table 18.9 guides the selection of a model for subsequent inference. Row indicators should be included in the model wherever R (row) appears in the table and column indicators where C (column) appears; for instance, the entry "R and C" suggests that both row and column indicators should be included. A row or column factor with an asterisk has been judged not to be important regardless of its position in the model sequence, but it still needs to be included in the sequence for the enhancement of the other factor.

Table 18.9 Guidelines for Selecting a Model

Rows|Columns Sequence	Columns|Rows Sequence			
	$I-I$	$N-I$	$I-N$	$N-N$
$I-I$	R and C	R and C	C	–
$N-I$	R and C	R and C	C	R^* and C
$I-N$	R	R	R (or C)	R
$N-N$	–	R and C^*	C	x_0

Note: I = importance; N = nonimportance; R = row; C = column.
*Judged not important but still needs to be included in the sequence.

With selection of a model we would then examine interesting contrasts among the levels for factors without an asterisk. Several ORTHO contrasts have already been evaluated in the two model sequences used to produce the table, which guided our model selection. Natural contrasts, or other ORTHO contrasts, may be examined by using ELIM or a program with an appropriate option. With both factors in the model the order of the row and column indicators in the model sequence does not affect the estimates and associated standard errors of natural contrasts from ELIM or a contrast option. If the choice between natural and ORTHO contrasts is not clear, a practical approach is to run both types of contrasts; any difference in significance levels becomes part of the overall interpretation of the data analysis.

Unit 19

Covariance Analysis and Comparison of Regression Lines

Objectives

- To combine indicator and measured x variables in the analysis of sample means.
- To reduce experimental error and adjust treatment means by covariance.
- To compare regression lines in factorial experiments.

In the past four units we have been comparing sample means. Starting with a general means model, data analyses by our regression methodology have utilized indicator variables. Previously, standard regression data sets with measured x variables were analyzed. Thus a reasonable extension to our past analyses would be an analysis of sample means combined with a measured x variable. Now both indicator and measured x variables will be included in the **X** matrix. One common example of such an analysis is called a *covariance analysis* of sample means.

SOYBEAN PHYSIOLOGICAL DATA

To develop a covariance analysis, we will use a soybean experiment. In this study the effect of two treatments, supplemental light and partial shading, on seed yield of soybeans were compared with a control. Fifteen replications of each treatment were grown in a greenhouse study. The data are given in Table 19.1.

The data in Table 19.1 reveal that the replicate-to-replicate variability within each of the three treatments is small compared with the mean-to-mean differences. Consequently, the ANOVA table (Table 19.2), based on a model sequence of $x_0|x_1\ x_2\ x_3$ and the model $y_{j\ell}=\mu_j+\varepsilon_{j\ell}$ ($j=1,2,3$; $\ell=1,2,\cdots,15$), shows a large mean square for treatments relative to the residual or experimental error mean square.

Despite the relatively large treatment mean square, the experimenter realized that the experimental error mean square of 0.2877 was large relative to other greenhouse experiments of this type. When setting up the experiment, the researcher recognized that the plants were not homogeneous. Consequently, the height of the plant, a measurable characteristic of plant vigor, was determined at the start of the experiment. When seed weight, the response variable, is plotted against initial plant height, it can be seen that part of the variability among the seed weights can be explained by the regression of seed weight on initial plant height. By introducing initial plant height into the model as a measured x variable, we will expect to reduce the residual mean square. The model can now be written as

$$y_{j\ell}=\mu_j+\beta(x_{j\ell}-\bar{x})+\varepsilon_{j\ell}$$

where $\bar{x}$ is the average of all the initial plant height measurements.

Table 19.1 Seed Yields (Grams) and Initial Plant Heights (Centimeters) for Soybean Study

Control		Supplemental Light		Partial Shading	
Yield	Height	Yield	Height	Yield	Height
12.2	48	16.6	63	9.5	52
12.4	52	15.8	50	9.5	54
11.9	42	16.5	63	9.6	58
11.3	35	15.0	33	8.8	45
11.8	40	15.4	38	9.5	57
12.1	48	15.6	45	9.8	62
13.1	60	15.8	50	9.1	52
12.7	61	15.8	48	10.3	67
12.4	50	16.0	50	9.5	55
11.4	33	15.8	49	8.5	40
12.3	48	15.0	35	8.6	41
12.2	51	16.2	50	10.4	67
12.6	56	16.7	62	9.4	55
13.2	65	15.8	49	10.2	66
12.3	51	15.9	52	9.3	56

Table 19.2 ANOVA for Soybean Study

Source of Variation	Degrees of Freedom	Sum of Squares	Mean Square
Mean	1	7063.79	—
Treatments	2	308.187	154.093
Residual	42	12.0853	0.2877

Initial plant height comes into the model as the deviation from the average plant height. In this form the model is the same as the mean (μ) plus slope (β) model of straight-line regression in Unit 5, except that we have μ_j in place of μ. There is only one slope, estimated from all the data, but instead of one line we can envision three lines, each with the same slope.

To examine the model with respect to the data, we set up the **X** matrix with four columns. The first three are the indicator columns for three

Soybean Physiological Data

treatments, followed by the initial plant height deviations, $x_{j\ell} - \bar{x}$, in the fourth column. The model sequence is the same as the sequence used when we fitted a four-x-variable model, including three intercepts and a common slope, with the electricity load data in Unit 13. For the soybean physiological data the ABDO calculations are shown in Table 19.3.

Table 19.3 ABDO Calculations for the Soybean Covariance Model

15	0	0	−28	(740)	183.90	
	15	0	−31	(737)	237.90	
		15	59	(827)	142.00	
			3695.2	(121,660)	−81.06	(28,785.50)
15	0	0	−28	(740)	183.90	
1	0	0	−1.87	(49.33 = $\bar{x}_1$)	12.26	= $\bar{y}_1$
	15	0	−31	(737)	237.90	
	1	0	−2.07	(49.13 = $\bar{x}_2$)	15.86	= $\bar{y}_2$
		15	59	(827)	142.00	
		1	3.93	(55.13 = $\bar{x}_3$)	9.47	= $\bar{y}_3$
			3346.8		195.35	
			1		0.0584 = b	

In practice, calculation of the $(x_{j\ell} - \bar{x})$ deviations is not needed for the fourth column of the **X** matrix. The numbers in parentheses in Table 19.3 are the values if the initial plant heights are used. The $x_{j\ell}$'s are then sequentially corrected in the first three stages so that the fourth-stage value of $b = 0.0584$ is the same.

The sum of the products of the pairs in the right-hand column for the first three stages gives 7371.9747, the general means model sum of squares with three degrees of freedom, leaving a residual sum of squares of 12.0853. Adding the initial plant height variable to the general means model accounts for an additional sum of squares equal to (195.346)(0.05837) = 11.402. We now have a residual sum of squares of 12.0853 − 11.4020 = 0.6833 and a residual mean square of 0.6833/41 = 0.0167, which is to be compared with the residual mean square of 0.2877 before considering the initial plant height.

One of the major objectives of a covariance analysis has now been achieved—namely, to remove from the residual sum of squares a source of variability that can be associated with a covariate x variable. The standard errors of the treatment means will be decreased; that is, the precision of the experiment will be increased.

Another part of most covariance analyses entails the adjustment of the treatment means for the measured x variable. If a treatment happens to have a larger initial plant height mean, then part of the mean seed weight is due to the higher initial plant height, since a positive relationship exists between seed weight and initial plant height for each treatment. The fairest comparison seems to be to adjust each of the treatment means to a common value of initial plant height. In Figure 19.1 an estimated regression line is drawn for each treatment.

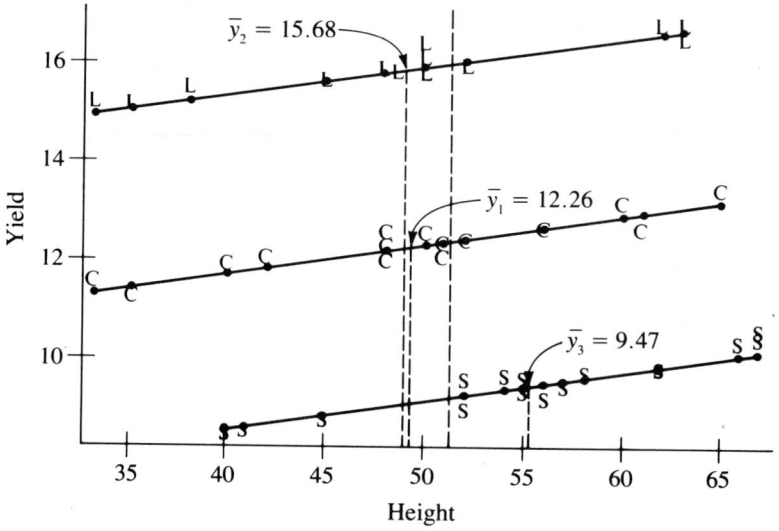

Figure 19.1 Graph of Yields (C=Control, L=Supplemental Light and S=Partial Shading), Predicted Values ($\cdot$) versus Initial Plant Height

Since the three treatments were allocated to the 45 plants at random, we would expect the $\bar{x}_j$'s to be the same. But because of natural variability, $\bar{x}_1 = 49.33$, $\bar{x}_2 = 49.13$, and $\bar{x}_3 = 55.13$. Practically speaking, then, treatment 1 (control, C) could be compared with treatment 2 (supplemental light, L) with the unadjusted $\bar{y}_1$ and $\bar{y}_2$, since $\bar{x}_1$ is close to $\bar{x}_2$ and the slope is a relatively small value. However, it is not appropriate to compare $\bar{y}_3$ with $\bar{y}_1$ or $\bar{y}_2$, because $\bar{x}_3$ is different from $\bar{x}_1$ and $\bar{x}_2$. Instead, ELIM is used to find a predicted value for each group, using a common initial plant height. If the model with deviations is used, then the common value of the deviation is usually 0.0. If the model with initial plant heights is used, then $\bar{x}$ is traditionally used as the common value; for example, $\mathbf{c} = [0\ 0\ 1\ \bar{x}]$ would be used to calculate the predicted value, the adjusted mean, for the third treatment group. The result for the jth group is denoted as $\bar{y}_j(\text{adj})$. Algebraically, we have $\bar{y}_j(\text{adj}) = \bar{y}_j - b(\bar{x}_j - \bar{x})$. Numerically we have the following results:

For treatment 1: $\bar{y}_1(\text{adj}) = 12.26 - 0.0584(49.33 - 51.2) = 12.37$
For treatment 2: $\bar{y}_2(\text{adj}) = 15.86 - 0.0584(49.13 - 51.2) = 15.98$
For treatment 3: $\bar{y}_3(\text{adj}) = 9.47 - 0.0584(55.13 - 51.2) = 9.24$

From Figure 19.1 it can be seen that the adjusted $\bar{y}_3$ will be less than the unadjusted $\bar{y}_3$.

Interestingly, the slope for regressing seed weight on initial plant height, without consideration of the three treatments, is $\Sigma(x_{j\ell} - \bar{x})y/\Sigma(x_{j\ell} - \bar{x})^2 = -0.0219$. The negative slope, evident from Figure 19.1 if the treatments are not considered, would not be a good candidate for adjusting the treatment means. Consequently, the b value used is based on the residuals after considering differences among the treatments; that is, we use the partial coefficient.

The adjusted treatment means are also partial regression coefficients that can be calculated from the ABDO in Table 19.3. From the even-numbered steps of the first three stages, we obtain

$$b_3 = 9.47 - (0.0584)(3.93) = 9.24 = \bar{y}_3(\text{adj})$$
$$b_2 = 15.86 - (0.0584)(-2.07) = 15.98 = \bar{y}_2(\text{adj})$$
$$b_1 = 12.26 - (0.0584)(-1.87) = 12.37 = \bar{y}_1(\text{adj})$$

Thus the adjusted treatment means and their standard errors can be part of the standard output from a regression program if the **X** matrix includes the indicator variables for the treatments and $x_{j\ell} - \bar{x}$ but does not include x_0.

If $x_{j\ell}$ is used in the fourth column of Table 19.3, then the estimated partials will be $\bar{y}_j - b\bar{x}_j$, the estimated intercepts for each of the three lines. However, contrasts of the μ_j are automatically adjusted for the covariable. For example, the **c** vector for estimating the difference between the control and the average of the two treatments is

$$\begin{array}{cccc} x_1 & x_2 & x_3 & x_4 \end{array}$$
$$\mathbf{c} = \begin{bmatrix} 1 & -1/2 & -1/2 & 0 \end{bmatrix}$$

This is verified by showing that any contrast among the adjusted means is identical to the same contrast among the estimated intercepts. Thus contrasts are easily obtained by using ELIM or a computer program with a contrast option.

The analysis of covariance table follows a format similar to the ANOVA table. For the physiological data the sources of variation and degrees of freedom for the model sequence x_0 x_4 x_1 x_2 x_3, where x_1, x_2, and

x_3 are the indicator variables for treatments and x_4 is the measured variable, are as shown in Table 19.4.

Table 19.4 Analysis of Covariance

Source of Variation	Degrees of Freedom
$R(0)$	1
$R(4\|0)$	1
$R(1,2,3\|0,4) = R(1\|0,4) + R(2\|0,4,1)$	2
Residual	41

The same residual sum of squares could also be calculated from the use of $R(0)$, $R(1,2,3|0) = R(1|0) + R(2|0,1)$, and $R(4|0,1,2,3)$. In general, the measured variable is not orthogonal to each of the treatment indicator variables. Then $R(1,2,3|0,4)$, commonly called the treatment sum of squares adjusted for the covariate, will not be equal to $R(1,2,3|0)$, the unadjusted treatment sum of squares. In practice, the adjusted treatment sum of squares is calculated from the sequential sums of squares output of a regression program, using a model sequence starting with x_0 and the measured x variable.

TESTING THE ASSUMPTION OF A COMMON SLOPE

An implicit assumption of the covariance analysis is now explicitly stated. We have assumed that the slope between the response variable, seed weight, and the x variable, initial plant height, is the same for each of the treatments. A slope common to all three treatments is calculated. If the assumption is not true, then a larger sum of squares of the remaining residual sum of squares, after fitting the general means model, can be accounted for by fitting a different slope for each treatment. Formally stated as a null hypothesis, we have

$$H_0: \beta_1 = \beta_2 = \cdots = \beta_t = \beta$$

and the alternate hypothesis is that at least one beta is different from the others.

We have already faced the equal-slope hypothesis for the electricity load data in Unit 13. And, indeed, we can use the same methodology here. For the soybean problem we set up a model sequence of three x variables for general intercepts, followed by one x variable for an equal slopes model and three x variables for a general slopes model. Consequently, the columns of the $\mathbf{X}$ matrix are

x_1 = indicator variable for treatment 1
x_2 = indicator variable for treatment 2
x_3 = indicator variable for treatment 3
x_4 = initial plant height
$x_5 = x_4$ for the first 15 rows (=0 for rows 16–45)
$x_6 = x_4$ for rows 16–30 (=0 for rows 1–15 and 31–45)
$x_7 = x_4$ for rows 31–45 (=0 for rows 1–30)

The estimated coefficient for x_7, the last variable, will be zero for the same reasons as stated for a general means model fitted after an equal means model. The calculations are not given here, but the additional sum of squares for x_5 and x_6 is equal to 0.0767 with two additional degrees of freedom.

Previously, we have used the F test statistic to test the adequacy of the equal means model relative to the general means model. The same methodology is used now, giving

$$F = \frac{[R(5|1,2,3,4) + R(6|1,2,3,4,5)]/2}{\text{residual mean square}} = \frac{(0.0046 + 0.0721)/2}{0.6066/39} = 2.46$$

with 2 and 39 degrees of freedom. From the magnitude of the F test statistic, we would continue to use a common slope; that is, the equal slopes model is adequate. The methodology for comparing regression lines will be more fully developed in the next section.

We conclude this covariance analysis part of the unit with several general remarks. The demonstrated methodology can be extended to include more than one x variable and to consider relationships other than straight lines (e.g., polynomials). More complex experimental designs and unbalanced data can also be handled. The calculations are more extensive but the basic methodology is the same as that developed in this and previous units.

COMPARING REGRESSION LINES IN FACTORIAL STUDIES

Evaluating the assumption of an equal slopes model in a covariance analysis is just one example of a larger group of data analysis problems. In Unit 13 the evaluation of a sequence of straight-line candidate models was illustrated by data from the electricity load observational study. In Units 17 and 18 a general means model for cross-classification data was used in the analysis of data with a factorial arrangement of treatments. The general approach was to estimate the $\mu_{j\ell}$ means and evaluate the adequacy of reduced models, starting with an evaluation of the need for the interaction.

When the factorial arrangement includes both qualitative and quantitative factors, the methodology of this section is appropriate; here we treat the quantitative factor as a measured x variable. Suppose we have an experiment with two factors, diet and polychlorinated biphenyl (PCB). If the diets differ in their ingredients, the diets are considered qualitative. The PCB factor has several levels and is considered quantitative. With two diets, diet A and diet B, and three levels of PCB, 0, 50, and 100 parts per million (ppm), the factorial experiment would have six treatments. Suppose animals were available for r replications for each treatment, and the response variable was liver weight. Then the most general model would be

$$y_{j\ell m} = \mu_{j\ell} + \varepsilon_{j\ell m} \qquad j=1,2;\; \ell=1,2,3;\; m=1,2,\ldots,r$$

Our factorial data analysis approach is to fit a general means model with six indicator variables, one for each treatment. The partition of the model degrees of freedom is reflected by one degree of freedom for the mean, one for diet, two for PCB, and two for the interaction between diet and PCB. More specifically, the PCB degrees of freedom can be expressed as two contrasts—for example, (1) the comparison between 0 and 100 ppm and (2) the comparison between 50 ppm and the average of 0 and 100 ppm, each with one degree of freedom. A corresponding subdivision for the interaction would also be appropriate. Five orthogonal contrasts, each in the format of a linear combination of the means ($c\beta$ in the notation of Unit 12), are evaluated, using the six indicator variables of the general means model and the c coefficients of Table 19.5.

Table 19.5 Contrast Coefficients (c) for Contrasts Among Means

Contrast	μ_{11}	μ_{12}	μ_{13}	μ_{21}	μ_{22}	μ_{23}
Diet (D)	$-1/3$	$-1/3$	$-1/3$	$1/3$	$1/3$	$1/3$
PCB (1)	$-1/2$	0	$1/2$	$-1/2$	0	$1/2$
PCB (2)	$1/4$	$-1/2$	$1/4$	$1/4$	$-1/2$	$1/4$
D×PCB (1)	$1/2$	0	$-1/2$	$-1/2$	0	$1/2$
D×PCB (2)	$-1/4$	$1/2$	$-1/4$	$1/4$	$-1/2$	$1/4$

As in previous units, the values of the c coefficients have divisors so that the estimates of the contrasts will be the desired contrasts among the sample means and not multiples of the contrasts. Remember, however, that the sum of squares and subsequent test statistics will not be affected if, for example, 1, −2, and 1 values are used for the PCB (2) contrast. Alternative sets of contrasts for PCB could be utilized—for instance, the zero PCB level compared with each of the other two levels. In general this set would not be used. Not only do the two contrasts lack orthogonality, but also contrasts among levels of a quantitative (continuous) factor should reflect a functional relationship with liver weight.

The five contrasts of Table 19.5 can also be envisioned as a comparison of two curves, one for each diet. Plotting the data with liver weight against PCB level is the first step. A quadratic polynomial will fit exactly the three means for each diet. Now our most general model would be two curves with different intercepts, different linear terms, and different quadratic terms. Note the similarity in approach to that of the electricity load data in Unit 13. Six parameters need to be estimated. Thus an **X** matrix with six columns (intercept, linear, and quadratic terms for each diet curve) would give the same model sum of squares as the general means model.

In Unit 13 we saw the advantages of using a model sequence approach to find a model as reduced as possible but still adequately describing the data. For example, if the two contrasts of PCB do not interact with diet, then common linear and quadratic parameters are estimated and the two curves differ in intercept only. Now the **X** matrix for the reduced model only needs four columns, one for each diet intercept, one for the common linear parameter, and one for the common quadratic parameter. The ANOVA table will be useful in comparing various candidate models.

If we view the data analysis as a regression problem, the relationship between y, the liver weight, and x, the PCB concentration, can be modeled as a quadratic polynomial regression, using orthogonal polynomial coefficients of -1, 0, and 1 for x_1 and 1, -2, and 1 for x_2 (three levels of PCB). Two curves are to be estimated, one for each diet. Then

$$y_{j\ell m} = \mu_j + \beta_{1j} x_{1j\ell m} + \beta_{2j} x_{2j\ell m} + \varepsilon_{j\ell m}$$

where μ_1, β_{11}, and β_{21} are the mean, linear, and quadratic parameters for diet A, and μ_2, β_{12}, and β_{22} the same parameters for diet B. Use of the orthogonal polynomial coefficients results in the diet mean, rather than the diet intercept, as the first parameter. The **X** matrix (one replication and in the order of 0, 50, and 100 ppm of PCB) for estimating the parameters for both curves is

$$\mathbf{X} = \begin{bmatrix} x_0 & x_1 & x_2 \\ 1 & -1 & 1 \\ 1 & 0 & -2 \\ 1 & 1 & 1 \end{bmatrix}$$

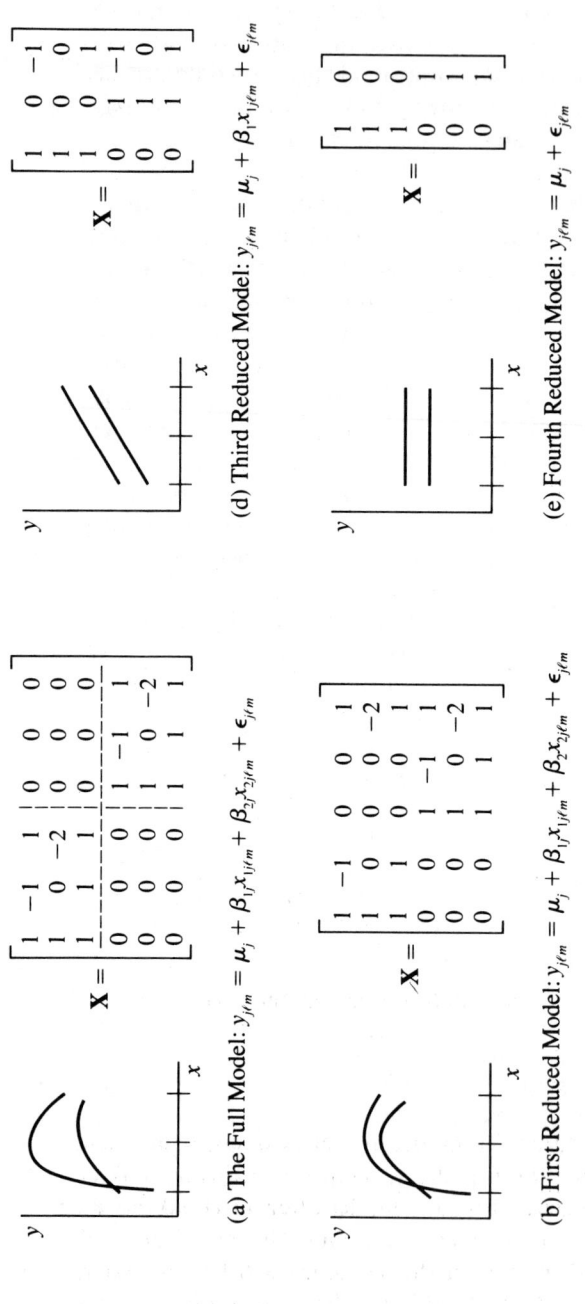

Figure 19.2 A Sequence of Candidate Models

With one factor quantitative and one factor qualitative, the various candidate models can be shown graphically, highlighting the sequence-of-models approach. Figure 19.2 presents one possible sequence, with each candidate model given algebraically and graphically. The **X** matrix is shown for each model using the diet and PCB example, with only the first replication written.

The dashed lines in the matrix in Figure 19.2(a) indicate that the six parameters could be estimated by running two regressions, one with the **X** matrix, as shown in the upper left-hand corner, and the other with the same **X** matrix, as shown in the lower right-hand corner. However, if we construct one **X** matrix with six columns, the individual parameters for each diet are preserved and the same six estimates can be found with one regression run. In this format it is also easier to see that the reduced models are subsets of the general model; for example, adding the third and sixth columns of the general model matrix gives the **X** matrix of the first reduced model with five columns. Now the estimate of the quadratic is forced to be the same for both curves, as denoted by β_2 in the first reduced model.

To evaluate the sequence of candidate models in Figure 19.2, we turn to our ANOVA methodology. Appropriate sums of squares can be calculated in one regression run by listing the **X** matrix for the most reduced model and augmenting it with additional variables for more general models. With the obviously redundant variables eliminated, we have

$$\mathbf{X} = \begin{array}{c} \phantom{\begin{bmatrix}}x_0 \ \ x_1 \ \ x_2 \ \ \ x_3 \ \ \ \ x_4 \ \ \ x_5 \ \ \ x_6 \ \ \ x_7 \ \ \ x_8 \\ \begin{bmatrix} 1 & 1 & 0 & -1 & 1 & -1 & 0 & 1 & 0 \\ 1 & 1 & 0 & 0 & -2 & 0 & 0 & -2 & 0 \\ 1 & 1 & 0 & 1 & 1 & 1 & 0 & 1 & 0 \\ 1 & 0 & 1 & -1 & 1 & 0 & -1 & 0 & 1 \\ 1 & 0 & 1 & 0 & -2 & 0 & 0 & 0 & -2 \\ 1 & 0 & 1 & 1 & 1 & 0 & 1 & 0 & 1 \end{bmatrix} \end{array}$$

If we write the sequence of six candidate models in the format of

$$x_0 | x_1 x_2 | x_3 | x_4 | x_5 x_6 | x_7 x_8$$

we will find that with each additional model in the sequence one new parameter will be estimated. For example, the sequence of fitting x_1 and x_2 after x_0 is the same as fitting a general means model after an equal means model and, as in Unit 15, x_2 is not needed. Consequently, the additional sum of squares for more general models in the sequence will be associated with one degree of freedom. The ANOVA table will have the outline shown in Table 19.6.

Table 19.6 ANOVA for the Sequence of Models

Source of Variation	Degrees of Freedom	Sum of Squares
Common diet mean	1	$R(0)$
Different diet means	1	$R(1,2\|0)$
Common straight line	1	$R(3\|0-2)$
Common curve	1	$R(4\|0-3)$
Different linears	1	$R(5,6\|0-4)$
Different quadratics	1	$R(7,8\|0-6)$
Residual	$6(r-1)$	Residual

We are now in a position to compare two data analysis approaches for a factorial with one factor (diet) qualitative and the other (PCB) quantitative. The conclusion is that both result in the same analysis. Specifically, the analysis of sample means uses a general means model with contrast and interaction terminology, whereas the classical regression approach is associated with a polynomial model and the terminology of functional relationships between the response variable and levels of the quantitative factor. The two approaches are summarized in Table 19.7 by comparing the ANOVA tables for each approach.

Table 19.7 Comparison of Factorial Data Analysis Approaches with Diet (D) Qualitative and PCB Quantitative

Polynomial Model	Degrees of Freedom	General Means Model
Common diet mean	1	Overall mean
Different diet means	1	Diet (D)
Common straight line	1	PCB (linear)
Common curve	1	PCB (quadratic)
Different linears	1	D×PCB (linear)
Different quadratics	1	D×PCB (quadratic)
Residual	$6(r-1)$	Residual

It would be interesting to go through a model sequence exercise if we had been willing to assume that straight lines were adequate for the relationship between liver weight and PCB level for each diet. One possible candidate model sequence,

common mean|different means|common slope|different slopes

flows directly from the polynomial model sequence without the straight-line assumption. Another realistic candidate model sequence is

common mean|common slope|different means|different slopes

In either case, six $\mu_{j\ell}$ parameters can be estimated from the data. The difference in the general means model sum of squares with six degrees of freedom and the straight-line candidate model sequence with four degrees of freedom is the sum of squares associated with the quadratics of our earlier sequence. We have only fitted four parameters because we made an assumption that straight lines were adequate for the relationship between liver weight and PCB level for each diet. That assumption could be evaluated by adding the general means model to the end of either straight-line sequence and including the additional sum of squares for the general means model in the ANOVA table, labeled as "lack-of-fit," with two degrees of freedom. Formal testing would be done by using F test statistics, starting with the lack-of-fit mean square divided by the residual mean square with 2 and 6 $(r-1)$ degrees of freedom. If the lack of fit is not important, we would then test the need for separate slopes and proceed up the ANOVA table until we found a reduced model that was judged to describe the data adequately.

SUMMARY

Models combining both indicator and measured variables arise in problems with different objectives. A measured x variable that is not affected (we hope) by the treatments can help explain variability in the response variable when analyzing sample means. Covariance analysis follows the same general flow of previous units, but the measured x variable usually is not orthogonal to the indicator variables for the treatments. Thus care has to be exercised in candidate model sequencing, treatment mean adjustment, and interpretation.

In a classical regression problem with one measured x variable, the data can sometimes be stratified into two or more categories, and a different slope can be envisioned for each category. Now the focus changes to comparing the slopes or curves for the relationship between the response variable and the measured x variable. One major group of data analysis problems with this general structure is the factorial experiment with one factor qualitative and one factor quantitative; only now the levels of the quantitative or continuous factor are the same for each state of the qualitative or categorical factor. Even though we now think in factorial terminology of treatment means and interaction between factors, the flow of the data analysis is that of comparing lines or curves.

Unit 20

Model Selection with Many Explanatory Variables

Objectives

- To identify the need for computer-assisted model selection.
- To present SWEEP, an algorithm for computer-assisted model selection.
- To discuss the difficulties associated with the residual sum of squares as a selection criterion and to propose PRESS as a supplementary criterion.

The past five units have emphasized the analysis of sample means. Our basic approach in those units, and also in earlier units, was to examine a limited number of candidate model sequences. In this unit model selection will be extended to many explanatory variables where the candidate model sequence is not clear. Now model selection is explored primarily for the purpose of finding a subset of the explanatory variables that will satisfy the objectives of the study.

At the design stage of a study, thought is given to identifying variables likely to be important, but the particular form of the relationship between the response and each explanatory variable usually receives less attention. At the time of the data analysis it may be observed that some relationships are not straight lines. It then becomes advantageous, or necessary, to construct additional variables by taking logarithms, powers, or products of the original variables, but creation of additional variables adds to the complexities of model selection. The measured and constructed variables are interrelated, and the number of explanatory variables makes the computing methods of earlier units too laborious.

Prerequisites for model selection, as in previous units, include (1) that a reasonable number of residual degrees of freedom be available after fitting the full model (all the variables), and (2) that the residuals from the estimated full model plotted against the predicted values reveal no unexpected patterns and the magnitude of the residuals be acceptably small. In short, we have sufficient variables to adequately describe the data but may have several of little value. Identification of superfluous variables is economically important information in designing the next study. Of more importance in the data analysis is the elimination of explanatory variables whose only effect is to increase the standard error of a predicted value or to introduce unnecessary complexities in the interpretation of the estimated parameters. In general, we continue in the spirit of the previous units; namely, we want to find the simplest model that is adequate for describing the data for inference.

The main objective of this unit is to illustrate potential problems encountered with many explanatory variables and to provide guidelines for data analysis approaches.

NEED FOR COMPUTER-ASSISTED MODEL SELECTION

The basic data analysis approach of previous units has been to use the Abbreviated Doolittle (ABDO) algorithm and perhaps two or three orderings of the explanatory variables. From the ABDO outputs we would try to select a model with as few variables as possible and with an acceptable residual sum of squares. Unfortunately, this technique becomes laborious when the number of variables is large.

To illustrate the amount of labor that is involved, 1023 candidate models can result from 10 explanatory variables. That is, this value gives the number of situations where each variable is either in or out ($2^{10} = 1024$) minus one, since a model without any variables usually would not be considered. From the sequential output of one ABDO we can glean the residual sum of squares from 9 models in addition to the full model. Even if we could judiciously choose different orderings of the variables so that each ABDO output would give the residual sum of squares for 9 distinct models, 114 ABDOs would be required. Clearly, we need an algorithm to examine many model sequences more efficiently. Specifically, we need a computer-assisted model selection approach.

THE SWEEP OPERATOR

Algorithms used for automatic model selection are based on the SWEEP operator. When the ordering of variables is fixed in advance, the partial regression coefficients are found by first doing ABDO and then solving the implicit set of equations found in the even steps. A sequence of SWEEP operators intermingles ABDO steps with solving the resulting equations. Precisely, a matrix **A** is said to have been swept on the rth (diagonal) element when it has been transformed into a matrix **C** such that

$$c_{rr} = 1/a_{rr}$$
$$c_{ir} = -a_{ir}/a_{rr} \qquad i \neq r$$
$$c_{rj} = a_{rj}/a_{rr} \qquad j \neq r$$
$$c_{ij} = a_{ij} - a_{ir}a_{rj}/a_{rr} \qquad i \neq r \text{ and } j \neq r$$

The SWEEP operator applied to the rth diagonal element of

$$\begin{bmatrix} X'X & X'Y \\ \hline Y'X & Y'Y \end{bmatrix}$$

brings x_r into the model. Variables may enter the model in any order. If x_r is already in the model, then a second sweep of the rth element will remove x_r from the model. When all x variables are swept, the partial regression

The SWEEP Operator

coefficients and the residual sum of squares for the full model are found in the right-most column. Intermediate stages give a variety of sequential and partial regression coefficients.

To illustrate, in Table 20.1 SWEEP is sequentially applied to the sums of squares and cross products matrix for the firefly data of Unit 10.

Table 20.1 Sequential Output of SWEEP Applied to the Sums of Squares and Cross Products Matrix of the Firefly Data

$$\begin{bmatrix} 17.0000 & 1{,}244.00 & 400.900 & 703.000 \\ 1244.00 & 109{,}328. & 30{,}228.9 & 49{,}557.0 \\ 400.900 & 30{,}228.9 & 9{,}555.65 & 16{,}363.1 \\ 703.000 & 49{,}557.0 & 16{,}363.1 & 30{,}211.0 \end{bmatrix}$$

Sweeping x_0 Gives

$$\begin{bmatrix} 0.058824 & 73.1765 & 23.5824 & 41.3529 & = \bar{y} \\ -73.1765 & 18{,}296.5 & 892.453 & -1886.06 & \\ -23.5824 & 892.453 & 101.484 & -215.293 & \\ -41.3529 & -1{,}886.06 & -215.293 & 1139.88 & = \sum [y - \hat{y}(0)]^2 \end{bmatrix}$$

Then Sweeping x_1 Gives

$$\begin{bmatrix} 0.351492 & -0.003999 & 20.0130 & 48.8962 & = b_{0 \cdot 1} \\ -0.003999 & 0.00005466 & 0.048777 & -0.103083 & = b_{1 \cdot 0} \\ -20.0130 & -0.048777 & 57.9529 & -123.296 & \\ -48.8962 & 0.103083 & -123.296 & 945.462 & = \sum [y - \hat{y}(0, 1)]^2 \end{bmatrix}$$

Then Sweeping x_2 Gives

$$\begin{bmatrix} 7.26262 & 0.012845 & -0.345332 & 91.4743 & = b_{0 \cdot 12} \\ 0.012845 & 0.00009571 & -0.000842 & 0.000692 & = b_{1 \cdot 02} \\ -0.345332 & -0.000842 & 0.017255 & -2.12752 & = b_{2 \cdot 01} \\ -91.4743 & -0.000692 & 2.12752 & 683.143 & = \sum [y - \hat{y}(0,1,2)]^2 \end{bmatrix}$$

Then Sweeping x_1 Again Removes It from Model

$$\begin{bmatrix} 5.53875 & -134.207 & -0.232374 & 91.3815 & = b_{0 \cdot 2} \\ 134.207 & 10{,}448.2 & -8.79399 & 7.22678 & \\ -0.232374 & 8.79399 & 0.009854 & -2.12144 & = b_{2 \cdot 0} \\ -91.3815 & 7.22678 & 2.12144 & 683.152 & = \sum [y - \hat{y}(0, 2)]^2 \end{bmatrix}$$

The SWEEP output provides everything needed for inference on $c\beta$, a linear combination of the elements of β. The predictor of the twelfth observation with x_0, x_1, and x_2 in the model is **cb**, with $c=[1\ 87\ 24.4]$. The vector **b** is in the upper right portion of the SWEEP output after sweeping x_0, x_1, and x_2. Then

$$\mathbf{cb} = \hat{y}_{12} = 91.4743 + (87 \times 0.000692) + (24.4 \times -2.12752) = 39.6$$

The residual sum of squares is in the lower right corner, and hence the residual mean square is $683.143/14 = 48.80$. From the upper left corner we calculate

$$v_c = [1\ \ 87\ \ 24.4] \begin{bmatrix} 7.26262 & 0.012845 & -0.345332 \\ 0.012845 & 0.00009571 & -0.000842 \\ -0.345332 & -0.000842 & 0.017255 \end{bmatrix} \begin{bmatrix} 1 \\ 87 \\ 24.4 \end{bmatrix}$$

$$= 0.0696$$

Thus the standard error of **cb** is $\text{SQRT}(0.0696 \times 48.80) = 1.843$.

The example illustrates that the SWEEP operator could have been used for all previous regression computations. However, it is less efficient than ABDO and is advantageous only when the order of the variables is not fixed. For example, the addition of a variable requires changing the full matrix with SWEEP, whereas only two partial rows are computed with ABDO. With regard to accuracy, SWEEP requires the same precautions as does ABDO (see Unit 3).

STEPWISE APPROACHES

In years past the procedures most used for automatic model selection have been stepwise. A forward selection scheme would fit an x_0 model first and then sequentially add variables. The x variable resulting in the biggest decrease in the residual sum of squares, given the x variables previously included, is added at each step. Another approach, called backward elimination, would start with all the variables and sequentially eliminate variables resulting in the smallest change in the residual sum of squares. Both of these procedures are stepwise in the sense of sequentially adding or removing one variable at a time. A commonly used approach, actually called stepwise, is the forward selection procedure with the additional provision at each step of eliminating a variable that had been selected at an earlier stage.

The SWEEP operator is well suited to stepwise procedures. At each stage it is easy to search the variables to find the one that will produce the maximum reduction in the residual sum of squares. Then that element is swept. With the firefly data suppose x_0 has been swept and we must decide to bring in x_1 or x_2. The decision to then sweep x_2 is based on possible reductions in the residual sum of squares as calculated by

$$\frac{(-1886.06)^2}{18296.5} = 194.42 = \text{reduction in residual sum of squares for } x_1$$

and

$$\frac{(-215.29)^2}{101.484} = 456.72 = \text{reduction in residual sum of squares for } x_2$$

Thus x_2 would be swept since its inclusion will give the smaller residual sum of squares.

Unfortunately, these various stepwise procedures have several disadvantages. The decision to stop at a certain step is arbitrary. If the number of variables to be included in the selected model is preset, then the three procedures will not necessarily yield the same variables to be included in the model. None of the procedures can guarantee that the selected model will be the one with the smallest residual sum of squares. Of more importance, however, is the complete reliance on the residual sum of squares as a criterion, a difficulty illustrated subsequently. One major change since the development of the stepwise approaches is the availability to many data analysts of improved programs and increased computing capacity. In addition, alternative criteria for comparing candidate models have also been developed. However, stepwise algorithms are fast and remain useful for preliminary screening of a very large number of variables.

ALL POSSIBLE REGRESSIONS

The SWEEP operator is used in algorithms to compute all possible candidate models. Variables known to be important and desired to be included in the model are swept and are not swept again. If there are k remaining variables, then there are $2^k - 1$ possible models, and all these models can be computed in $2^k - 1$ sweeps. For example, suppose x_0 has been swept and we want to inspect combinations of three other variables. Then we have the following process:

Element Swept	Effect	Resulting Model
1	Add x_1	x_1
2	Add x_2	x_1, x_2
1	Remove x_1	x_2
3	Add x_3	x_2, x_3
1	Add x_1	x_1, x_2, x_3
2	Remove x_2	x_1, x_3
1	Remove x_1	x_3

The pattern is easy to see. If there were four variables, the above sequence would then generate all models, not including x_4. We could sweep x_4 and then repeat the above sequence of elements swept and generate all models containing x_4. In practice, there are sophisticated checks that can be made that would facilitate finding the best models without doing all $2^k - 1$ sweeps.

Several programs can identify the m models of each size (number of variables) having the smallest residual sums of squares. The word SELECT is used to indicate any one of the several available. If there are 10 variables and we choose 3 for m, SELECT would identify 28 models for closer scrutiny (there is only one 10-variable model). These all-possible-regressions programs are more reasonable in cost if the total number of measured and constructed variables is limited. The specific value depends on the particular program. When the total number of variables exceeds the number recommended or necessitated by the program, a procedure is needed to screen variables for entry in an all-possible-regressions program. The stepwise regression program described in the previous section is effective, easy to use, and relatively inexpensive. With each of several runs to stepwise regression forcing different combinations of known important variables in the first step, a large number of variables can be efficiently screened. Coupled with subject matter knowledge, the variables to enter the SELECT program are picked from the total available pool of variables.

Unlike the stepwise procedures, SELECT guarantees the subset for each size having the smallest residual sum of squares. However, as mentioned earlier, the residual sum of squares must be used with caution. The next section describes the nature of the problem.

DIFFICULTIES WITH THE RESIDUAL SUM OF SQUARES AS A SELECTION CRITERION

From previous units we know that the variance of a predicted value is $v_c\sigma^2$, where v_c depends on the $\mathbf{X}$ matrix. It can be shown that v_c will not decrease and in general will increase with each additional variable in $\mathbf{X}$. The polynomial models introduced in Unit 11 with five levels of one explanatory variable illustrate the changes in the variance of a predicted value. If the correct model includes only x_0, then the estimator of $\hat{y}$ is $\bar{y}$, and we have the results shown in Table 20.2.

In general, each predicted value is a linear combination of the estimated partial regression coefficients, simply $\bar{y}$ in this case. Remember that each estimated coefficient is a linear combination of the observations. Therefore each predicted value can be expressed as a linear combination of the observations and the variance can be calculated by the Variance Rule or by ELIM.

Table 20.2 Variance of Predicted Value When Model Includes Only x_0

c	$\hat{y} = cb$	Variance $(\hat{y}) = v_c \sigma^2$
[1]	$\hat{y}_1 = 0.2 y_1 + 0.2 y_2 + 0.2 y_3 + 0.2 y_4 + 0.2 y_5$	$0.2\sigma^2$
[1]	$\hat{y}_2 = 0.2 y_1 + 0.2 y_2 + 0.2 y_3 + 0.2 y_4 + 0.2 y_5$	$0.2\sigma^2$
[1]	$\hat{y}_3 = 0.2 y_1 + 0.2 y_2 + 0.2 y_3 + 0.2 y_4 + 0.2 y_5$	$0.2\sigma^2$
[1]	$\hat{y}_4 = 0.2 y_1 + 0.2 y_2 + 0.2 y_3 + 0.2 y_4 + 0.2 y_5$	$0.2\sigma^2$
[1]	$\hat{y}_5 = 0.2 y_1 + 0.2 y_2 + 0.2 y_3 + 0.2 y_4 + 0.2 y_5$	$0.2\sigma^2$

If the correct model includes both x_0 and x_1, then we have the results given in Table 20.3.

Table 20.3 Variance of Predicted Value When Model Includes x_0 and x_1

c	$\hat{y} = cb$	Variance $(\hat{y}) = v_c \sigma^2$
[1 1]	$\hat{y}_1 = 0.6 y_1 + 0.4 y_2 + 0.2 y_3 + 0 y_4 - 0.2 y_5$	$0.6\sigma^2$
[1 2]	$\hat{y}_2 = 0.4 y_1 + 0.3 y_2 + 0.2 y_3 + 0.1 y_4 + 0 y_5$	$0.3\sigma^2$
[1 3]	$\hat{y}_3 = 0.2 y_1 + 0.2 y_2 + 0.2 y_3 + 0.2 y_4 + 0.2 y_5$	$0.2\sigma^2$
[1 4]	$\hat{y}_4 = 0 y_1 + 0.1 y_2 + 0.2 y_3 + 0.3 y_4 + 0.4 y_5$	$0.3\sigma^2$
[1 5]	$\hat{y}_5 = -0.2 y_1 + 0 y_2 + 0.2 y_3 + 0.4 y_4 + 0.6 y_5$	$0.6\sigma^2$

Calculation of the $\hat{y}$'s as a linear combination of the observations again is straightforward since b_0 and b_1 are linear combinations of the observations, as shown in Unit 12.

If the correct model includes x_0, x_1 and $x_2 = x_1^2$, then we have the results shown in Table 20.4. Comparing the three models, we see that one

Table 20.4 Variance of Predicted Value When Model Includes x_0, x_1, and $x_2 = x_1^2$

c	$\hat{y} = cb$	Variance $(\hat{y}) = v_c \sigma^2$
[1 1 1]	$\hat{y}_1 = 0.89 y_1 + 0.26 y_2 + 0.086 y_3 - 0.14 y_4 + 0.086 y_5$	$0.89\sigma^2$
[1 2 4]	$\hat{y}_2 = 0.26 y_1 + 0.37 y_2 + 0.34 y_3 + 0.17 y_4 - 0.14 y_5$	$0.37\sigma^2$
[1 3 9]	$\hat{y}_3 = -0.086 y_1 + 0.34 y_2 + 0.49 y_3 + 0.34 y_4 - 0.086 y_5$	$0.49\sigma^2$
[1 4 16]	$\hat{y}_4 = -0.14 y_1 + 0.17 y_2 + 0.34 y_3 + 0.37 y_4 + 0.26 y_5$	$0.37\sigma^2$
[1 5 25]	$\hat{y}_5 = 0.086 y_1 - 0.14 y_2 - 0.086 y_3 + 0.26 y_4 + 0.89 y_5$	$0.89\sigma^2$

feature stands out:

- The variance of $\hat{y}$ increases as additional variables are added; for instance, the variance of $\hat{y}_1$ triples by adding x_1 and increases from $0.6\sigma^2$ to $0.89\sigma^2$ by adding x_2.

Viewed from another perspective, the coefficient associated with the recorded observation increases when predicting that observation as additional variables are added. With x_0 alone each y is used with a coefficient of 0.2 for the prediction of itself. If y_1 were based on the mean of five observations not including y_1, the deviation, $y_1 - \hat{y}_1$, would be expected to be larger than the deviation when the predictor is based on a mean including the observed value. Consequently, the results obtained from including y_1 in calculating $\hat{y}_1$ will be overly optimistic. This optimism becomes more pronounced with additional variables added to the predictor. For $\hat{y}_1$ the coefficient associated with y_1 increases from 0.2 to 0.6 to 0.89 by adding x_1 and then x_2. Note that these are the same values as the v_c coefficient for the variance of $\hat{y}_1$.

One overall conclusion from comparing the predicted values for the three models is apparent. If a reduced model is the correct model, then the data analysis will be adversely affected by including additional variables. Of course, the correct model usually is not known in practice. Then the deviation $y_i - \hat{y}_i$ cannot be completely explained by the variance argument, but part of the deviation is possibly due to the use of an incorrect model. For example, if the x_0 model is not adequate, the deviations, $y_i - \bar{y}$ can be made smaller by including important additional variables. Consequently, in model selection there is a trade-off between including variables that will reduce the residual sum of squares from the use of an insufficient model, and including variables that have little effect on the residual sum of squares but will increase the variance of a predicted value. A criterion to offset this difficulty is the prediction sum of squares.

THE PREDICTION SUM OF SQUARES (PRESS)

A criterion more realistic than the residual sum of squares for comparing models is the prediction sum of squares (PRESS). For a particular set of variables PRESS is obtained by predicting each observation using all the *other* observations. The resulting residuals (PRESS residuals) are squared and summed to form PRESS. The criterion PRESS is appealing because it simulates prediction; it does not use an observation to aid in the prediction of itself. The suggested procedure is to calculate PRESS for selected models identified by SELECT. Those few models with small values of PRESS are examined carefully in the subject matter context. Then models with acceptable standard errors for predicted values—and, it is hoped, with interpretable regression coefficients—are reported.

The Prediction Sum of Squares (PRESS)

Computation of PRESS seems a formidable task, since n sets of regression coefficients, each based on $(n-1)$ observations, appear to be required for each model. Actually, the computations are not extensive and depend only on the results from fitting the model to all the observations. To see this, we examine the second observation predicted by the quadratic polynomial model in the previous section,

$$\hat{y}_2 = 0.26y_1 + 0.37y_2 + 0.34y_3 + 0.17y_4 - 0.14y_5$$

We know that both y_2 and $\hat{y}_2$ estimate $\beta_0 + 2\beta_1 + 4\beta_2$, the population mean response for $x_1 = 2$ and $x_2 = 4$. Hence

$$\hat{y}_2 - 0.37y_2 = 0.26y_1 + 0.34y_3 + 0.17y_4 - 0.14y_5$$

estimates $(1-0.37)(\beta_0 + 2\beta_1 + 4\beta_2)$.

Thus

$$\frac{0.26y_1 + 0.34y_3 + 0.17y_4 - 0.14y_5}{1-0.37} \quad \text{estimates} \quad \beta_0 + 2\beta_1 + 4\beta_2$$

and does not depend on y_2. This is the predictor of y_2 that does not depend on y_2.

The PRESS residual is the difference between y_2 and the above predictor:

$$y_2 - \frac{0.26y_1 + 0.34y_3 + 0.17y_4 - 0.14y_5}{1-0.37}$$

$$= \frac{y_2 - 0.26y_1 - 0.37y_2 - 0.34y_3 - 0.17y_4 + 0.14y_5}{1-0.37}$$

$$= \frac{y_2 - \hat{y}_2}{1-0.37}$$

The numerator is just the ordinary residual and the denominator is just one minus v_c, the coefficient of σ^2 in the variance of $\hat{y}_2$. Hence a PRESS residual can be computed by using an algorithm, for instance, ELIM from Unit 12, which outputs $\hat{y}$ and the associated variance. To calculate PRESS, we apply ELIM to each observation; then the PRESS residual is formed and the sum of squares of the residuals are accumulated. Indeed, if we follow the excellent practice of examining the residuals, predicted values, and their variances, the main calculations have been done and all that remains is to form the PRESS residuals and sum their squares.

Table 20.5 Finger Lakes Phosphorus Pollution Data

Lake	Total Phosphorus (mg/m^3)	Lake Surface (km^2)	Effective Mixing Depth (m)	Sewered Population	Unsewered Population	Active Agriculture Land Area (km^2)	Total Land Area (km^2)	Runoff (100m)^3/year
Conesus	17.6	13.7	8.8	1,591	2,589	83.1	158.2	76.23
Hemlock	10.9	7.2	8.5	0	1,186	24.2	96.2	36.63
Canadice	9.2	2.6	7.7	0	291	4.5	31.8	10.08
Honeoye	16.2	7.0	4.9	0	1,276	9.1	95.0	27.83
Canandaigua	10.1	42.3	9.6	0	6,633	158.2	407.2	115.06
Keuka	11.7	47.0	10.0	0	12,125	132.2	404.6	147.62
Seneca	17.9	175.4	11.5	32,842	36,717	501.5	1180.6	651.84
Cayuga	21.1	172.1	11.6	36,334	43,319	904.0	1807.5	749.74
Owasco	14.7	26.7	11.2	2,668	10,530	227.6	470.0	254.95
Skaneateles	7.7	35.9	10.7	0	3,931	66.6	151.4	81.65
Otisco	8.4	7.6	10.0	0	1,536	43.4	93.8	33.53
Oneida	32.8	206.7	6.8	29,800	61,400	1253.0	3293.4	2,183.19
Canadarago	39.5	7.6	6.7	1,527	2,000	81.1	165.0	97.27

Source: R. T. Oglesby and W. R. Schaffner, "Phosphorus Loadings to Lakes and Some of Their Responses. Part 2: Regression Models of Summer Phyloplankton Standing Crops, Winter Total P, and Transparency of New York State Lakes with Known Phosphorus Loadings," *Limnology and Oceanography* 23 (no. 1, 1978): 135–145.

FINGER LAKES DATA

In this section our developed methodology is applied to a set of data concerning phosphorus pollution in 13 New York State lakes. The dependent variable is total phosphorus, which is regarded as a pollutant. The main objective is to identify variables or combinations of variables that contribute to pollution. We are not particularly concerned with inference, since the sample includes all the lakes for which a prediction equation is likely to be valid. The first step is to identify the potentially important variables, starting with the measured data in Table 20.5.

As data analysts, we need to consider the nature of the relationships between these variables. Note that total phosphorus is recorded in terms of a concentration. It therefore seems reasonable to scale the independent variables by the effective volume of the lake. That is, we divide sewered population, unsewered population, active agriculture land area, total land area, and runoff by the product of lake surface and effective mixing depth. This gives us five basic independent variables:

$$PS/LV = (\text{sewered population})/(\text{lake volume})$$
$$PU/LV = (\text{unsewered population})/(\text{lake volume})$$
$$AA/LV = (\text{active agriculture})/(\text{lake volume})$$
$$TL/LV = (\text{total land})/(\text{lake volume})$$
$$R/LV = (\text{runoff})/(\text{lake volume})$$

where

$$\text{lake volume} = (\text{lake surface}) \times (\text{effective mixing depth})$$

Now let's consider active agriculture. Fertilizer applied to the soil could not pollute a lake without runoff to carry phosphorus from the land to the lake. Hence a variable that is large, when there is a combination of large agriculture and large runoff, would seem appropriate. The variable

$$RAA/LV = (\text{runoff}/\text{total land}) \times (AA/LV)$$

is such a variable. The use of (runoff/total land) instead of just runoff or some other function of runoff is somewhat arbitrary.

Similar points apply to other variables as well. Runoff is needed to carry pollutants into a lake. Therefore the variables

$$RPS/LV = (\text{runoff}/\text{total land}) \times (PS/LV)$$

and

$$RPU/LV = (\text{runoff}/\text{total land}) \times (PU/LV)$$

seem to be reasonable. This brings the number of potential variables to nine (including x_0), so that there are $2^9-1=511$ possible models.

The SELECT procedure, treating x_0 as an additional variable, identified the three best models for each size; they are shown in Table 20.6. For each model PRESS was calculated and is listed in the last column.

Table 20.6 Variables Identified for the Finger Lakes Data by the SELECT Procedure

Model	Size	Residual SS	PRESS
RPU/LV	1	372	526
R/LV	1	380	405
AA/LV	1	491	600
PS/LV, PU/LV	2	84	120
PU/LV, RPS/LV	2	92	130
X(0), RPS/LV	2	104	154
X(0), TL/LV, RPS/LV	3	35	54
X(0), PS/LV, R/LV	3	44	66
X(0), PS/LV, TL/LV	3	45	65
X(0), PU/LV, TL/LV, RPS/LV	4	32	68
X(0), PS/LV, R/LV, RAA/LV	4	32	63
X(0), TL/LV, RPS/LV, RAA/LV	4	33	77
X(0), PU/LV, R/LV, RPS/LV, RPU/LV	5	24	74
X(0), PU/LV, TL/LV, RPS/LV, RPU/LV	5	27	77
X(0), PU/LV, TL/LV, RPS/LV, RAA/LV	5	30	113

As expected, the residual sum of squares decreases as variables are added to the model. The PRESS criterion decreases at first and then starts to increase. There is a clear favorite. The model with PRESS=54 has the prediction equation

$$\hat{y} = 5.98 + 3.25(TL/LV) + 1.30(RPS/LV)$$

From a subject matter point of view, this equation is somewhat of a surprise if active agriculture were conjectured as the sole or major contribution of phosphorus pollution. Total land, including both agricultural and nonagricultural, is positively related to phosphorus concentration, perhaps vindicating agriculture.

SUMMARY

Model selection with many x variables depends on computer assisted procedures for screening of variables and for identification and evaluation of a large number of candidate models. On the other hand, model selection also depends heavily on the data analyst's knowledge of the subject matter of the data for judgments with respect to a usable, interpretable model. An estimated regression equation with noninterpretable estimated regression coefficients and unreasonable predicted values, both associated with large standard errors, is discounted in scientific worth. Obviously, the ultimate success or failure of methodology is related to the original choice of possible useful variables. Past experience and reasoning, as well as plotting, is important in examining original, transformed, and constructed variables.

Unit 21

Coping with Large Standard Errors and Unrealistic Estimates

Objectives

- To show how prior information can be incorporated into an analysis by using additional created observations.
- To give a rationale for weighting these additional observations.
- To use prior information to cope with large standard errors and unrealistic estimates.

The basic methodology for estimation and prediction is given in Unit 12. It was emphasized that calculation of an estimate or a predicted value should always be accompanied by calculation of its standard error. A large standard error indicates that our estimate is, practically, of little value. In some instances the estimate itself may have an unrealistic value, signaling that a problem exists. However, an unrealistic estimate is usually accompanied by a large standard error, and hence the problems of unrealistic estimates and large standard errors are very much related. This unit poses suggestions for coping with these problems.

A brief outline of the basic computations for estimating a linear combination of the regression coefficients $c\beta$ includes the following steps:

1. Calculate the ABDO matrix.
2. Apply the ELIM algorithm to c to obtain the estimator cb and v_c, the coefficient of σ^2 in the variance of the estimator.
3. Calculate the standard error of the estimator, SQRT($v_c s^2$). Note that the standard error is composed of two parts:
   a. s^2, a measure of experimental error, usually the residual mean square.
   b. v_c, a factor depending on X and c.

If the standard error is large because the experimental error is large, very little can be done with the existing data. More data should be collected, perhaps by using improved experimental or laboratory techniques. If the standard error is large because v_c is large, then the data analyst may provide insight and a partial solution to the problem.

THE BARLEY RESPONSE DATA

To illustrate the problem and to explore possible solutions, we will use a set of Tunisian data for barley yield with four levels of applied nitrogen fertilizer and four replications of each nitrogen level; the data were collected by Francillon (1973). See Table 21.1.

Suppose we want to predict the yield when 12 units of nitrogen (120 kg/ha) are applied. Of course, the treatment design of the experiment without an actual applied level of 12 is not good for this objective. However, it is not uncommon to attempt an extrapolation one level beyond the data.

Table 21.1 Barley Yield Data

Nitrogen (10kg/ha)	Yield (kg/ha)	Mean
0	23, 22, 19, 21	21.25
3	31, 35, 31, 34	32.75
6	37, 40, 38, 38	38.25
9	40, 42, 40, 40	40.50

Source: F. Francillon, *Notes on Descriptive Statistics of Two Dimensions* (Tunisia: National Agronomic Institute, 1973).

The quadratic polynomial model, which was developed in Unit 12, can be used for prediction, that is, $\hat{y} = b_0 + b_1 x_1 + b_2 x_2$, where $x_2 = x_1^2$ and one unit of x_1 is 10 kg/ha of applied nitrogen.

Our objective of predicting yield at an applied level of 12 can be expressed as estimating $c\beta$, where $c = [1\ 12\ 144]$. This linear combination will be called the objective linear combination. Figure 21.1 shows the data and the fitted curve. The right-most point of the curve represents the estimate of the objective linear combination.

There are two distressing features of this estimate. First, the variance of the estimator is $1.94\sigma^2$. This value is large, because a single observation at

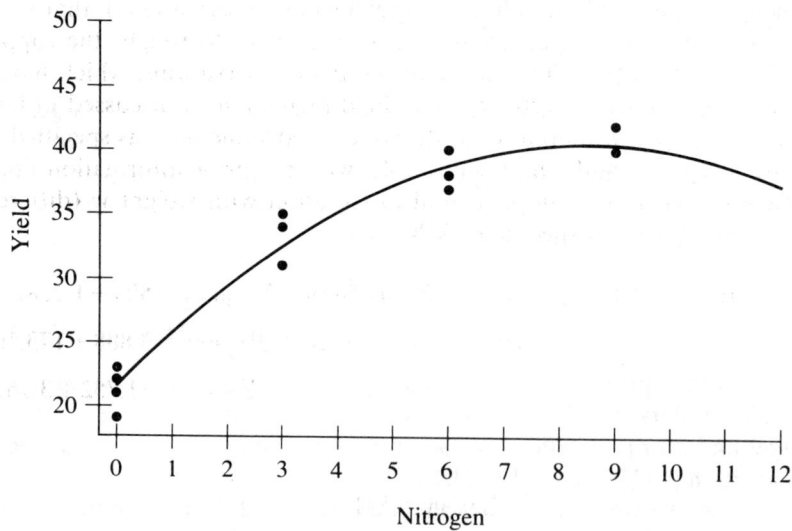

Figure 21.1 Data and Fitted Curve for Barley Yield

an applied nitrogen level of 12 would have a variance of one times σ^2. Said another way, our predictor is about twice as variable as a single direct observation. This result is due to the extrapolation. Second, the downward turn of the prediction curve is contrary to our knowledge of fertilizer response within this interval. In fact, we are reasonably certain that the response at 12 should be slightly higher than the response at 9, say, 2% higher. This prior knowledge can be expressed in terms of a linear combination of the parameters as follows:

$$(\text{response at } 12) = 1.02 \, (\text{response at } 9)$$
$$\beta_0 + \beta_1 12 + \beta_2 144 = 1.02 \, (\beta_0 + \beta_1 9 + \beta_2 81)$$
$$0 = 0.02\beta_0 - 2.82\beta_1 - 61.38\beta_2$$

Or $0 = \mathbf{c}\boldsymbol{\beta}$, where $\mathbf{c} = [0.02 \; -2.82 \; -61.28]$. This linear combination will be called the prior knowledge linear combination.

Having identified the problem and quantified the objective and prior information, we attempt to cope with the problem. The technique is to augment the data with supplemental "observations" that reflect our prior knowledge.

DATA AUGMENTATION TO ACHIEVE A SPECIFIED v_c

Each row of the data matrices $\mathbf{X}$ and $\mathbf{Y}$ represents an item of information about $\boldsymbol{\beta}$. Our prior knowledge can be incorporated by including the prior knowledge linear combination as a supplemental observation(s), that is, by augmenting the data. In all likelihood we will want to weight the supplemental observation(s) differently than the data observations, which have a weight (w) of one, and perform a weighted regression as discussed in Unit 7. The weight will be chosen to produce a desired value of v_c as specified by the data analyst. For the barley example, with the prior information linear combination used as a supplemental observation with weight w (different from one but as yet unspecified), $\mathbf{X}'\mathbf{X}$ is

$$\Sigma w x_0^2 = 16 + 0.0004w \quad \Sigma x_0 w x_1 = 72 - 0.0564w \quad \Sigma x_0 w x_2 = 504 - 1.228w$$
$$\Sigma w x_1^2 = 504 + 7.952w \quad \Sigma x_1 w x_2 = 3{,}888 + 173.1w$$
$$\Sigma w x_2^2 = 31{,}752 + 3{,}767w$$

and $\mathbf{X}'\mathbf{Y}$ is

$$\Sigma x_0 w y = 531$$
$$\Sigma x_1 w y = 2{,}769$$
$$\Sigma x_2 w y = 19{,}809$$

An exact formula for the specific value of w that would achieve a specified value of v_c is not available in general. However, if we try a zero weight (no data augmentation) and two other trial values, then we can apply the interpolation formula:

$$w \doteq \frac{w_1 w_2 (r_0 - r)(r_1 - r_2)}{w_2(r - r_2)(r_0 - r_1) + w_1(r_1 - r)(r_0 - r_2)}$$

where

$r_0 =$ the value of v_c with no data augmentation or weight zero
$r_1 =$ the value of v_c for the trial weight w_1
$r_2 =$ the value of v_c for the trial weight w_2
$r =$ the desired value of v_c

When the augmentation consists of an additional single row to **X** and **Y**, the formula for w is exact and only one application is required. A negative value for the denominator of the formula indicates that the desired value of v_c is not obtainable. When more than one row is augmented, the formula is not exact. However, repeated application of the formula, using the w given by the immediately preceding application in conjunction with the best of the other preceding w's, will produce a w associated with the desired value of v_c. When there are multiple rows, a negative denominator is not absolute proof that the desired value of v_c is not obtainable. If a negative denominator does occur, a larger trial value should be used.

A reasonable rule of thumb for the trial values is to choose w_1 and w_2 so that the affected diagonal elements of **X'X** are changed by about 1% and 10%. In this case, considering only the third diagonal element, $31{,}752 + 3{,}767w$, this rule of thumb suggests using 0.08 (0.01 of 31,752 is approximately 0.08 of 3,767) and 0.8. Using the trial values of 0, 0.08, and 0.8 and the ELIM algorithm of Unit 12, we determine $r_0 = 1.9375$, $r_1 = 1.8001$, and $r_2 = 1.1243$. In this example we set our objective to be having an estimator that is no more variable than the average of two direct observations. That is, we want $v_c = 0.5$. It is obtainable and to require a smaller v_c would increase the chances of an unacceptable fit of the model. Applying the interpolation formula gives

$$w = \frac{(0.08)(0.8)(1.9375 - 0.5)(1.8001 - 1.1243)}{0.8(0.5 - 1.1243)(1.9375 - 1.8001) + 0.08(1.8001 - 0.5)(1.9375 - 1.1243)}$$

$$= \frac{0.06217}{-0.06862 + 0.08458} = 3.90$$

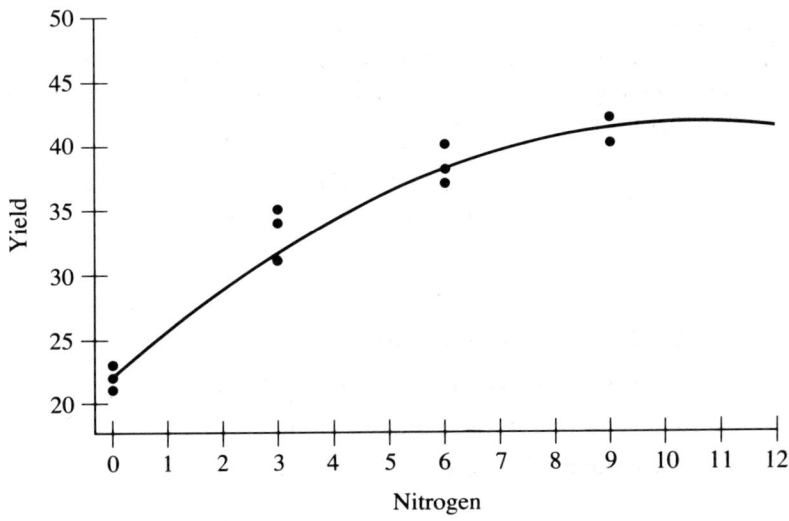

Figure 21.2 Prediction Curve for the Augmented Model

The prediction curve for the augmented model with $w=3.90$ is shown in Figure 21.2. While this curve has a slight turn downward at an applied nitrogen level of 12, it is generally more appealing than Figure 21.1. Some summary statistics for augmented and nonaugmented models are given in Table 21.2.

Table 21.2 Summary Statistics for the Barley Data for Augmented and Nonaugmented Models

Model	cb	v_c	Residual Sum of Squares	Standard Error of cb
Nonaugmented	37.44	1.94	30.76	2.14
Augmented	41.37	0.50	39.40	1.23

The results from the augmented model are more satisfying. The predictor is consistent with prior knowledge in that the predicted yield for 12 is slightly larger than the predicted yield for 9. The v_c component of the variance is about a fourth of the original value. The negative aspect is that the residual sum of squares has increased by about 28%. However, since the fitted curve still seems consistent with the data, there seems little doubt that augmentation has been advantageous.

In general, the reduction in v_c cannot be regarded as a gift. It comes at the expense of a poorer fit of the curve to the data, which is manifested in an increased residual sum of squares. Thus one should subjectively weigh the value of a decreased v_c with an increased residual sum of squares. There is no guarantee that data augmentation will always be successful. Indeed, there are situations where no inference is warranted.

It should be mentioned that the residual sum of squares that comes from ABDO applied to the augmented data is too large because it includes the additional observation(s). To adjust for this, subtract the weight multiplied by the sum of squares of the residuals associated with the supplemental observation(s). For this example the sum of squares from ABDO is 41.53, the residual for the supplemental observation is -0.74, and the weight is 3.90. Hence the sum of squares in Table 21.2 is $41.53 - 3.90(-0.74)^2 = 39.40$.

RIDGE REGRESSION

In models containing x_0 and k other variables, a commonly used set of supplemental observations is given below.

$$\begin{matrix} x_0 & x_1 & x_2 & & x_{k-1} & x_k & & y \\ \begin{bmatrix} 0 & d_1 & 0 & \cdots & 0 & 0 \\ 0 & 0 & d_2 & \cdots & 0 & 0 \\ \vdots & \vdots & \vdots & & \vdots & \vdots \\ 0 & 0 & 0 & \cdots & d_{k-1} & 0 \\ 0 & 0 & 0 & \cdots & 0 & d_k \end{bmatrix} & & = & \begin{bmatrix} 0 \\ 0 \\ \vdots \\ 0 \\ 0 \end{bmatrix} \end{matrix}$$

where $d_j = \text{SQRT}[\Sigma(x_{ij} - \bar{x}_j)^2]$. Augmentation with supplemental observations of this type is called *ridge regression*. The effect of ridge regression on any **cb**, where the first component of **c** is one, is to pull it toward $\bar{y}$. This is often a reasonable thing to do but should not be done indiscriminantly. The d_j are used rather than ones to make an approximate adjustment for x's having different scales so that the same weight may be given each supplemental observation.

While the expression for the d_j above is consistent with classical ridge regression, another choice would be to let d_j equal the range of x_j, that is, the largest x_j minus the smallest x_j. This value, in some instances, has a nice interpretation, as will be illustrated with an example later.

FIREFLY DATA

The two-variable firefly data introduced in Unit 10 will be used here to further illustrate augmentation to achieve a specified v_c. Attention will be confined to estimating $\mathbf{c}\boldsymbol{\beta}$, where $\mathbf{c}$ is of the form $[1 \ c_1 \ c_2]$.

If c_1 and c_2 are possible values of x_1 and x_2, $\mathbf{cb}$ is a predicted value. Predicted values for various combinations of x_1 and x_2 for the firefly data are given in Unit 12. Figure 21.3 shows the combinations of x_1 and x_2 for the firefly data. The three ellipses represent those values of c_1 and c_2 (plotted on the x_1 and x_2 axes) that give a constant value for v_c. Predicting y for the measured combination $\mathbf{c}=[1\ 130\ 25.5]$ gives the same v_c as other combinations of x_1 and x_2 that fall on the inner ellipse. The value of v_c is 0.25 for the inner ellipse, 0.50 for the middle ellipse, and 1.0 for the outer ellipse.

Any attempted prediction outside the largest ellipse would have a v_c greater than one. Therefore, a single observation taken at any point in this region would be a better predictor than that based on the existing data. A prediction on the inner ellipse would be equivalent to the average of four direct observations. Of course, additional direct observations may not be possible. The analogy is mentioned to provide insight about what is a reasonable interval of values for v_c.

Figure 21.3 demonstrates the obvious fact that v_c is small in the area of high concentration of data and large when the data are sparse. It is hoped that the $\mathbf{c}$'s of interest to the experimenter are consistent with the data and hence yield reasonable values of v_c.

In the firefly data suppose it is realistic to predict the time of first flash for high values of light intensity (x_1) and low values of temperature (x_2). No measured data points fall in this new region of interest. Ideally, more data would be collected. If this is not feasible, we might want to proceed with the data analysis by using data augmentation. Three $\mathbf{c}$'s of interest are selected and shown in Table 21.3. The first $\mathbf{c}$ is the approximate average of the variables. The second $\mathbf{c}$ is on the tip of the elongated concentration of

Table 21.3 Selection of Three c's

c	v_c	Indicator in Figure 21.3
[1 73 24]	0.06196	Z
[1 135 28]	0.3016	X
[1 135 18]	1.5433	Y

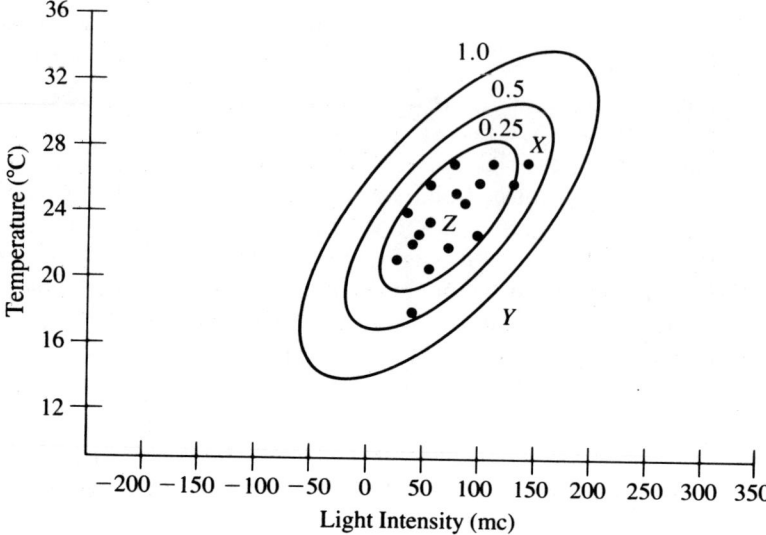

Figure 21.3 Variance Contours for Least Squares Predictors

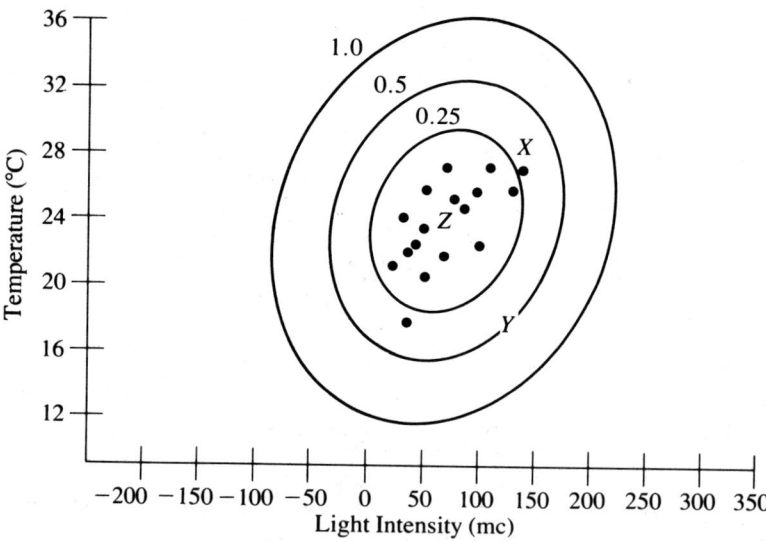

Figure 21.4 Variance Contours with the First Supplemental Observation

Firefly Data

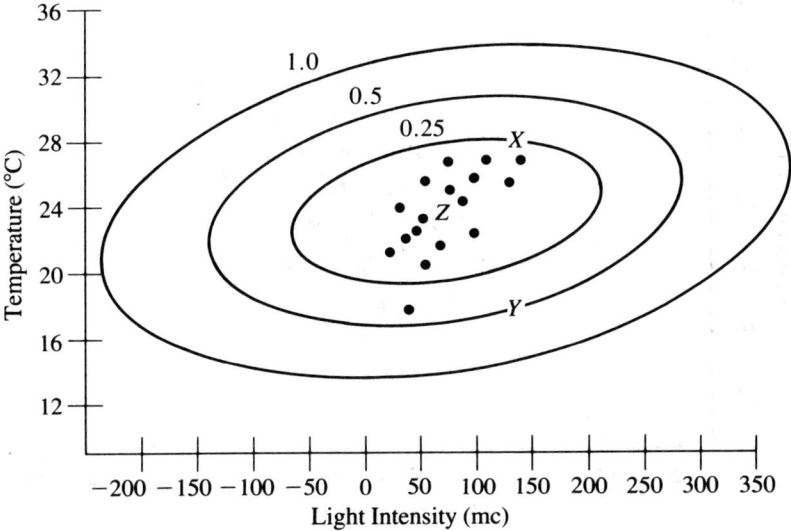

Figure 21.5 Variance Contours with the Second Supplemental Observation

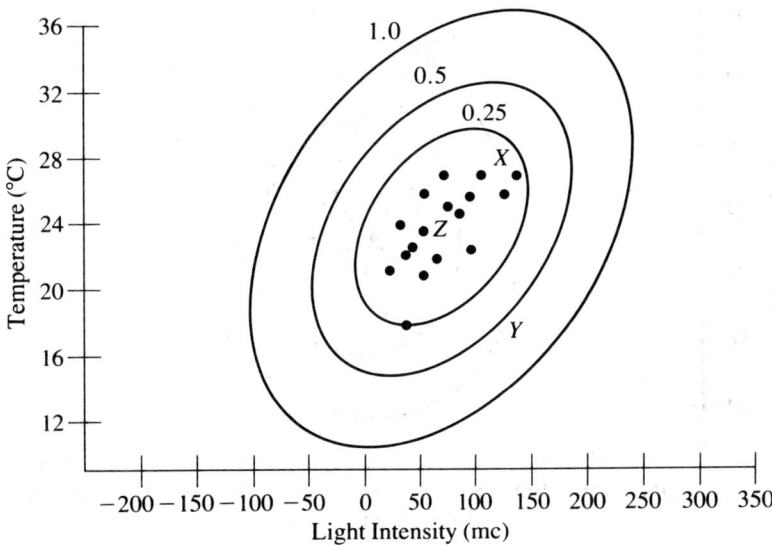

Figure 21.6 Variance Contours with the Third Supplemental Observation

the data. The third is outside the concentration of data but still within limits of each respective x variable. The techniques of Unit 12 are not only adequate but are recommended for the first two c's. However, $v_c = 1.5433$, associated with the third c, is regarded as large. All three are included here so that the effect of data augmentation on a variety of different c's may be seen.

Three ways of augmenting the data are considered. For each set of supplemental observations we choose a weight such that v_c for $c = [1\ 135\ 18]$ will become 0.5. The sets of supplemental observations, their rationale, and the computation of the weights are discussed first. The discussion of the results is deferred until all the results have been obtained.

The first supplemental observation results from the supposition that

$$\beta_0 + 135\beta_1 + 18\beta_2 \quad \text{equals} \quad \beta_0 + 30\beta_1 + 27\beta_2$$

In other words, we think each of the "corners" in Figure 21.3 with sparse data should give an equal response. To incorporate this conjecture, we set the two responses equal to each other, and thus

$$0 = 0\beta_0 + 105\beta_1 - 9\beta_2$$

is the augmented observation. With this additional observation having weight w (as yet unspecified), $\mathbf{X'X}$ and $\mathbf{X'Y}$ are

$$\mathbf{X'X} = \begin{bmatrix} 17 & 1{,}244 & 400.9 \\ & 109{,}328 + 11{,}025w & 30{,}228.9 - 945w \\ & & 9{,}555.65 + 81w \end{bmatrix}$$

$$\mathbf{X'Y} = \begin{bmatrix} 703 \\ 49{,}557 \\ 16{,}363.1 \end{bmatrix}$$

Using trial weights of 0, 1, and 10 gives $r_0 = 1.5433$, $r_1 = 0.3533$, and $r_2 = 0.0948$. Application of the interpolation formula gives

$$w = \frac{(1)(10)(1.5433 - 0.5)(0.3533 - 0.0948)}{10(0.5 - 0.0948)(1.5433 - 0.3533) + 1(0.3533 - 0.5)(1.5433 - 0.0948)}$$

$$= 0.5851$$

Since there is only one supplemental observation, $w = 0.5851$ will produce a v_c of exactly 0.5.

A second possibility is to conjecture that the response does not change as x_1 changes. Said another way, $\beta_1 = 0$ or $\mathbf{c}\beta = 0$, where $\mathbf{c} = [0\ 1\ 0]$. The sums of squares and cross products with this additional observation having weight w are

$$\mathbf{X'X} = \begin{bmatrix} 17 & 1{,}244 & 400.9 \\ & 109{,}328 + w & 30{,}228.9 \\ & & 9{,}555.65 \end{bmatrix} \quad \mathbf{X'Y} = \begin{bmatrix} 703 \\ 49{,}557 \\ 16{,}363 \end{bmatrix}$$

With trial weights of 0, 1,000, and 10,000, the interpolation formula gives

$$w = \frac{(1{,}000)(10{,}000)(1.5433 - 0.5)(1.4405 - 0.9675)}{10{,}000(15 - 0.9675)(1.5433 - 1.4405) + 1{,}000(1.4405 - 0.5)(1.5433 - 0.9675)}$$

$$= 80{,}965$$

where $r_0 = 1.5433$, $r_1 = 1.4405$, $r_2 = 0.9675$, and $r = 0.5$.

The third conjecture is that the response is flat. That is, $\beta_1 = 0$ and $\beta_2 = 0$. Hence we use ridge regression and add two supplemental observations, $0 = [0\ 135.26\ 0]\beta$ and $0 = [0\ 0\ 10.05]\beta$. The nonzero numbers in these observations are square roots of sums of squares of the x's about their respective means and not one. This is an attempt to compensate for different scales of the x's. The sums of squares and cross products with the same unspecified weight applied to each additional observation are

$$\mathbf{X'X} = \begin{bmatrix} 17 & 1{,}244 & 400.9 \\ & 109{,}328 + 18{,}296.47w & 30{,}228.9 \\ & & 9{,}555.65 + 101.48w \end{bmatrix}$$

$$\mathbf{X'Y} = \begin{bmatrix} 703 \\ 49{,}557 \\ 16{,}363.1 \end{bmatrix}$$

Application of the interpolation formula with trial weights of 0, 1, and 10 gives

$$w = \frac{(1)(10)(1.5433 - 0.5)(0.4408 - 0.1087)}{10(0.5 - 0.1087)(1.5433 - 0.4408) + 1(14{,}408 - 0.5)(1.5433 - 0.1087)}$$

$$= 0.8193$$

where $r_0 = 1.5433$, $r_1 = 0.4408$, $r_2 = 0.1087$, and $r = 0.5$. This weight is used with each of the augmented observations and gives $v_c = 0.4743$. This value would ordinarily be close enough. However, since we are interested in comparing different sets of supplemental observations, we will reapply the formula iteratively. First replace the weight of 10 with 0.8193. The formula gives $w = 0.7146$, which in turn gives $v_c = 0.5434$. Then replace the weight of one with 0.7146, giving a new $w = 0.7789$ and $v_c = 0.5158$. We accept this value as sufficiently close to 0.5.

A slightly different way of rationalizing ridge regression is to conjecture

$$\beta_0 + 30\beta_1 + x_2\beta_2 = \beta_0 + 135\beta_1 + x_2\beta_2$$

and

$$\beta_0 + x_1\beta_1 + 18\beta_2 = \beta_0 + x_1\beta_1 + 27\beta_2$$

This conjecture indicates a belief that the response does not change as x_1 takes values from 30 to 135 or as x_2 goes from 18 to 27. Transposing and collecting terms gives

$$0 = 0\beta_0 + 105\beta_1 + 0\beta_2$$

and

$$0 = 0\beta_0 + 0\beta_1 + 9\beta_2$$

Here the d_j are the approximate ranges of the x's. This is similar to assuming that the response is flat over the rectangular x space rather than over the elongated space in which the data actually lie. In this case the d_j are similar enough to those previously used that we will not process them.

EXAMINATION OF RESULTS

Table 21.4 gives the results of estimating three linear combinations of the parameters by using the two-variable model, and this model is augmented with three different sets of supplemental observations. In each case the weight used with the supplemental observation was chosen so that for $\mathbf{c} = [1\ 135\ 18]$ the value of v_c is approximately 0.5.

Table 21.4 Summary of Results for Three Linear Combinations

Supplemental Observation	Weight	Residual Sum of Squares	$\mathbf{c}=[1\ 73\ 24]$ cb	v_c	$\mathbf{c}=[1\ 135\ 28]$ cb	v_c	$\mathbf{c}=[1\ 135\ 18]$ cb	v_c
None		683	40.46	0.062	32.00	0.302	53.27	1.543
$[0\ 105\ -9]$	0.5851	728	40.81	0.060	31.95	0.302	45.09	0.500
$[0\ 1\ 0]$	80,965.	683	40.47	0.061	31.98	0.257	53.20	0.500
$\begin{bmatrix} 0 & 135.26 & 0 \\ 0 & 0 & 10.07 \end{bmatrix}$	0.7789	760	40.92	0.060	34.91	0.224	45.40	0.516

For $c=[1\ 73\ 24]$ the results for the estimates and v_c are similar for all cases considered. For all but ridge regression the results for the estimates and v_c are similar for $c=[1\ 135\ 28]$, because only ridge regression dampens the slope in the direction of the elongated concentration of the data. For $c=[1\ 135\ 18]$ the supplemental observation $[0\ 1\ 0]$ is a clear favorite. It has reduced v_c by more than two-thirds and has not increased the residual sum of squares. This supplemental observation is used to pull b_1 toward zero. As our analysis in Unit 13 has shown, b_1 is already almost zero, and thus pulling it toward zero has almost no effect on the residual sum of squares. The use of the supplemental observation $c=[0\ 1\ 0]$ is very similar to leaving x_1 out of the model.

Figures 21.4, 21.5, and 21.6 show the variance contours that result from using the respective sets of supplemental observations. They are all considerably expanded from the least square contours in Figure 21.3. For the two sets having a single observation, the direction of principal expansion is related to the supplemental observation in a predictable way. Weighting b_1 toward zero expands the contours primarily in the direction of the x_1 axis. Weighting equality of corners expands the contours toward the corners.

SUMMARY

In this unit statistical inference is reviewed and extended. Now we assume that we have a model that is appropriate for the data and s^2, the residual mean square, has a reasonable value. Interest then centers on linear combinations of the parameters ($c\beta$) and on making inferences. Problems can arise with the magnitude of the estimates and standard errors, and the following paragraphs summarize correctional procedures discussed in this unit.

1. Estimate $c\beta$ and its standard error, using the methods of Unit 12. If the estimate is reasonable and the standard error is small, we are done. A judgment about what is small is often facilitated by looking at a confidence interval. If the interval is small enough to be informative, then we have been successful. If the interval is not informative, we may give up, collect more data, or go to the next step.

2. Augment the data with additional observations reflecting prior knowledge of the subject matter. These observations should be selected to complement and not duplicate the existing data. But the additional observations may be related to the c's of the prediction. We might want to assume $c\beta$ is equal to the average of our observations. That is,

$$\beta_0 + c_1\beta_1 + \cdots + c_k\beta_k = \beta_0 + \bar{x}_1\beta_1 + \cdots + \bar{x}_k\beta_k,$$

or

$$(c_1 - \bar{x}_1)\beta_1 + \cdots + (c_k - \bar{x}_k)\beta_k = 0$$

is a possible choice for a supplemental observation. Ridge regression is a popular technique to use here, but its influence is to pull the entire prediction toward $\bar{y}$, which is not always appropriate. If there is some doubt, more than one set of supplemental observations may be tried.

3. Choose a value of v_c such that the standard error is small enough to be meaningful. However, excessive greed here increases the chance that the model will no longer adequately fit the data. Select two trial values for the weight to be used with the supplemental observations. Apply the interpolation formula to obtain a weight that should yield a value of v_c close to the desired value. If the value obtained is not close enough, use it in conjunction with the better of the original two trial values and apply the interpolation formula again. The formula is quite accurate, and it is unlikely that many applications will be required. This procedure can be automated on a computer. The model with the appropriate weight will have been fitted while verifying that it has the specified v_c.

4. Check to see if the model still adequately fits the data. The usual techniques of looking at the residual sum of squares and plots should be used. If the model doesn't fit, try a less demanding value of v_c or give up. If the model fits, proceed with the usual methods.

Unit 22

Split Experimental Units and Repeated Measures

Objectives

- To present a split-unit design, a design that is used extensively in clinical trials and agricultural experiments.
- To show the effects of having two random components in each of the observations.
- To demonstrate the complications that arise with missing data in the split-unit design and to show how to deal with them.

A factorial experiment with one factor assigned to subdivisions of an experimental unit is called a split-experimental-unit experiment, or simply a split-unit experiment. The principle has been historically applied in agricultural experiments with plots of land as units, but is also important in clinical trials where a drug is administered to a subject. Subjects are the experimental units with regard to drugs and, in addition, each level of another factor is given to each subject. The experimental unit for this second factor is a subject at a particular time. The usefulness of this type of design goes far beyond field plots. Therefore, the term *split-unit* is used instead of *split-plot*, the traditional term.

Split-unit experiments require two separate randomization procedures. First, levels of one factor are allocated to the experimental units, commonly called whole units. Then the levels of the second factor are allocated to the split units within each whole unit. With two types of experimental units, each with its own randomization, we need to have two residual components in the model. With no missing data the methods of Unit 18 are easily modified to handle the additional random component. In fact, the only difference here is that the whole-unit mean square is used for testing equality of levels of the whole-unit factor.

Two general situations motivate the use of split-unit experiments. At times the physical application of a treatment requires a larger experimental unit for one factor than for another. For instance in field trials land can be divided into different-sized units, enabling the application of a sprinkler irrigation factor to whole units and a fertilizer factor to splits of each whole unit. In addition to physical needs, experimenters might want different levels of precision for each factor. For example, with the contiguous nature of soil, the random variability among the split units within a whole unit is expected to be less than the variability among whole units. The variance for split units would then be expected to be less than the variance for whole units.

A situation similar in appearance to the split-unit design is one in which the split-unit factor is some unit of time. In this case the experiment is called a repeated-measurements experiment. Since time cannot be randomized, we do not have randomization as a safeguard to ensure independence (apart from the common whole-unit component) of observations on the same whole unit. If lack of independence is not considered a problem, the repeated-measures experiment is analyzed as a split-unit experiment.

The next section is developed in the context of clinical trials. A small data set is presented first to demonstrate basic calculations, including the handling of different residual terms. The basic inferences to be made are very similar to those in Unit 18. We are interested in the presence or absence of interaction and in the comparison of treatment means. Subsequent sections discuss problems resulting from missing data, and practical suggestions for handling these difficulties are presented.

DRUG AND ALCOHOL DATA

Tranquilizers have become one of the most prescribed classes of drugs. Unfortunately, the combination of tranquilizers and alcohol can compromise a driver's ability to operate a motor vehicle. Pharmaceutical companies are trying to develop new tranquilizers that serve their intended purpose but do not combine with alcohol to give an undesirable effect. A trial to compare drug A (an experimental drug), drug B (a currently popular drug), and a placebo (control) is planned. Each drug is to be taken with and without alcohol. Twelve subjects have volunteered to participate in the study, and each subject can receive three different experimental conditions well separated in time. The response is the subject's performance on a simulated driving test. While multiple responses usually are recorded, only the mean deviation (in feet) from the center of the driving lane is discussed here.

One approach to carrying out such an experiment would be to use a completely randomized design with the six combinations of (drug A, drug B, control) with (alcohol, no alcohol), assigned at random to the 12 subjects. Such a design assumes equal importance of alcohol and drugs and also allows an adjustment for time through replicating each treatment-subject combination. However, in this trial time is not believed to be a factor. Furthermore, interest lies more in the drugs, including the interaction with alcohol, than in the difference between alcohol and no alcohol.

The split-unit trial is a natural for this situation, with the subject being the whole unit. The alcohol factor, which requires less precision, is randomly allocated to the whole units. The split unit is a subject at a particular time. Obviously the subjects cannot be "split," but a practical split is the subject's time of availability for the trial. For each whole unit the drug factor, for which more precision is desired, is randomly allocated to the split units. In the present context this simply means that the order in which each subject receives the drug is assigned at random.

The data from this trial are shown in Table 22.1.

Drug and Alcohol Data

To analyze these data, we construct a model sequence with x_0, indicator variables for alcohol and no alcohol (x_1 and x_2), indicator variables for subjects (x_3-x_{14}), indicator variables for drugs ($x_{15}-x_{17}$), and indicator variables for the general means model ($x_{18}-x_{23}$). The sequence of candidate models is similar to those in Units 17–19, with the addition of the indicator variables for subjects. The resulting **X** matrix is

x_0	$x_1 x_2$	$x_3 - x_{14}$	$x_{15} - x_{17}$	$x_{18} - x_{23}$
1	1 0	1 0 0 0 0 0 0 0 0 0 0 0	1 0 0	1 0 0 0 0 0
1	1 0	1 0 0 0 0 0 0 0 0 0 0 0	0 1 0	0 1 0 0 0 0
1	1 0	1 0 0 0 0 0 0 0 0 0 0 0	0 0 1	0 0 1 0 0 0
1	1 0	0 1 0 0 0 0 0 0 0 0 0 0	1 0 0	1 0 0 0 0 0
1	1 0	0 1 0 0 0 0 0 0 0 0 0 0	0 1 0	0 1 0 0 0 0
1	1 0	0 1 0 0 0 0 0 0 0 0 0 0	0 0 1	0 0 1 0 0 0
1	1 0	0 0 1 0 0 0 0 0 0 0 0 0	1 0 0	1 0 0 0 0 0
1	1 0	0 0 1 0 0 0 0 0 0 0 0 0	0 1 0	0 1 0 0 0 0
1	1 0	0 0 1 0 0 0 0 0 0 0 0 0	0 0 1	0 0 1 0 0 0
1	1 0	0 0 0 1 0 0 0 0 0 0 0 0	1 0 0	1 0 0 0 0 0
1	1 0	0 0 0 1 0 0 0 0 0 0 0 0	0 1 0	0 1 0 0 0 0
1	1 0	0 0 0 1 0 0 0 0 0 0 0 0	0 0 1	0 0 1 0 0 0
1	1 0	0 0 0 0 1 0 0 0 0 0 0 0	1 0 0	1 0 0 0 0 0
1	1 0	0 0 0 0 1 0 0 0 0 0 0 0	0 1 0	0 1 0 0 0 0
1	1 0	0 0 0 0 1 0 0 0 0 0 0 0	0 0 1	0 0 1 0 0 0
1	1 0	0 0 0 0 0 1 0 0 0 0 0 0	1 0 0	1 0 0 0 0 0
1	1 0	0 0 0 0 0 1 0 0 0 0 0 0	0 1 0	0 1 0 0 0 0
1	1 0	0 0 0 0 0 1 0 0 0 0 0 0	0 0 1	0 0 1 0 0 0
1	0 1	0 0 0 0 0 0 1 0 0 0 0 0	1 0 0	0 0 0 1 0 0
1	0 1	0 0 0 0 0 0 1 0 0 0 0 0	0 1 0	0 0 0 0 1 0
1	0 1	0 0 0 0 0 0 1 0 0 0 0 0	0 0 1	0 0 0 0 0 1
1	0 1	0 0 0 0 0 0 0 1 0 0 0 0	1 0 0	0 0 0 1 0 0
1	0 1	0 0 0 0 0 0 0 1 0 0 0 0	0 1 0	0 0 0 0 1 0
1	0 1	0 0 0 0 0 0 0 1 0 0 0 0	0 0 1	0 0 0 0 0 1
1	0 1	0 0 0 0 0 0 0 0 1 0 0 0	1 0 0	0 0 0 1 0 0
1	0 1	0 0 0 0 0 0 0 0 1 0 0 0	0 1 0	0 0 0 0 1 0
1	0 1	0 0 0 0 0 0 0 0 1 0 0 0	0 0 1	0 0 0 0 0 1
1	0 1	0 0 0 0 0 0 0 0 0 1 0 0	1 0 0	0 0 0 1 0 0
1	0 1	0 0 0 0 0 0 0 0 0 1 0 0	0 1 0	0 0 0 0 1 0
1	0 1	0 0 0 0 0 0 0 0 0 1 0 0	0 0 1	0 0 0 0 0 1
1	0 1	0 0 0 0 0 0 0 0 0 0 1 0	1 0 0	0 0 0 1 0 0
1	0 1	0 0 0 0 0 0 0 0 0 0 1 0	0 1 0	0 0 0 0 1 0
1	0 1	0 0 0 0 0 0 0 0 0 0 1 0	0 0 1	0 0 0 0 0 1
1	0 1	0 0 0 0 0 0 0 0 0 0 0 1	1 0 0	0 0 0 1 0 0
1	0 1	0 0 0 0 0 0 0 0 0 0 0 1	0 1 0	0 0 0 0 1 0
1	0 1	0 0 0 0 0 0 0 0 0 0 0 1	0 0 1	0 0 0 0 0 1

Table 22.1 Alcohol-Drug Data

	Subject	Drug A	Drug B	Control
Alcohol	1	3.56	4.04	3.26
	2	3.79	3.88	3.49
	3	4.09	5.32	3.79
	4	3.10	4.38	2.80
	5	3.33	3.63	3.03
	6	3.35	3.63	3.05
No alcohol	7	2.83	2.55	2.63
	8	2.93	2.42	2.73
	9	3.58	3.99	3.38
	10	2.98	3.07	2.78
	11	2.32	2.15	2.12
	12	2.73	3.23	2.53

Table 22.2 ANOVA for Data When y Is Regressed on x_0 Through x_{23}

Source of Variation	Sum of Squares	Degrees of Freedom	Mean Square
x_0	376.813	1	376.813
Alcohol (A)	5.8968	1	5.8968
Subjects	5.8340	10	0.5834
Drugs (D)	1.8772	2	0.9386
AD Interaction	0.8687	2	0.4343
Residual	1.4126	20	0.0706

When y is regressed on x_0 through x_{23}, the ANOVA table (Table 22.2) can be calculated from the sequential sums of squares. The mean responses are displayed in Figure 22.1. The first point of interest is examination for interaction between alcohol and drugs. Graphically, parallel lines in Figure 22.1, within the limits allowed by chance, would indicate no interaction. This is checked by an F test from the ANOVA table, $F = 0.4343/0.0706 = 6.15$. With evidence of an interaction between alcohol and drugs, the inappropriateness of comparing drug means and comparing alcohol with no alcohol is emphasized. For example, it is difficult to compare drug A with drug B when at least part of the comparison depends on the presence or

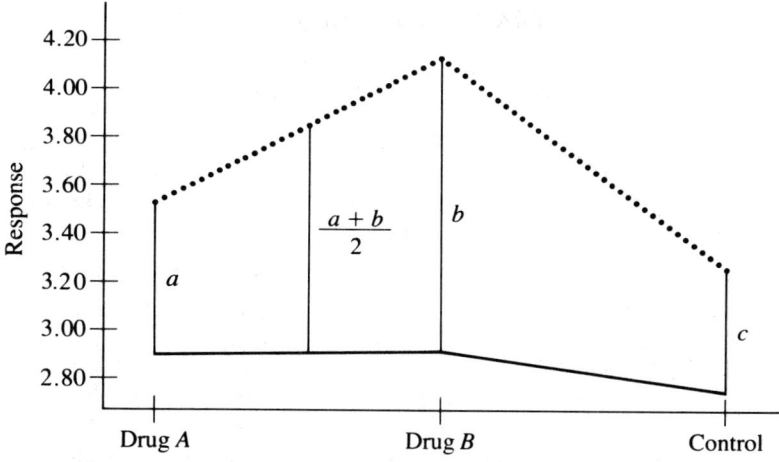

Figure 22.1 Alcohol (Dotted) and No Alcohol (Solid) Responses

absence of alcohol. Had there been no interaction (parallel lines), the comparisons among drugs would not be dependent on presence or absence of alcohol, and both alcohol and drugs would have been examined as in Units 17–19. The one difference would be in using the subjects' mean square as the appropriate residual term for alcohol, because subjects are the experimental units with respect to alcohol.

Returning to the present situation where interaction is evident, we need to fit the general means model, followed by contrasts of interest among drugs for both alcohol and no alcohol. The results are shown in Table 22.3.

Table 22.3 Results from Fitting the General Means Model

Contrast	Estimate	v_c	Standard Error
11+12−2*13−21−22+2*23	0.8033	2.0000	0.3758
11−12−21+22	−0.6033	0.6667	0.2170
11+12−2*13	1.2100	1.0000	0.2658
11−12	−0.6100	0.3333	0.1534
21+22−2*23	0.4067	1.0000	0.2658
21−22	−0.0067	0.3333	0.1534

Note: Two-digit numbers identify combination of alcohol and drugs; single-digit numbers are constants; the asterisk (*) denotes multiplication.

In the table the two-digit numbers identify the combination of alcohol and drugs (e.g., 13 denotes alcohol and control drug), single digits are constants, and * denotes multiplication. The first two contrasts subdivide the two degrees of freedom for interaction in the previous ANOVA table, as discussed in Unit 18. With reference to Figure 22.1, the first contrast compares the average length of lines a and b with the length of line c. The t statistic is $0.8033/0.3758 = 2.14$. The second contrast compares the length of line a to the length of line b. The t statistic is $-0.6033/0.2170 = -2.78$. The next two contrasts involve drug comparisons within the alcohol group, while the last two contrasts are within the no-alcohol group. It is noteworthy that drug A and drug B are very different in the presence of alcohol but not different in the absence of alcohol. Had there not been an interaction, the general means portion of the sequence would have been omitted, and contrasts among drug means and between means for alcohol and no alcohol would have been made. Subsequent sections will consider the no-interaction situation in more detail.

THE REASONS FOR DIFFERENT RESIDUAL TERMS

In order to explicitly show why the whole-unit mean square is the appropriate error term for the whole-unit factor (when there is no interaction), and to establish a basis for handling missing data in the next section, we will examine a small configuration in detail. It is shown in Table 22.4.

Table 22.4 Data from a Drug Trial

	Subject	Stress 1	Stress 2	Stress 3
Drug A	1	y_{111}	y_{112}	y_{113}
	2	y_{121}	y_{122}	y_{123}
	3	y_{131}	y_{132}	y_{133}
Drug B	4	y_{241}	y_{242}	y_{243}
	5	y_{251}	y_{252}	y_{253}
	6	y_{261}	y_{262}	y_{263}

The layout in the table represents data from a drug trial comparing two drugs (whole-unit factor) in conjunction with three stresses allocated to each subject: rest, mild exercise, and strenuous exercise. Numerical values of the observations are not given because the comments apply in general. A specification of the model is

$$y_{j\ell k} = \mu_{j\ell} + S_k + \varepsilon_{j\ell k}$$

where $\mu_{j\ell}$ is the mean for the jth drug and ℓth stress combination, S_k is a random subject component, and $\varepsilon_{j\ell k}$ is a random error. It is assumed that the S_k and $\varepsilon_{j\ell k}$ are normally and independently distributed with mean zero and respective variances of σ_s^2 and σ^2.

When there is no interaction, we have

$$\mu_{j\ell} = \mu + \alpha_j + \beta_\ell$$

where μ is the regression coefficient of x_0, the α_j's are the regression coefficients of drug indicator variables, and the β_ℓ's are the regression coefficients of the stress indicator variables. While the α_j's and β_ℓ's are not uniquely determined, any contrast among them is uniquely determined and equal to the same contrast among the corresponding means.

In the following demonstration no interaction between drugs and stresses is assumed. The **X** matrix is

x_0 Constant	$x_1 - x_3$ Stresses	$x_4 - x_5$ Drugs	$x_6 - x_{11}$ Subjects
1	1 0 0	1 0	1 0 0 0 0 0
1	0 1 0	1 0	1 0 0 0 0 0
1	0 0 1	1 0	1 0 0 0 0 0
1	1 0 0	1 0	0 1 0 0 0 0
1	0 1 0	1 0	0 1 0 0 0 0
1	0 0 1	1 0	0 1 0 0 0 0
1	1 0 0	1 0	0 0 1 0 0 0
1	0 1 0	1 0	0 0 1 0 0 0
1	0 0 1	1 0	0 0 1 0 0 0
1	1 0 0	0 1	0 0 0 1 0 0
1	0 1 0	0 1	0 0 0 1 0 0
1	0 0 1	0 1	0 0 0 1 0 0
1	1 0 0	0 1	0 0 0 0 1 0
1	0 1 0	0 1	0 0 0 0 1 0
1	0 0 1	0 1	0 0 0 0 1 0
1	1 0 0	0 1	0 0 0 0 0 1
1	0 1 0	0 1	0 0 0 0 0 1
1	0 0 1	0 1	0 0 0 0 0 1

With an equal number of observations for each combination of stress and drug, the data layout is balanced, and the ordering of stresses, drugs, and subjects is immaterial with one exception. The subject variables can explain the drug variables exactly. If subjects were put before drugs in the model sequence, then drugs would have a sum of squares of zero. If drugs are put before subjects, then the maximum possible sum of squares is attributed to drugs, and the balance of the total subject sum of squares is attributed to subjects. The latter is the desired partitioning of the sequential sums of squares, and therefore drugs should always be before subjects. The particular ordering given here was chosen to provide a comparison when missing data are discussed in a subsequent section.

The ABDO matrix (odd rows of the ABDO calculations) is as follows:

Constant	Stresses			Drugs				Subjects						z
18	6	6	6	9	9	3	3	3	3	3	3	3		z_1
	4	−2	−2	0	0	0	0	0	0	0	0	0		z_2
		3	−3	0	0	0	0	0	0	0	0	0		z_3
			0	0	0	0	0	0	0	0	0	0		0
				4.5	−4.5	1.5	1.5	1.5	1.5	−1.5	−1.5	−1.5		z_5
					0	0	0	0	0	0	0	0		0
						2	−1	−1	0	0	0	0		z_7
							1.5	−1.5	0	0	0	0		z_8
								0	0	0	0	0		0
									2	−1	−1			z_{10}
										1.5	−1.5			z_{11}
											0			0

The z's in the right-most column are analogous to the z's in Unit 12, the only difference being that the subject components contribute to the variances instead of the means. We will examine the properties of z_5 in detail and simply state the properties of the other z's.

From the ABDO matrix we can state that z_5, a linear combination of the observations, is

$$4.5\alpha_1 - 4.5\alpha_2 + 1.5S_1 + 1.5S_2 + 1.5S_3 - 1.5S_4 - 1.5S_5 - 1.5S_6$$
$$+ (\text{a linear combination of the } \varepsilon_{j\ell k})$$

Since the S_k and $\varepsilon_{j\ell k}$ have mean zero, the mean of z_5 is simply the fixed portion,

$$\text{mean}(z_5) = 4.5\alpha_1 - 4.5\alpha_2 = 4.5(\alpha_1 - \alpha_2)$$

The variance of z_5 is the sum of one part due to the $\varepsilon_{j\ell k}$ and another part due to the S_k. The part due to $\varepsilon_{j\ell k}$ is $4.5\sigma^2$, where 4.5 is the fifth diagonal element of the ABDO matrix. (Unit 11 shows why this is true.) The part due to the S_k is found by the equal variance rule,

$$\left[1.5^2 + 1.5^2 + 1.5^2 + (-1.5)^2 + (-1.5)^2 + (-1.5)^2\right]\sigma_s^2 = 13.5\sigma_s^2$$

Thus

$$\text{variance}(z_5) = 4.5\sigma^2 + 13.5\sigma_s^2 = 4.5(\sigma^2 + 3\sigma_s^2)$$

Similar manipulations on the other z's give the results shown below.

z	Mean(z)	Variance(z)
z_1	$18\mu + 6(\beta_1 + \beta_2 + \beta_3) + 9(\alpha_1 + \alpha_2)$	$18(\sigma^2 + 3\sigma_s^2)$
z_2	$2(2\beta_1 - \beta_2 - \beta_3)$	$4\sigma^2$
z_3	$3(\beta_2 - \beta_3)$	$3\sigma^2$
z_4	0	0
z_5	$4.5(\alpha_1 - \alpha_2)$	$4.5(\sigma^2 + 3\sigma_s^2)$
z_6	0	0
z_7	0	$2(\sigma^2 + 3\sigma_s^2)$
z_8	0	$1.5(\sigma^2 + 3\sigma_s^2)$
z_9	0	0
z_{10}	0	$2(\sigma^2 + 3\sigma_s^2)$
z_{11}	0	$1.5(\sigma^2 + 3\sigma_s^2)$
z_{12}	0	0

In addition, the z's are mutually independent. The discussion in Unit 11 shows why the $\varepsilon_{j\ell k}$ portion of the z's are independent. The subject components of the z's are shown to be independent by noting that rows of the subject block of the ABDO matrix are orthogonal and by applying the independence rule for linear combinations.

Apart from the adjustments required for differences in diagonal elements of the ABDO matrix, the z's associated with drugs and subjects all have the same variance, $\sigma^2 + 3\sigma_s^2$. The mean of z_5 (associated with drugs) is $4.5(\alpha_1 - \alpha_2)$ and the means of the remaining z's are zero. Hence the subjects mean square is the appropriate error term for drugs. The z's associated with stresses do not depend on σ_s^2 and the residual term is appropriate for error.

MISSING DATA IN THE SPLIT-UNIT TRIAL

Here we consider the same situation as that of the preceding section except that three values, indicated by —, are missing. The configuration is given in Table 22.5.

As discussed in Unit 18, two model sequences need to be fitted in the analysis of a two-factor experiment with no interaction and unequal numbers of observations. With stresses coming after drugs and subjects in the model sequence, stresses are adjusted for drugs and subjects and are tested against the residual mean square with no new complications. With the ordering of stresses, drugs, and subjects, drugs and subjects are adjusted for stresses, but we have the problem of violating the assumptions underlying the F test, as shown by the ABDO matrix:

Constant	Stresses			Drugs			Subjects					z
	5	5	5	7	8	2	2	3	2	3	3	
15	3.33	−1.67	−1.67	−0.33	0.33	0.33	−0.67	0	0.33	0	0	z_1
		2.50	−2.50	0.50	−0.50	0.50	0	0	−0.50	0	0	z_2
			0	0	0	0	0	0	0	0	0	z_3
				3.60	−3.60	1	1	1.60	−0.80	−1.40	−1.40	0
					0	0	0	0	0	0	0	z_5
						1.32	1.32	0	0	0	0	0
							−0.48	−0.84	0.02	−0.01	−0.01	z_7
							1.15	−1.15	0.03	−0.02	−0.02	z_8
								0	0	0	0	0
									1.42	−0.71	−0.71	z_{10}
										1.50	−1.50	z_{11}
											0	0

Table 22.5 Data from a Drug Trial with Three Missing Values

	Subject	Stress 1	Stress 2	Stress 3
	1	y_{111}	y_{112}	—
Drug A	2	—	y_{122}	y_{123}
	3	y_{131}	y_{132}	y_{133}
	4	y_{241}	—	y_{243}
Drug B	5	y_{251}	y_{252}	y_{253}
	6	y_{261}	y_{262}	y_{263}

Since the present interest is in the test for drugs, only the properties of z_5 through z_{12} are examined. Using the same techniques as in the previous section, we have

z	Variance(z)
z_5	$3.60(\sigma^2 + 2.53\sigma_s^2)$
z_6	0
z_7	$1.32(\sigma^2 + 2.03\sigma_s^2)$
z_8	$1.15(\sigma^2 + 2.30\sigma_s^2)$
z_9	0
z_{10}	$1.42(\sigma^2 + 2.13\sigma_s^2)$
z_{11}	$1.5(\sigma^2 + 3.00\sigma_s^2)$
z_{12}	0

Note that the variances adjusted for the diagonal elements are not equal. Furthermore, the rows of the subject block of the ABDO matrix are not orthogonal, and hence the z's are not independent. The consequences of the assumptions not being satisfied depend on the amount and nature of imbalance and the relative magnitude of σ_s^2 to σ^2. There is no exact way to overcome this problem. Previous research has compared several methods in conjunction with several different patterns of missing data and ratios of σ_s^2 to σ^2. A general recommendation, provided that the missing observations comprise no more than 25% of the data, is to multiply the ratio of drug mean square to the subject mean square by a constant and then proceed in the usual way. The constant is

$$\frac{\text{average multiplier of } \sigma_s^2 \text{ from the variances of the subject } z\text{'s}}{\text{average multiplier of } \sigma_s^2 \text{ from the variances of the drug } z\text{'s}}$$

For this example we have

$$[(2.03+2.30+2.13+3.00)/4]/2.53/1 = 2.37/2.53 = 0.93$$

Using this adjusted F statistic for the test of drugs and the previously mentioned test for stresses, we can then follow the general strategy for the analysis of a two-factor experiment given in the summary of Unit 18.

MORE ON THE INTERPRETATION OF INTERACTION

The previous section showed how to test the whole-unit factor when there are missing data. A certain amount of additional work was required. This section demonstrates that if one level of each drug is a control or a baseline, then the objective of the analysis is almost fulfilled by examining the interaction. In these situations we need not be concerned with testing the whole-unit factor.

Consider a situation such as the one shown in Figure 22.2. Each line represents the average of all subjects on the particular drug. The important aspect is that response to the control drug is almost constant, while the response to the test drug is a substantial decrease from its initial value. A response of this type would be delightful if the response is blood pressure and the test drug is one proposed to reduce blood pressure. Such a response would be detected by the test for interaction and not by the test for drugs.

Now consider the situation shown in Figure 22.3. The response for each drug remains at about its initial level, an uninteresting result. This situation

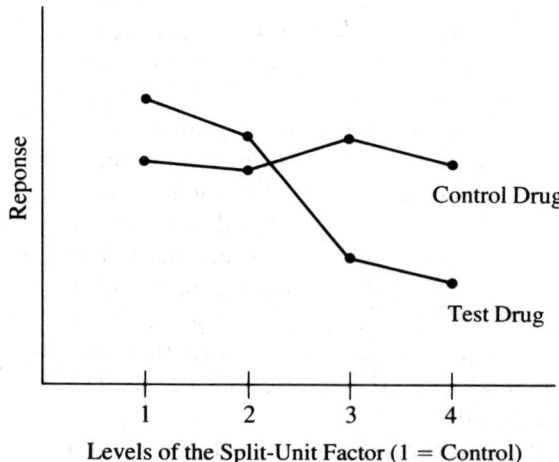

Figure 22.2   Response to the Test Drug Shows a Decrease

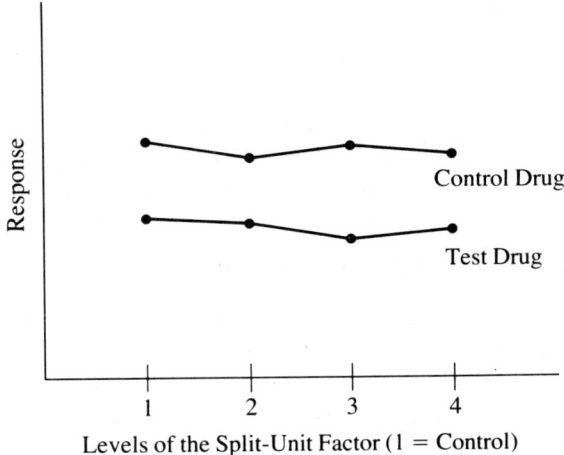

Figure 22.3 No Change in Either Response

depicts no interaction but a significant difference between drugs. In view of the difficulty in testing drugs with missing data, the detection of a particular type of drug response by testing the interaction is preferred.

MISSING DATA IN THE REPEATED-MEASURES TRIAL

The data layout for a repeated-measures experiment looks identical to that of the split unit. The difference is that the second factor is time. This design is very common for drug trials where the response of each subject is observed periodically. In the split-unit design the order in which the stresses were applied was random for each subject, and this randomization is a safeguard to ensure independence of the $\varepsilon_{j\ell k}$. But it is impossible to randomize the order of time. Hence the repeated-measures experiment does not have the safeguard that the split-unit experiment does. Lack of the safeguard does not necessarily mean that the $\varepsilon_{j\ell k}$ are not independent, for repeated-measures data are often analyzed in exactly the same manner as split-unit data. If there is doubt concerning the independence of the $\varepsilon_{j\ell k}$, though, tests for independence requiring multivariate methods and no missing data are available. The procedure presented here allows for both missing data and nonindependence, provided that the pattern of nonindependence is the same for each subject, and is carried out by using methodology already developed.

Consider the data layout shown in Table 22.6. The first concern is interaction between drugs and weeks. A specification of the model is $y_{j\ell k} = \mu_{j\ell} + S_k + \varepsilon_{j\ell k}$, as previously defined. The assumptions about S_k and

$\varepsilon_{j\ell k}$ are the same as for the split-unit design, except that $\varepsilon_{j\ell k}$ in the same row need not be independent provided that the pattern of nonindependence is the same for each row.

The test for interaction begins by forming a table of successive differences in which we subtract each column of numbers from the following column. If in any subtraction one number is missing, then the result is considered missing. The data layout for the differences is shown in Table 22.7.

To see the effect of taking these differences, we apply the same operations to the right-hand side of the model specification (omitting the $\varepsilon_{j\ell k}$), as shown in Table 22.8. Taking each new column separately, we can test for equality of drugs by using the methods of Unit 15. For the first

Table 22.6 Data from a Drug Trial with Three Missing Values; Repeated-Measures Trial

	Subject	Weeks		
		1	2	3
Drug A	1	y_{111}	y_{112}	—
	2	—	y_{122}	y_{123}
	3	y_{131}	y_{132}	y_{133}
Drug B	4	y_{241}	—	y_{243}
	5	y_{251}	y_{252}	y_{253}
	6	y_{261}	y_{262}	y_{263}

Table 22.7 Successive Differences for Data in Table 22.6

	Subject	Successive Difference	
		1	2
Drug A	1	$y_{112} - y_{111}$	—
	2	—	$y_{123} - y_{122}$
	3	$y_{132} - y_{131}$	$y_{133} - y_{132}$
Drug B	4	—	—
	5	$y_{252} - y_{251}$	$y_{253} - y_{252}$
	6	$y_{262} - y_{261}$	$y_{263} - y_{262}$

Table 22.8 Successive Differences for Means

	Subject	Successive Difference 1	Successive Difference 2
Drug A	1	$\mu_{12}-\mu_{11}$	—
	2	—	$\mu_{13}-\mu_{12}$
	3	$\mu_{12}-\mu_{11}$	$\mu_{13}-\mu_{12}$
Drug B	4	—	—
	5	$\mu_{22}-\mu_{21}$	$\mu_{23}-\mu_{22}$
	6	$\mu_{22}-\mu_{21}$	$\mu_{23}-\mu_{22}$

column we would be testing the hypothesis

$$\mu_{12}-\mu_{11}=\mu_{22}-\mu_{21} \quad \text{or} \quad \mu_{11}-\mu_{12}-\mu_{21}+\mu_{22}=0$$

which is the condition of no interaction of drugs with the first two weeks. Similarly, the comparison of drugs on the second column is testing the hypothesis of no interaction of drugs with the second and third weeks. While the technique does not give one overall test for interaction, it is actually more informative than an overall test in that it pinpoints the nature of the interaction.

If there is no interaction, the data can be represented (omitting the $\varepsilon_{j\ell k}$) as shown in Table 22.9. A comparison of drugs can be done separately for each week. While having three tests is not as satisfying as having one test, each test should give similar results since no interaction was detected.

Table 22.9 Data When There Is No Interaction

	Subject	Week 1	Week 2	Week 3
Drug A	1	$\mu+\alpha_1+\beta_1+S_1$	$\mu+\alpha_1+\beta_2+S_1$	—
	2	—	$\mu+\alpha_1+\beta_2+S_2$	$\mu+\alpha_1+\beta_3+S_2$
	3	$\mu+\alpha_1+\beta_1+S_3$	$\mu+\alpha_1+\beta_2+S_3$	$\mu+\alpha_1+\beta_3+S_3$
Drug B	4	$\mu+\alpha_2+\beta_1+S_4$	—	$\mu+\alpha_2+\beta_3+S_4$
	5	$\mu+\alpha_2+\beta_1+S_5$	$\mu+\alpha_2+\beta_2+S_5$	$\mu+\alpha_2+\beta_3+S_5$
	6	$\mu+\alpha_2+\beta_1+S_6$	$\mu+\alpha_2+\beta_2+S_6$	$\mu+\alpha_2+\beta_3+S_6$

Table 22.10 Successive Differences When There Is No Interaction

	Subject	Difference 1	Difference 2
Drug A	1	$\beta_2 - \beta_1$	—
	2	—	$\beta_3 - \beta_2$
	3	$\beta_2 - \beta_1$	$\beta_3 - \beta_2$
Drug B	4	—	—
	5	$\beta_2 - \beta_1$	$\beta_3 - \beta_2$
	6	$\beta_2 - \beta_1$	$\beta_3 - \beta_2$

To test for time effects, we return to the successive differences. The data representation of the differences (omitting the $\varepsilon_{j\ell k}$) with no interaction is shown in Table 22.10. Thus using a means model on the first difference provides the test for equality of week 1 and week 2. Similarly, the second difference provides a test for equality of week 2 and week 3.

COMMENTS ON USING STATISTICAL PACKAGE PROGRAMS

Examination of ABDO output shows the effects on the analysis of the observations containing more than one random component. If the analysis of balanced data is done with a statistical package program, then we must form appropriate F ratios, which are to be used instead of the default statistics that use the residual mean square in the denominator. Some programs do have an option to specify the desired error term, while other programs have an option to print expected (average value) mean squares that show the appropriate error terms for the various tests. With unbalanced data the coefficients of σ_s^2 in the expected mean squares are the average multipliers of σ_s^2 from the variances of the z's and can be used for forming approximate F statistics. Most programs provide transformations that may be used to compute successive differences with automatic handling of missing values.

SUMMARY

This unit has described one of the simplest but most often used of the split-unit designs. This design is appropriate when experimental units are naturally subdivided or when more precision on one factor is desired at the expense of less precision on the other factor. With balanced designs the analysis is no more complicated than the analysis of cell means models discussed in Units 15 and 17. When data are missing, some of the ratios of mean squares no longer have F distributions, and hence special techniques are required, as was discussed. A repeated-measures experiment, having the same data layout as the split-unit design, was also presented.

Unit 23

Nonlinear Models

Objectives

- To show the distinction between linear and nonlinear models.
- To show that nonlinear models are fit by iteratively fitting linear models.
- To illustrate the fitting of nonlinear models to fertilizer response data and to drug concentration data.

Throughout the previous units we have considered linear regression models of the general form

$$y = \beta_0 + \beta_1 x_1 + \cdots + \beta_k x_k + \varepsilon$$

Linear refers to the betas. Models are called linear even if some of the x's are powers or logarithms of other x's. An example of a nonlinear model is

$$y = \beta_1 - \beta_2 \exp(-\alpha_1 x) + \varepsilon$$

This model is nonlinear when regarded as a function of the betas and alpha and no transformation will give a linear model. Note that if α_1 were a known constant, the model would be linear. Hence β_1 and β_2 are called linear parameters, and α_1 is called a nonlinear parameter. Throughout, β will denote linear parameters, and α will denote nonlinear parameters. This distinction is helpful with computer programs that can exploit the fitting of linear parameters in a nonlinear model.

WHERE NONLINEAR MODELS ARE USED

In previous units models were fitted to data to find a satisfactory relationship between a response variable and the explanatory variables. Logarithms, powers, or products of variables were also used when they helped provide a simpler explanation. But almost no attention was given to the underlying mechanisms that generate the data. Indeed, in most situations the complexity of underlying mechanisms prohibits reasonable alternatives. However, there are situations where a dominant mechanism is sufficiently simple for a model to be derived. The actual derivation may be mathematically sophisticated; whether we, as data analysts, derive the model or obtain it from the literature, considerable effort is likely to be involved. Nevertheless, as the underlying mechanisms of drug metabolism, fertilizer response, and other processes become better understood, the use of nonlinear models will increase.

There are rewards to be found in using derived models. The parameters are usually much more interpretable, and the estimated response has a "built in" pleasing form. Derived models also tend to extrapolate much better than polynomial models. Finally, the objectives of a nonlinear regression analysis are similar to those of linear regression—namely, model formulation, estimation, and prediction.

HOW NONLINEAR MODELS ARE FITTED

Nonlinear models are more difficult to fit than linear models. The ideas behind the methods for fitting nonlinear models are presented in a subsequent section. For the moment it suffices to say that we begin with crude estimates of the parameters and then approximate the nonlinear model by a linear model. Estimates of the parameters of the approximating linear model are found by standard techniques. These estimates are used to improve our initial estimates and the process is repeated. However, since the initial estimates of the parameters may be far from the final estimates, a sizable number of approximations may be required. Fortunately, computer programs are available for the computations.

When a small amount of fertilizer is applied to soils with poor fertility, a substantial increase in crop yield is usually obtained. However, successive increments of fertilizer result in smaller marginal yields. A nonlinear model, called the Mitscherlich equation, has been used by soil scientists to relate crop yield to fertilizer application and can be expressed as

$$y = \beta_1 - \beta_2 \exp(-\alpha_1 x) + \varepsilon$$

where y is yield and x is the amount of fertilizer applied. But before this model is fit to data, further development of the model in terms of parameter interpretation is given.

Notice that β_1 and β_2 are linear parameters and α_1 is a nonlinear parameter. The parameters of the Mitscherlich nonlinear model can be interpreted more directly in terms of the subject matter compared with a polynomial model. Specifically:

- β_1 is the upper bound of the response to increasing amounts of applied nitrogen, that is, the asymptote.
- β_2 is the difference between the upper bound and the response with no nitrogen, that is, the potential increase in yield.
- α_1 is the ratio of the change in response per unit change of x to the remaining possible increase of the response.

To better understand the interpretation of the parameters and α_1 in particular, consider a situation with observations at each integer value of x.

At $x=0$ the response is $(\beta_1-\beta_2)$ and the potential increase is β_2. The response at $x=1$ is the response at $x=0$ plus a fraction (α) of the potential increase at $x=0$; that is,

$$(\text{response at } x=1) = (\text{response at } x=0) + \alpha(\text{potential increase at } x=0)$$
$$= \beta_1 - \beta_2 + \alpha\beta_2 = \beta_1 - \beta_2(1-\alpha)$$

At $x=2$,

$$(\text{response at } x=2) = (\text{response at } x=1) + \alpha(\text{potential increase at } x=1)$$
$$= \beta_1 - \beta_2(1-\alpha) + \alpha[\beta_2(1-\alpha)] = \beta_1 - \beta_2(1-\alpha)^2$$

Continuing this procedure would show that the response at any x is $\beta_1 - \beta_2(1-\alpha)^x$. If this entire process is repeated with smaller divisions of the x units and with α adjusted accordingly, the response at any x is $\beta_1 - \beta_2(1-\alpha/k)^{kx}$, where k is the number of intervals in an original unit of x. As these intervals become smaller and smaller (k becomes larger and larger), the response function becomes $\beta_1 - \beta_2 e^{-\alpha x}$. This approach represents a derivation of the Mitscherlich model by taking the limit of a discrete x model to a continuous x model. The interpretation of α is the same in either model but is easier to understand in terms of the discrete model.

The use of a nonlinear regression program for fitting the Mitscherlich model can be illustrated with the barley response data from Unit 21. Any computer program will require the analyst to write an expression for the model in terms of a common programming language such as FORTRAN, PL-1, or PASCAL. In addition, the initial estimates of the parameters must be provided. Some information about available computer programs is given in an appendix.

The analysis presented here uses the BMDP program PAR. PAR does not exploit linear parameters and stores the parameters in an array P. The correspondence is $P(1)=\beta_1$, $P(2)=\beta_2$, and $P(3)=\alpha_1$. The initial estimates provided are $P(1)=10.0$, $P(2)=10.0$, and $P(3)=0.1$. Nitrogen is designated by the variable X(1), and yield is designated as the response variable. The expected response is specified for the program by the FORTRAN statement

"F=P(1)−P(2)∗DEXP(−P(3)∗X(1))"

The pertinent output is as follows.

THE RESIDUAL SUM OF SQUARES (= 29.3444) WAS SMALLEST WITH THE FOLLOWING PARAMETER VALUES

```
   1 P(1)        2 P(2)        3 P(3)
  42.6213       21.3845       0.260075
```

ESTIMATE OF ASYMPTOTIC CORRELATION MATRIX

```
          P(1)        P(2)        P(3)
           1           2           3
P(1)  1   1.0000
P(2)  2   0.8927      1.0000
P(3)  3  -0.9242     -0.7536      1.0000
```

THE ESTIMATED MEAN SQUARE ERROR IS   2.257

ESTIMATES OF ASYMPTOTIC STANDARD DEVIATIONS OF PARAMETER ESTIMATES WITH   13 DEGREES OF FREEDOM ARE

```
   1 P(1)        2 P(2)        3 P(3)
  1.58854       1.63597      4.96425D-02
```

CASE NO. NAME	RESIDUAL	OBSERVED 2 YIELD	PREDICTED 2 YIELD	STD. DEV. PREDICTED	1 NITROGEN
1	1.763186	23.000000	21.236814	0.748420	0.0
2	0.763186	22.000000	21.236814	0.748420	0.0
3	-2.236814	19.000000	21.236814	0.748420	0.0
4	-0.236814	21.000000	21.236814	0.748420	0.0
5	-1.820731	31.000000	32.820731	0.667121	3.000000
6	2.179269	35.000000	32.820731	0.667121	3.000000
7	-1.820731	31.000000	32.820731	0.667121	3.000000
8	-1.179269	34.000000	32.820731	0.667121	3.000000
9	-1.129675	37.000000	38.129675	0.466751	6.000000
10	1.870325	40.000000	38.129675	0.466751	6.000000
11	-0.129675	38.000000	38.129675	0.466751	6.000000
12	-0.129675	38.000000	38.129675	0.466751	6.000000
13	-0.562780	40.000000	40.562780	0.685496	9.000000
14	1.437220	42.000000	40.562780	0.685496	9.000000
15	-0.562780	40.000000	40.562780	0.685496	9.000000
16	-0.562780	40.000000	40.562780	0.685496	9.000000

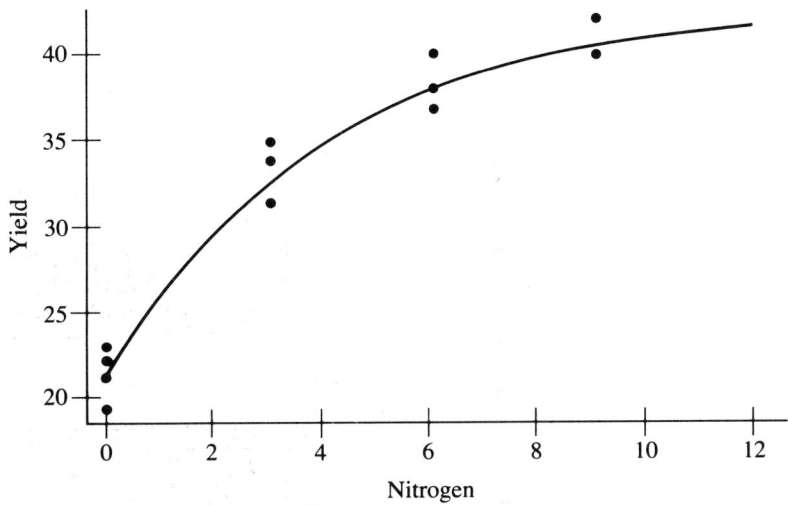

Figure 23.1 Data and Fitted Curve for Barley Yield

A graph of the data and fitted curve is shown in Figure 23.1. Comparing our nonlinear model fit with the quadratic polynomial model used in Unit 21, we note the following:

- The residual sum of squares of 29.34 is smaller than the 30.76 and 39.40 found with the nonaugmented and augmented fits of the quadratic polynomial.
- The curve does not turn down with increasing levels of nitrogen.
- For the predicted yield at nitrogen equal to 12, v_c (not available from the BMDP output) is 0.47, compared with 1.94 and 0.5 for the nonaugmented and augmented polynomial models. The value of v_c is computed from the approximating linear model. With linear models v_c depends only on **X** and **c**. Since the approximating model depends on the estimated parameter, v_c also depends on the estimated parameter. This dependence should not present any problems unless the sample size is small.

DRUG ASSIMILATION DATA

The kinetic diagram

$$\text{drug injection} \rightarrow \boxed{\text{muscle tissue}} \xrightarrow{KA} \boxed{\text{blood plasma}} \xrightarrow{K} \text{urine}$$

with KA and K representing first-order rate constants, gives rise to a model that is often appropriate for drug assimilation studies. A drug is injected into muscle tissue; it diffuses into the blood plasma and is eliminated after diffusing into the urine. The constants KA and K characterize the rates of diffusion, and inference on these two parameters is a likely objective of an experiment. This kinetic diagram is an oversimplification, but it often provides an excellent vehicle for analysis.

Deriving a model from a kinetic diagram can be approached through simple rationalization. Let $f(t)$ represent the amount of drug in muscle tissue at time t. First-order kinetics implies that [change in $f(t)$ per unit change in t]/$f(t)$ is proportional to $-KA$, which in turn implies $f(t)$ is of the form $c\exp(-KAt)$. Observations are made on the concentration of the drug in the plasma. Since there is diffusion into and out of the plasma, the response function for the plasma drug levels is the difference between two exponential terms. Specifically, the model for concentration of the drug in the plasma is

$$y = R\left(\frac{KA}{KA-K}\right)[\exp(-Kt) - \exp(-KAt)] + \varepsilon$$

The parameter R depends on the amount of drug administered and the volume of the system. While R is generally not interpretable, it is a necessary part of the model. This model is fitted with the data shown in Table 23.1, using the Statistical Analysis System (SAS). A feature of SAS is the use of meaningful names for parameters and variables in specifying the model. The input for the present model is

Table 23.1 Drug Assimilation Data

Time	Concentration
0.5	27.3
1.0	34.1
1.5	33.7
2.0	27.0
4.0	10.8
6.0	5.8
9.0	0.6

PARAMETERS: R = 50, 60, 70
  KA = 1, 1.5, 2
  K = .25, .5, .75;

MODEL CONC =
  R * (KA/(KA − K)) *
  (EXP(− K * TIME) − EXP(− KA * TIME))

NON-LINEAR LEAST SQUARES SUMMARY STATISTICS
DEPENDENT VARIABLE CONC

SOURCE	DF	SUM OF SQUARES	MEAN SQUARE
REGRESSION	3	3917.96617729	1305.98872576
RESIDUAL	4	5.46382271	1.36595568
UNCORRECTED TOTAL	7	3923.43000000	
(CORRECTED TOTAL)	6	1151.36000000	

PARAMETER	ESTIMATE	ASYMPTOTIC STD. ERROR	ASYMPTOTIC 95 % CONFIDENCE INTERVAL LOWER	UPPER
R	58.06707376	4.87537137	44.53105594	71.60309158
KA	1.58016985	0.22222430	0.96318462	2.19715508
K	0.49267592	0.05777470	0.33226981	0.65308202

ASYMPTOTIC CORRELATION MATRIX OF THE PARAMETERS

	R	KA	K
R	1.000000	-0.926007	0.953956
KA	-0.926007	1.000000	-0.865216
K	0.953956	-0.865216	1.000000

NOTE: ALL ASYMPTOTIC STATISTICS ARE APPROXIMATE. REFERENCE: RALSTON AND JENNRICH, TECHNOMETRICS, FEBRUARY 1978, P 7-14.

OBS	TIME	CONC	Y_HAT	RES
1	0.5	27.3	27.6621	-0.36205
2	1.0	34.1	34.1755	-0.07548
3	1.5	33.7	32.4103	1.28974
4	2.0	27.0	27.9190	-0.91895
5	4.0	10.8	11.6064	-0.80644
6	6.0	5.8	4.3830	1.41700
7	9.0	0.6	1.0011	-0.40112

The program will compute the residual sum of squares for all possible combinations of the initial estimates of the parameters to find the best starting set.

The pertinent output is shown on the previous page.

NONLINEAR LEAST SQUARES ALGORITHM

The principles of an algorithm for fitting nonlinear models are best illustrated graphically. To establish an analogy, let us consider a one-variable linear regression model with two observations, y_1 and y_2, and values of x equal to 4 and 5. In matrix notation

$$\mathbf{Y} = \begin{bmatrix} y_1 \\ y_2 \end{bmatrix} = \begin{bmatrix} 5 \\ 4 \end{bmatrix} \beta + \varepsilon$$

and the predicted values are of the form

$$\hat{\mathbf{Y}} = \begin{bmatrix} \hat{y}_1 \\ \hat{y}_2 \end{bmatrix} = \begin{bmatrix} 5 \\ 4 \end{bmatrix} b$$

Possible values of $\hat{\mathbf{Y}}$ include

$$\begin{bmatrix} 0 \\ 0 \end{bmatrix} \quad \begin{bmatrix} 0.5 \\ 0.4 \end{bmatrix} \quad \begin{bmatrix} 1.0 \\ 0.8 \end{bmatrix} \quad \begin{bmatrix} 2.5 \\ 2.0 \end{bmatrix} \quad \begin{bmatrix} 5. \\ 4. \end{bmatrix}$$

Note that all these values lie on a straight line of a plot with y_1 and y_2 as the axes. Indeed, if the line in Figure 23.2 was extended indefinitely in each direction, it would represent all possible value of $\hat{\mathbf{Y}}$. Since the $\mathbf{X}$ matrix determines the possible predicted values, it is reasonable to refer to $\mathbf{X}$ as the model. In the same sense the line in Figure 23.2 is referred to as the model.

Suppose $y_1 = 4$ and $y_2 = 1$. Then $\mathbf{Y}$ is one point in Figure 23.2 and is denoted as ⊙. We want to find b such that the residual sum of squares, $(4-5b)^2 + (1+4b)^2$, is as small as possible. Graphically, the residual sum of squares for a given b value is the square of the distance between $\mathbf{Y}$ and $\hat{\mathbf{Y}}$. For an arbitrarily selected value of b, for example, $b = 0.4$, then $\hat{y}_1 = 2.0$ and $\hat{y}_2 = 1.6$, and $\hat{\mathbf{Y}}$ is denoted by △ in Figure 23.2. The residual sum of squares is $(4.0-2.0)^2 + (1.0-1.6)^2 = 4.36$, and the distance between ⊙ and △ is SQRT(4.36) = 2.09. The value of b that gives the minimum residual sum of squares is easily found from ABDO calculations. The results are

$$b = 0.5854 \quad \text{residual sum of squares} = 2.9512 \quad \hat{\mathbf{Y}} = \begin{bmatrix} 2.93 \\ 2.34 \end{bmatrix}$$

This $\hat{\mathbf{Y}}$ is denoted by + in Figure 23.2. It is the value of $\hat{\mathbf{Y}}$ that is closest to $\mathbf{Y}$.

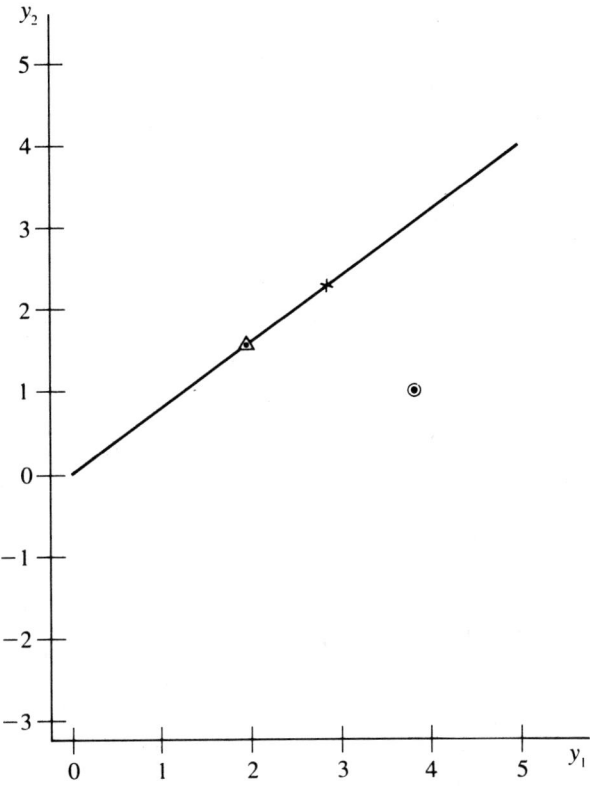

Figure 23.2 One-Variable Linear Regression Model

An example of a nonlinear model is

$$y = \alpha^x/(4x-3) + \varepsilon$$

where $x = 1$ and 2. There are no linear parameters in this model. In matrix notation the model is

$$\mathbf{Y} = \begin{bmatrix} y_1 \\ y_2 \end{bmatrix} = \begin{bmatrix} \alpha \\ \alpha^2/5 \end{bmatrix} + \varepsilon$$

The parameter vector on the right-hand side contains the mean values of the elements of **Y**. This corresponds to $\mathbf{X}\boldsymbol{\beta}$ in linear models, but nonlinear models cannot be written in terms of a matrix product. The predicted values are of the form

$$\hat{\mathbf{Y}} = \begin{bmatrix} a \\ a^2/5 \end{bmatrix}$$

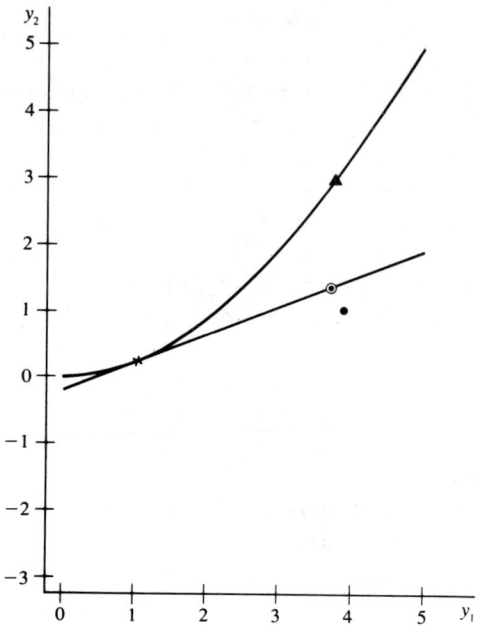

Figure 23.3 First Approximating Linear Model

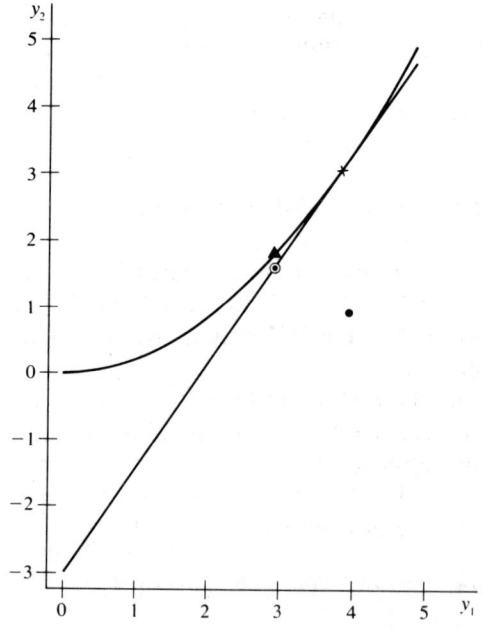

Figure 23.4 Second Approximating Model

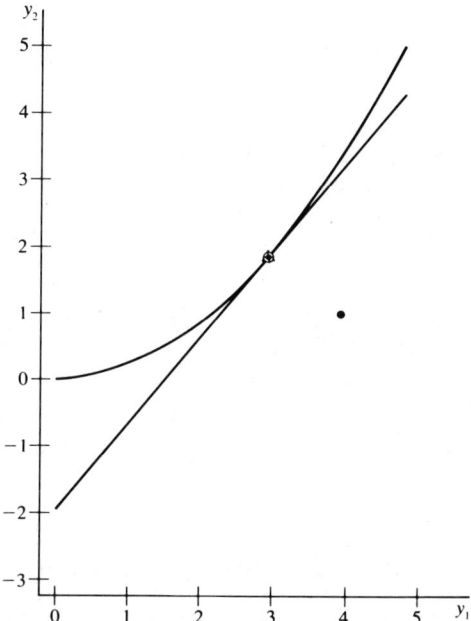

Figure 23.5 Last Approximating Model

and possible values of $\hat{\mathbf{Y}}$ include

$$\begin{bmatrix} 0 \\ 0 \end{bmatrix} \quad \begin{bmatrix} 1.0 \\ 0.2 \end{bmatrix} \quad \begin{bmatrix} 2.0 \\ 0.8 \end{bmatrix} \quad \begin{bmatrix} 4.0 \\ 3.1 \end{bmatrix} \quad \begin{bmatrix} 5.0 \\ 5.0 \end{bmatrix}$$

In Figure 23.3 all these values lie on the smooth curve called the nonlinear model.

Again we will suppose $y_1 = 4$ and $y_2 = 1$. Analogous to what we did with the linear model, we want to find the point on the curve that is closest to **Y**. Unfortunately, a one-step solution to this problem is not available; it is necessary to successively approximate the nonlinear model with linear models. To do this, we must have a provisional estimate of α, and for this example $a = 1.0$ is arbitrarily selected. The $\hat{\mathbf{Y}}$ for $a = 1.0$ is denoted by X on Figure 23.3. The approximating linear model is the line drawn through the X and the tangent to the curve.

Note on Figure 23.3 that the approximating linear model is good near zero but not very good near **Y** (denoted by ●). However, if this approximating linear model is analyzed in the standard way, we obtain $a = 3.86$. (The details of obtaining this value are given in the next section.) The value of a was chosen so that $\hat{\mathbf{Y}}$ from the approximating linear model (denoted by ⊙) is as close as possible to **Y**. However, the $\hat{\mathbf{Y}}$ of the nonlinear model (denoted by ▲) is obtained by substituting $a=3.86$ into the $\hat{Y}$ formula. Clearly, ▲ is not the point on the curve that is closest to **Y**. However, ▲ is closer to **Y** than **X**.

Now we repeat the whole process again, except that the approximating linear model is a line through ▲ and the tangent to the curve. The results are shown in Figure 23.4. Figures 23.3 through 23.5 display the results of successively approximating the nonlinear model with linear models in the manner just discussed. Note on Figure 23.5 that the final approximating linear model is a good approximation to the nonlinear model in the region nearest **Y**. Approximate standard errors for estimated parameters are obtained by applying standard formulas for linear models to final approximation. The residual sums of squares for the nonlinear model and for the approximating linear model are equal. Some programs use the ratio of these two sums of squares as a criterion for stopping the successive approximations.

CONSTRUCTING THE APPROXIMATING LINEAR MODEL

The preceding section conveyed the ideas behind fitting nonlinear models. Now the specific details needed to construct the approximating linear model are presented and demonstrated in terms of the same two-observation example. Graphically, an approximating linear model is a straight line that touches but does not pass through the nonlinear model. To construct such a model, we begin by finding the equation of the line that passes through two points of the nonlinear model and then we move the two points together. To illustrate, the approximating linear model for $a=1.0$ is constructed. Values of $\hat{y}_1$ and $\hat{y}_2$ when $a=1.0$ and when $a=1+\delta$, where δ is an unspecified number, are

$$\begin{bmatrix} 1.0 \\ 0.2 \end{bmatrix} \quad \text{and} \quad \begin{bmatrix} 1+\delta \\ (1+\delta)^2/5 \end{bmatrix}$$

If the value of d in the expression

$$[1-(d/\delta)]\begin{bmatrix} 1.0 \\ 0.2 \end{bmatrix} + (d/\delta)\begin{bmatrix} 1+\delta \\ (1+\delta)^2/5 \end{bmatrix}$$

is changed, all the points on the line through these two values of $\hat{Y}$ are generated. This line is shown in Figure 23.6 for $\delta=1.0$.

The variable d is divided by δ to give it the proper scale. This is manifested by the fact that $d=\delta$ gives the same point as does evaluating $\hat{Y}$ with the value of a increased by δ. Thus it is reasonable to use d for interpolation. Algebraic simplification of the expression for the line results in

Constructing the Approximating Linear Model

$$\begin{bmatrix} 1.0 \\ 0.2 \end{bmatrix} + \begin{bmatrix} 1.0 \\ 0.4+0.2\delta \end{bmatrix} d$$

and is shown in Figure 23.7 for several different values of δ.

This line is the approximating linear model when $\delta=0$. Explicitly, the approximating linear model is

$$\begin{bmatrix} 4.0 \\ 1.0 \end{bmatrix} = \begin{bmatrix} 1.0 \\ 0.2 \end{bmatrix} + \begin{bmatrix} 1.0 \\ 0.4 \end{bmatrix} d + \varepsilon$$

The first vector to the right of the equal sign is $\hat{Y}$ for $a=1.0$. Since it does not have an unknown multiplier, it is transposed to the left, and

$$\begin{bmatrix} 3.0 \\ 0.8 \end{bmatrix} = \begin{bmatrix} 4.0 \\ 1.0 \end{bmatrix} - \begin{bmatrix} 1.0 \\ 0.2 \end{bmatrix} = \begin{bmatrix} 1.0 \\ 0.4 \end{bmatrix} d + \varepsilon$$

The Y vector in this model is actually the residual. The X matrix is derived from the vector of mean values. (In mathematics this matrix would actually be called the derivative of the mean value vector evaluated at one.) The quantity d takes the role of **b**. Because d is set up to have a scaling comparable to δ and because the second point on the curve is $\hat{Y}$ evaluated at $a+\delta$, the new provisional estimate of a is $a+d$, where d is found by applying standard methods to the approximately linear model. Specifically,

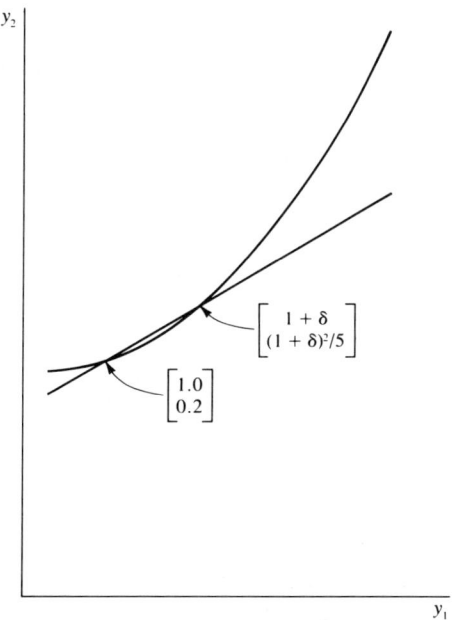

Figure 23.6 Approximating Line for $\delta=1.0$

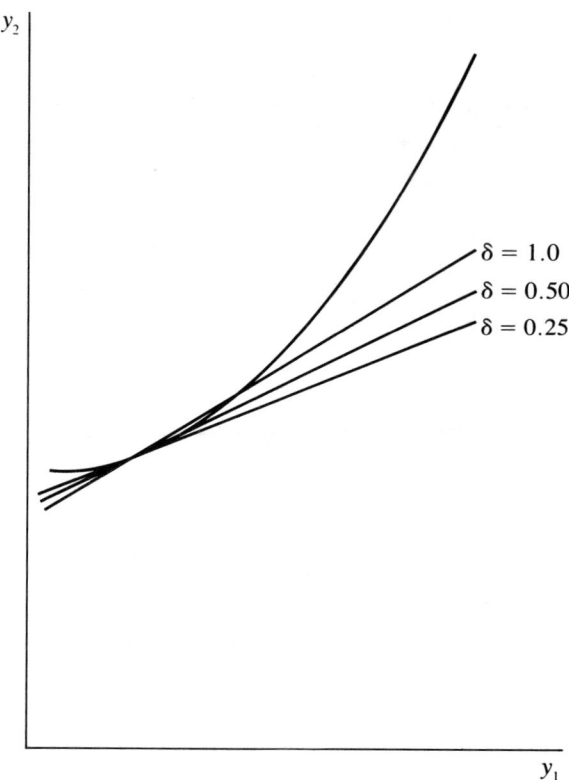

Figure 23.7 Approximating Lines for Various Values of δ

the ABDO computations are as follows:

X'X|X'Y:	1.16	3.32
		9.64
Step 1:	1.16	3.32
Step 2:	1.00	2.86
Step 3:		0.14
Step 4:		1.0

Hence $d=2.86$ and the new a is $1.0+2.86=3.86$, as given in the preceding section. The residual sum of squares from the approximating linear model is 0.14, which is much smaller than the true residual sum of squares of $3^2+0.8^2=9.64$. This result indicates that we have not obtained the solution and the process must be repeated.

SUMMARY

In this unit nonlinear models were shown to arise in two situations where the underlying mechanics are considered. The analysis of data from nonlinear models is accomplished by fitting a sequence of approximating linear models. The use of two widely distributed computer programs to analyze data from nonlinear models was also demonstrated.

Unit 24

A Regression Program for Microcomputers

STAN is an interactive statistical analysis program for microcomputers and is exceedingly easy to use. Prompts are given at every stage to show possible options. Variables are referenced by mnemonic names given by the user and not by their positions in the data set or a model. STAN tries to recover user input errors. Should a variable be specified that is not in the data set, then the names of the legal variables are displayed so that a substitute may be entered. STAN checks for illegal characters in numerical input and requests reentry when not of proper form.

STAN is modular, with related procedures grouped together to make an efficient system. The first version has five modules, described below.

1. *Data management*: Data sets are created, corrected, and updated by using the system editor. STAN has procedures for summarizing, transforming, sorting, and neatly printing selected rows and columns of the data set.

2. *Linear models*: Standard least squares or Bayesian techniques may be used. Qualitative variables are automatically expanded into indicator variables. Analysis of variance, regression coefficients, linear combinations of parameters, and calculation of predicted values, residuals, and related variances are some of the procedures available. Models may be fit to selected portions of the data. The algorithm for fitting models is among the most accurate available.

3. *Nonlinear models*: Models may have different functional forms for different portions of the data. Models with restrictions may be fit, and implicit functions of the parameters may be estimated.

4. *Graphics*: Graphic procedures are very flexible. Different patterns of dotted lines and different symbols are selected by the user. Graphs may use the entire data set or selected portions of the data. Multiple variables may be placed on the same graph.

5. *Text formatting*: Since the result of a statistical analysis is usually presented in a report or paper, STAN has a text-formatting procedure to merge discussion and selected tables and graphs. A paper or report is created by using the system editor. Simple commands control margins, indentations, centering, underlining, spacing, top and bottom running titles, page numbering, justification, and several other features. A simple code indicates where the tables and figures produced by the other modules are to be inserted.

A STAN user needs no knowledge of any programming language (except for nonlinear models). However, when selecting a statistical analysis system, the user should consider the following point. STAN is written in PASCAL, which is compiled into an object code. This object code is executed much more rapidly than a similar program would be if written in an interpreted language such as most versions of BASIC.

The graphic portion of STAN uses the features of modern printers and terminals, such as the NEC Spinwriter or the Anderson Jacobson 832, which allow a much higher-quality figure than is possible with traditional printer plots. The sample runs at the end of this unit show the resolution of these devices.

STAN is available for Apple and North Star computers, using their respective versions of the University of California, San Diego (UCSD), PASCAL system. The development of STAN is a continuing process; more modules will be added and present modules will be expanded. The direction of development will be influenced by the needs and suggestions of users. Information on obtaining STAN may be obtained from the publisher of this textbook or from the authors.

The remainder of this unit consists of sample runs of STAN. The distinction between prompts and user responses should be clear from the context. Notes in braces { } are annotations inserted after the run. This unit is not intended to be a user's manual for STAN, since some of the prompts and responses might change slightly by the publication date. This unit merely demonstrates some of the capabilities of STAN and its ease of use, and provides another look at two of the data sets introduced in earlier units.

Analysis of the Arsenic Data Using STAN

{This section is a demonstration of STAN using the Arsenic Data from Unit 5. Examination of a graph of concentration with miles gives a hint that the relationship is curved. Here the logarithmn of the concentration is used as the response variable. It is interesting to compare the results of this analysis to that done in Units 4, 5, and 6.}

{Before execution of the MODELS branch of STAN a data set is named ARSENIC containing variables MILES and LNCONC is created on disk. The following results from executing the MODELS branch.}

TYPE OF FIT: L(east squares, B(ayes? L

{The Bayes option would be used to apply the methods of Unit 22.}

Enter name of data set to be used.
ARSENIC
What name shall be given to the implicit column of 1's?
CONSTANT

{To save time and space the data set does not need to have a column of 1's. STAN creates the column when it is needed but asks the analyst to name it.}

Does every observation receive a weight of one? Y

{A "N" response would give the analyst the opportunity to specify a variable that contains weights.}

Enter names of expanatory variables in the model.
Follow list by ";".
CONSTANT MILES;
Enter name of response variable.
LNCONC
Use A(ll or P(art of the observations? A
Model is fit.
Error degrees of freedom: 8.
Standard deviation: 0.3170

{At this point STAN has computed a matrix equivalent to the ABDO matrix and is ready for the analyst to select the appropriate subsequent computations.}

MODEL:A(NOV,B(eta,C(onsole,L(C,P(rint S,R(esidual Q(uit, ? A
Name file where table is to be stored:
ANOV

ANALYSIS OF VARIANCE

SOURCE	SUM OF SQUARES	D.F.	MEAN SQUARE	P-VALUE
CONSTANT	0.7457	1	0.7457	0.0261
MILES	4.2083	1	4.2083	0.0002
RESIDUAL	0.8041	8	0.1005	

Save? Y

{This table is now on a disk file and can be electronically merged into a report without retyping.}

MODEL:A(NOV,B(eta,C(onsole,L(C,P(rint S,R(esidual Q(uit, ? B
Name file where table is to be stored:
BETA

REGRESSION COEFFICIENTS

VARIABLE	ESTIMATE	Vc	STANDARD ERROR	P-VALUE
CONSTANT	1.2569	0.3300	0.1821	0.0001
MILES	-0.0611	0.0009	0.0094	0.0002

Save? Y
MODEL:A(NOV,B(eta,C(onsole,L(C,P(rint S,R(esidual Q(uit, ? L
Name file where table is to be stored:
LC
Enter title.
MILES=5
       CONSTANT: 1
         MILES: 5
        LNCONC: 0.0
Another linear combination? Y
Enter title.
MILES=15
       CONSTANT: 1
         MILES: 15
        LNCONC: 0.0
Another linear combination? Y
Enter title.
MILES=25
       CONSTANT: 1
         MILES: 25
        LNCONC: 0.0
Another linear combination? Y
Enter title.
MILES=35
       CONSTANT: 1
         MILES: 35
        LNCONC: 0.0
Another linear combination? N

LINEAR COMBINATIONS

TITLE	ESTIMATE	Vc	STANDARD ERROR	P-VALUE
MILES=5	0.9514	0.2093	0.1451	0.0002
MILES=15	0.3403	0.1011	0.1008	0.0097
MILES=25	-0.2708	0.1703	0.1308	0.0722
MILES=35	-0.8819	0.4170	0.2047	0.0026

```
Save? Y
MODEL:A(NOV,B(eta,C(onsole,L(C,P(rint S,R(esidual Q(uit, ? R
A new file consisting of ARSENIC with residuals,
predicted values, and Vc added will be created.
Name this file.
ARSRES
Name the predicted values.
PREDICT
Name the residuals.
RESIDUAL
Name the Vc.
VC
ARSRES has been created.
MODEL:A(NOV,B(eta,C(onsole,L(C,P(rint S,R(esidual Q(uit, ? Q
```

{We now leave the MODEL branch and go to the GRAPHICS branch in order to graph the data and the predicted values just computed.}

```
GRAPHIC: G(raph, H(istogram, Q(uit, ? G
Line density sampler? N
```

{A "Y" response would show the available patterns of broken lines.}

```
Name file where plot is to be stored:
ARSPLOT
Enter name of data set to be used.
ARSRES          {Created by MODEL branch}
What name shall be given to the implicit column of 1's?
Z         {Name is immaterial since it will not be used}
Enter height of graph.
3
Enter width of graph.
5
Enter title.
Logarithm of concentration and estimated values
Enter name of variable on horizontal axis.
MILES
Enter names of variables on the vertical axis.
Follow list by ";".
LNCONC PREDICT;
For the variable LNCONC is graph
over the E(ntire data set or S(egmented? E
Plot LNCONC with L(ine, S(ymbol, or sK(ip? S
```

```
Enter symbol. o
Make legend entry? N
```

{STAN will label lines or symbols using variable names, but in this case they are not needed.}

```
For the variable PREDICT is graph
over the E(ntire data set or S(egmented? E
Plot PREDICT with L(ine, S(ymbol, or sK(ip? L
Enter line density.
Ø           {Solid line}
Make legend entry? N
Horizonal axis? Y
Vertical axis? Y
Send plot to R(emout: or to C(onsole:? C
```

{Console: and Remout: are the UCSD Pascal names for the primary and secondary input-output ports, respectively. Plot output must go to the port to which the printer is connected.}

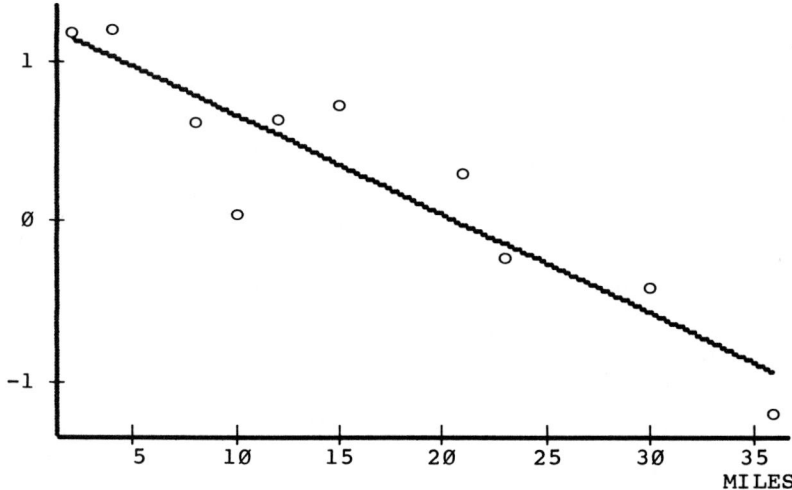

LNCONC   Logarithm of concentration and estimated values

```
Graph takes 22 lines.
Save? Y
GRAPHIC: G(raph, H(istogram, Q(uit, ? Q
```

{This completes the production of tables and figures. The next section shows what a report processed by STAN's text formatting branch and using these tables and figures might look like.}

The Relationship Between the Arsenic Determinations on Hair Samples and the Distance from a Power Plant Burning Coal

This is an alternate analysis of the Arsenic Data given in Unit 5. In Unit 8, ln(y) is suggested as a transformation to straighten a decreasing curved relationship. The ln(y) transformation is also appropriate when a fixed change in the explanatory variable causes a fixed percentage change rather than a fixed absolute change in the response variable. For these reasons, the use of the logarithmn of concentration as the response variable seems worthy of examination. Since this report contains computer generated tables and figures it is convenient to use the computer labels throughout. The variable labels are:

    LNCONC = Natural logarithmn of the community average concentration of arsenic (parts per million) found in the hair of ten year old boys.

    MILES = Distance in miles from the plant to community.

LNCONC is assumed to be a straight line function of MILES.

To decide if the transformation is helpful a graph of the data and the fitted equation is compared with Figure 6.1, and an analysis of variance is compared to Table 6.1.

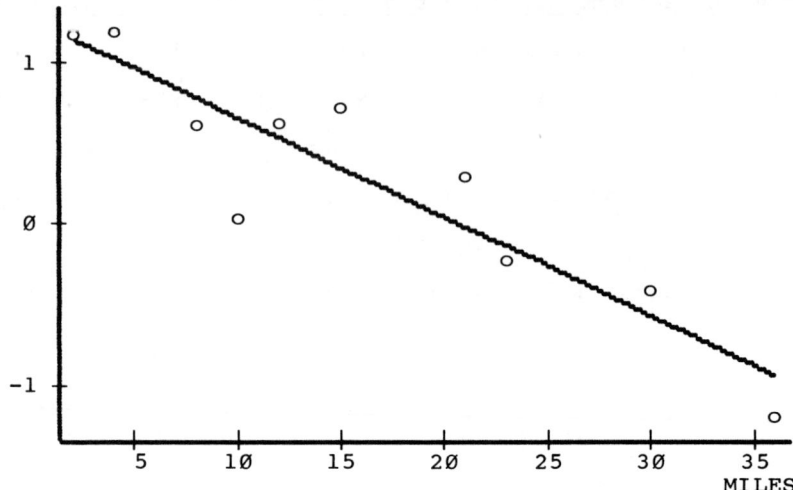

LNCONC Logarithm of concentration and estimated values

This seems a better fit than Figure 6.1. Figure 6.1 shows that predicted values for distances over 36 miles would be negative but it is impossible for a concentration to be negative. The present model can never predict negative values when transformed back into the original units.

ANALYSIS OF VARIANCE

SOURCE	SUM OF SQUARES	D.F.	MEAN SQUARE	P-VALUE
CONSTANT	0.7457	1	0.7457	0.0261
MILES	4.2083	1	4.2083	0.0002
RESIDUAL	0.8041	8	0.1005	

The F statistic for MILES is 41.87. The analogous statistic computed from Table 6.1 is 23.72. The graph and F statistic both suggest that the logarithmn of concentration is more nearly linearly related to distance than is concentration.

The specific details of fitted model are now presented.

REGRESSION COEFFICIENTS

VARIABLE	ESTIMATE	Vc	STANDARD ERROR	P-VALUE
CONSTANT	1.2569	0.3300	0.1821	0.0001
MILES	-0.0611	0.0009	0.0094	0.0002

The prediction equation is

$$LNCONC = 1.2568 - 0.0611\ MILES.$$

Predictions for selected distances are given in:

LINEAR COMBINATIONS

TITLE	ESTIMATE	Vc	STANDARD ERROR	P-VALUE
MILES=5	0.9514	0.2093	0.1451	0.0002
MILES=15	0.3403	0.1011	0.1008	0.0097
MILES=25	-0.2708	0.1703	0.1308	0.0722
MILES=35	-0.8819	0.4170	0.2047	0.0026

Confidence limits are obtained by finding confidence limits for the transformed values and then back transforming the limits. For example, a 95% confidence interval for the predicted value when MILES=25 is

$$(-0.2708-(2.31)(0.1308), -0.2708+(2.31)(0.1308)) =$$

$$(-0.5729, 0.0313)$$

where 2.31 is the value from the t-distribution with eight degrees of freedom associated with 0.95 central area. Taking the exponential function of these limits give a confidence interval in the original units of (0.5639, 1.0318).

Analysis of Electricity Load Data Using STAN

{This section is a demonstration of STAN using the Electricity Load Data from Unit 13. Four features of STAN not demonstrated with the Arsenic Data are shown here: the structure of data set, qualitative variables, segmenting data, and adjusted treatment means.}

{The data set named "KU" prepared for STAN is

```
DAY TEMP ELOAD;
WEEK 69 1061  WEEK 71 1066  WEEK 71 1081  WEEK 72 1045
WEEK 72 1105  WEEK 73 1062  WEEK 73 1068  WEEK 75 1073
WEEK 75 1130  WEEK 75 1111  WEEK 75 1063  WEEK 77 1156
WEEK 77 1136  WEEK 77 1121  WEEK 77 1174  WEEK 78 1078
WEEK 78 1126  WEEK 79 1179  WEEK 79 1155  WEEK 81 1213
WEEK 85 1273  SAT 68 880    SAT 74 893    SAT 74 961
SAT 82 1034   SAT 83 1092   SUN 74 866    SUN 77 889
SUN 77 825    SUN 78 818    SUN 82 899
```

First, the variable names are listed followed by a semicolon. These are followed by the data values separated by one or more blanks. Typically, one observation of each variable is put on each line. However, if the number of variables is small, then multiple observations may be put on the same line as shown here. If the number of observations is large, then several lines may be used for one observation. A non-numeric variable is automatically assumed to be qualitative. A numeric variable may be designated as qualitative. A qualitative variable entered into a model is automatically expanded into a set of indicator variables.}

{We now enter the MODEL branch.}

```
TYPE OF FIT: L(east squares, B(ayes? L
Enter name of data set to be used.
KU
What name shall be given to the implicit column of 1's?
CONSTANT
Does every observation receive a weight of one? Y
Enter names of expanatory variables in the model.
Follow list by ";".
CONSTANT DAY TEMP;
Enter name of response variable.
ELOAD
Use A(ll or P(art of the observations? A
Model is fit.
Error degrees of freedom: 27.
Standard deviation:     32.9510
```

{Compare the following analysis of variance with that of Unit 13.}

```
MODEL:A(NOV,B(eta,C(onsole,L(C,P(rint S,R(esidual Q(uit, ? A
```

Name file where table is to be stored:
ANOV

ANALYSIS OF VARIANCE

SOURCE	SUM OF SQUARES	D.F.	MEAN SQUARE	P-VALUE
CONSTANT	3.43520E7	1	3.43520E7	0.0000
DAY	308672.	2	154336.	0.0000
TEMP	77555.1	1	77555.1	0.0000
RESIDUAL	29315.8	27	1085.77	

Save? Y

{Adjusted treatment (group) means are discussed in Unit 19. Adjusted group means are predicted values for the respective groups with the covariable equal to its average value. It is immaterial if the model contains a column of 1's or not. However, if the column of 1's is present, the first stage of the ABDO matrix contains the means of all the variables. An "M" entered into a linear combination will cause STAN to fetch the mean for that variable. This is a convenience for obtaining adjusted group means because the mean of the covariable need not be known in advance.}

MODEL:A(NOV,B(eta,C(onsole,L(C,P(rint S,R(esidual Q(uit, ? L
Name file where table is to be stored:
LC
Enter title.
WEEKDAYS
```
        CONSTANT: 1
        DAY.WEEK: 1
        DAY.SAT:  0.0
        DAY.SUN:  0.0
           TEMP:  M =       76.0645
          ELOAD:  0.0
```
Another linear combination? Y
Enter title.
SATURDAYS
```
        CONSTANT: 1
        DAY.WEEK: 0.0
        DAY.SAT:  1
        DAY.SUN:  0.0
           TEMP:  M =       76.0645
          ELOAD:  0.0
```
Another linear combination? Y
Enter title.
SUNDAYS
```
        CONSTANT: 1
        DAY.WEEK: 0.0
        DAY.SAT:  0.0
        DAY.SUN:  1
           TEMP:  M =       76.0645
          ELOAD:  0.0
```
Another linear combination? N

{We have produced the following table of adjusted group means.}

LINEAR COMBINATIONS

TITLE	ESTIMATE	Vc	STANDARD ERROR	P-VALUE
WEEKDAYS	1122.98	0.0480	7.2155	0.0000
SATURDAYS	970.272	0.2000	14.7376	0.0000
SUNDAYS	839.814	0.2049	14.9172	0.0000

Save? Y

{The predicted values are computed for subsequent graphic display.}

```
MODEL:A(NOV,B(eta,C(onsole,L(C,P(rint S,R(esidual Q(uit, ? R
A new file consisting of KU with residuals,
predicted values, and Vc added will be created.
Name this file.
KURES
Name the predicted values.
PREDICT
Name the residuals.
RESIDUAL
Name the Vc.
VC
KURES has been created.
MODEL:A(NOV,B(eta,C(onsole,L(C,P(rint S,R(esidual Q(uit, ? Q
```

{We now leave the MODEL branch and go to the GRAPHIC branch.}

```
GRAPHIC: G(raph, H(istogram, Q(uit, ? G
Line density sampler? N
Name file where plot is to be stored:
KUPLOT
Enter name of data set to be used.
KURES
What name shall be given to the implicit column of 1's?
Z
Enter height of graph.
3
Enter width of graph.
5
Enter title.
Electricity Load and Temperature
Enter name of variable on horizontal axis.
TEMP
Enter names of variables on the vertical axis.
Follow list by ";".
ELOAD PREDICT;
For the variable ELOAD is graph
over the E(ntire data set or S(egmented? S
```

{The data is segmented so that each group can have its own symbol and line.}

```
Segmented by values of the variable:
DAY
Enter name of legend variable:
DAY
Plot ELOAD for DAY = WEEK with L(ine, S(ymbol, or sK(ip? S
Enter symbol. .
Make legend entry? Y
Plot ELOAD for DAY = SAT with L(ine, S(ymbol, or sK(ip? S
Enter symbol. o
Make legend entry? y
Plot ELOAD for DAY = SUN with L(ine, S(ymbol, or sK(ip? s
Enter symbol. x
Make legend entry? y
For the variable PREDICT is graph
over the E(ntire data set or S(egmented? s
Segmented by values of the variable:
DAY
Enter name of legend variable:
DAY
Plot PREDICT for DAY = WEEK with L(ine, S(ymbol, or sK(ip? L
Enter line density.
Ø
Make legend entry? N
Plot PREDICT for DAY = SAT with L(ine, S(ymbol, or sK(ip? L
Enter line density.
Ø
Make legend entry? N
Plot PREDICT for DAY = SUN with L(ine, S(ymbol, or sK(ip? L
Enter line density.
Ø
Make legend entry? N
Horizonal axis? Y
Vertical axis? Y
Send plot to R(emout: or to C(onsole:? C
```

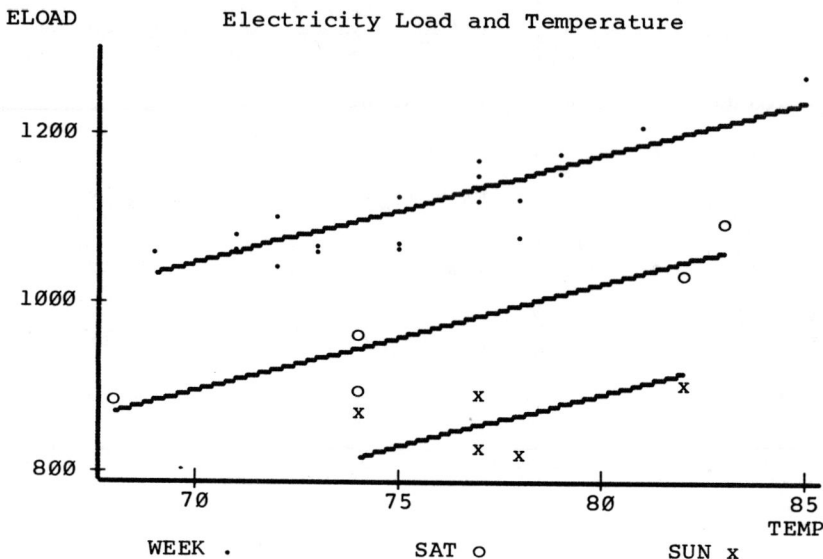

```
Graph takes 23 lines.
Save? Y
GRAPHIC: G(raph, H(istogram, Q(uit, ? Q
```

Appendix A

Exercises

Unit 2

1. Write a brief research proposal in your field. The following items may be used as a guide:
   - What is the general research objective?
   - How will data be collected to help meet this objective? Is this an experiment or a survey?
   - What variables will be recorded? Which of the variables are quantitative and which are qualitative? Which of the variables are explanatory or predictor and which are response variables?
   - Restate the research objective; be very specific and relate it to the variables recorded.
2. Review papers in which data are analyzed; determine which variables are explanatory variables and which are response variables.

Unit 3

1. Refer to your brief research proposal prepared for Exercise 1 of Unit 2. Give models that would be appropriate if various research hypotheses were true.
2. One way to transform a model from observational equation format into matrix notation is given by the following steps:
   - Write down in a column the observations on the response variable. This gives **y**.
   - Write down the parameters in a row at the top. This gives β'. At this point the display might look like this.

	β_0	β_1
y_1		
y_2		
y_3		
y_4		
y_5		
y_6		
y_7		

- In each row write down the multiplier for each parameter below that parameter. This gives **X**.

For the model $y = \beta_0 + \beta_1 x + \varepsilon$ and the (x, y) pairs $(1, 4.9)$, $(2, 4.7)$, $(2, 3.8)$, $(3, 4.2)$, $(3, 4.1)$, and $(4, 3.9)$, write the model in matrix notation.

3. The elements of $\boldsymbol{\beta}$ need not always be denoted by subscripted β's. For example, the greek letter μ is a traditional symbol used to represent a mean. Thus if we had a sample of size four from each of four populations, then we might write the model as $y_{jl} = \mu_j + \varepsilon_{jl}, j = 1, 2, 3; l = 1, 2, 3, 4$. Suppose that the observed values of the response variable are as follows:

Population 1:	18.13	19.14	17.37	18.54
Population 2:	13.45	18.36	14.29	17.84
Population 3:	12.21	13.63	14.36	12.98

Put the model in matrix notation.

4. For the matrices

$$\mathbf{A} = \begin{bmatrix} 4 & 2 & 1 \\ -1 & 2 & 8 \end{bmatrix} \quad \mathbf{B} = \begin{bmatrix} 3 & 2 & 1 \\ 6 & 8 & 2 \end{bmatrix} \quad \mathbf{C} = \begin{bmatrix} 3 & 2 \\ 1 & 4 \end{bmatrix}$$

find each of the following (if it is defined): $\mathbf{A} + \mathbf{B}$; $\mathbf{A} + \mathbf{C}$; $\mathbf{AC}$; $\mathbf{CA}$.

5. Suppose that

$$\mathbf{Y} = \begin{bmatrix} 1.10 \\ 1.01 \\ 1.90 \\ 1.00 \\ 1.99 \\ 1.42 \end{bmatrix} \quad \text{and} \quad \mathbf{X} = \begin{bmatrix} 1 & 26 & 63 \\ 1 & 29 & 57 \\ 1 & 54 & 26 \\ 1 & 2 & 84 \\ 1 & 3 & 26 \\ 1 & 19 & 61 \end{bmatrix}$$

Calculate $\mathbf{X'X}$ and $\mathbf{X'Y}$ and solve the equations $\mathbf{X'Xb} = \mathbf{X'Y}$ for **b**.

6. Verify that

$$(a - b)^2 / 2 = a^2 + b^2 - (a + b)^2 / 2$$

Pretend you are a computer that carries four significant digits (that is, after every operation you replace those digits four places beyond the first nonzero digit with zeros). For $a = 10, 100, 1000$, and $10,000$ and $b = a + 5$, evaluate both sides of the above expression. Moral: Algebraic identities are not equal when evaluated by using finite precision arithmetic.

Unit 4

1. Suppose that the observed values of a random sample are 12.2, 8.1, 16.0, 8.6, 5.8, and 8.2. Fit the Mean Model to these data and plot the residuals with the order in which the data are listed.

2. Verify that

$$\Sigma(y_i - \mu)^2 = \Sigma(y_i - \bar{y})^2 + n(\bar{y} - \mu)^2.$$

The left-hand side is a simple measure of how far the observations are from μ. If we were to choose an estimator of μ to minimize this measure, then we see from the right side that the estimator would be $\bar{y}$ and the minimum value would be $\Sigma(y_i - \bar{y})^2$.

Unit 5

1. The accompanying table provides data on an explanatory variable x and a response variable y.

x	y
13.2	9.0
14.6	9.5
9.8	7.7
30.4	14.2
16.9	10.2
14.5	9.5
19.7	10.7
12.3	8.1
14.4	9.6
13.6	9.1
14.8	9.9
18.0	10.6

   a. Fit the Mean and Slope model.
   b. Fit the Intercept and Slope Model.
   c. Assess the similarities and differences between the two models.

Unit 6

1. Reproduce the analysis of variance in Table 6.2 by using ABDO.

2. Ten observations of three explanatory variables x_1, x_2, x_3 and a response variable y are given in the accompanying table. Append an x_0 column and form the sum of squares and cross products matrix.

x_1	x_2	x_3	y
5	9	7	42.6
1	3	1	16.0
8	0	7	8.2
6	6	4	55.4
10	4	8	97.1
4	9	6	50.1
2	2	3	49.7
1	9	2	32.1
4	10	7	40.3
8	5	9	55.3

  a. Find $\hat{y}(0)$.
  b. Find $\hat{y}(0,1)$ and $\hat{y}(0,1) - \hat{y}(0)$.
  c. Find $\hat{y}(0,1,2)$ and $\hat{y}(0,1,2) - \hat{y}(0,1)$.
  d. Find $\hat{y}(0,1,2,3)$ and $\hat{y}(0,1,2,3) - \hat{y}(0,1,2)$. (Hint: All these can be computed from the ABDO output using all the variables.)

3. Refer to Exercise 2. Verify that the sums of squares of the differences between successive $\hat{y}$'s are the same as those calculated from the ABDO output.

Unit 7

1. The accompanying table gives a set of data and its sum of squares and cross product matrix. Display the analysis of variance table for the data.

	x_0	x_1	x_2	x_3	y
	1	4	9	37	184.48
	1	5	10	43	211.22
	1	5	11	43	215.57
	1	5	11	43	215.00
	1	3	8	30	152.47
	1	4	9	35	177.22
	1	3	7	27	133.86
	1	4	8	34	167.92
	1	3	8	31	153.18
	1	4	9	37	182.18
	1	2	4	17	85.52
	1	4	9	37	185.80
	1	4	9	37	186.73
	1	4	8	33	165.61
	1	5	11	42	212.77
x_0	15	59	131	526	2,629.55
x_1		243	536	2,153	10,762.2
x_2			1,189	4,766	23,829.4
x_3				19,136	95,652.0
y					478,175.

2. In the accompanying table the values of y are averages of plot yields at a particular location. The number of plots at a location is denoted by n. The variables x_1 and x_2 are environmental variables that are constant for a given location.

n	x_0	x_1	x_2	y
3	1	0	1	3.3
6	1	3	8	39.3
9	1	3	9	43.1
3	1	1	2	11.4
3	1	4	11	52.5
6	1	4	12	56.3
12	1	2	8	36.8
3	1	3	12	54.8
6	1	5	13	62.9
6	1	2	7	33.1
3	1	6	17	80.9
3	1	4	11	52.7
9	1	6	14	68.9
12	1	5	15	71.2

a. Find the weighted sum of squares and cross products matrix.
b. Display the analysis of variance table. How would this table differ from the analysis of variance table computed by using all the individual plot yields?

Unit 8

1. Draw a graph of the following relationships:

$$y = 2^{2x-1} \qquad y = 3x^2 \qquad y = \sqrt{x+1} \qquad y = \ln(x+2)$$

2. Decipher the nature of the following relationships:
   a.

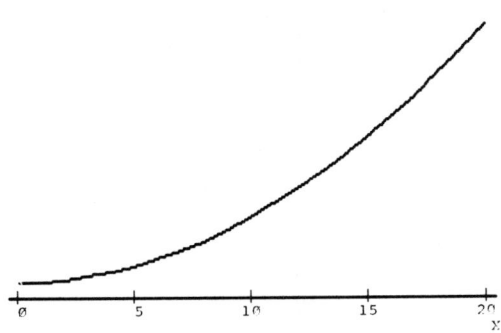

b.

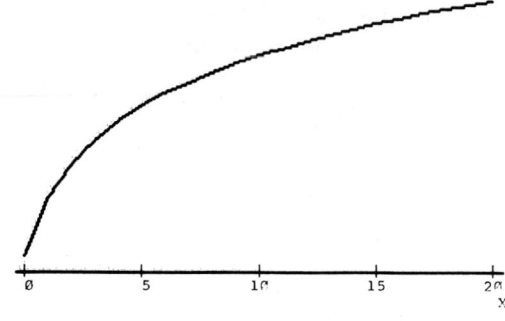

c.

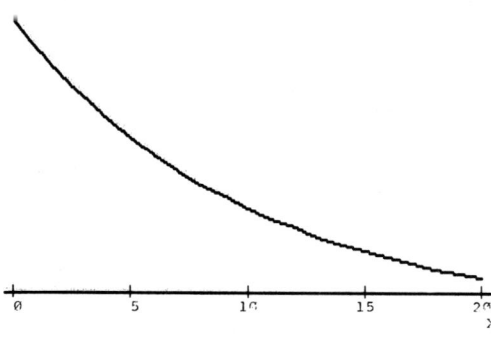

d.

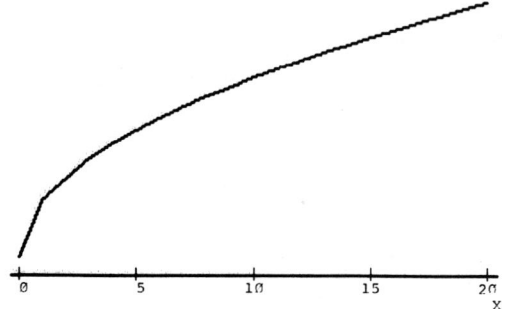

3. Find the value of c such that $y = b_0 + b_1 \ln(x+c) + \varepsilon$ for the accompanying data.

x	y
1	4.46
2	6.47
3	7.88
4	9.05
5	9.95
6	10.75
7	11.41
8	12.02
9	12.57
10	13.00

Unit 10

1. Use the accompanying data to find the expressions for $\hat{y}(0,1)$ and $\hat{y}(0,1,2)$. Note that the coefficients for x_1 in the two expressions are of opposite sign. Explain. [Hint: Since the regression coefficients under discussion are also coefficients of $x_1 - \hat{x}_1(0)$ and $x_1 - \hat{x}_1(0,2)$ in single-variable regressions, a look at $x_1 - \hat{x}_1(0)$ and $x_1 - \hat{x}_1(0,2)$ might aid the interpretation.]

x_0	x_1	x_2	y
1	0	1	6.93
1	−3	3	8.46
1	0	0	5.45
1	3	−4	−0.55
1	−2	1	7.11
1	3	−4	−0.67
1	−1	1	6.34
1	−3	3	10.46
1	−2	2	6.63
1	0	0	5.15
1	−1	1	5.04
1	2	−1	4.84
1	−1	0	5.63
1	1	−2	2.82
1	0	1	6.72

2. Find the regression coefficients for the accompanying data. Note that $x_2 = x_1^2$. How does this affect interpretation of the regression coefficients?

x_0	x_1	x_2	y
1	2	4	7.3
1	3	9	11.4
1	2	4	4.3
1	2	4	8.2
1	1	1	3.4
1	4	16	13.5
1	4	16	12.1
1	4	16	10.5
1	2	4	6.9
1	4	16	10.9
1	3	9	10.4
1	3	9	12.4
1	5	25	17.2
1	0	0	1.6
1	3	3	10.0

Unit 11

1. Orthogonality of a finite number of observations of Y_1, Y_2, and Y_3 is analogous to Y_1, Y_2, and Y_3 being uncorrelated. The accompanying table gives ten observations of Y_1, Y_2, and Y_3 that are orthogonal.

Y_1	Y_2	Y_3
−3	−24	−6
0	−28	11
−5	−8	14
4	−52	1
−6	18	−1
9	−36	−14
−6	34	−11
15	60	6
−5	30	−6
−3	6	6

    a. Calculate the estimated variances for Y_1, Y_2, and Y_3 and denote them as s_1^2, s_2^2, and s_3^2.
    b. Calculate another column $Y = 4Y_1 - 3Y_2 + 2Y_3$ and the estimated variance of Y (denoted by s^2).
    c. Verify that $s^2 = 4^2 s_1^2 + (-3)^2 s_2^2 + 2^2 s_3^2$.
    d. Repeat parts a–c by using any three constants in place of 4, −3, and 2.

2. Apply ABDO to the sums of squares and cross products matrix of the data in the accompanying table.

x_0	x_1	x_2	y
1	1	1	y_1
1	1	2	y_2
1	2	1	y_3
1	2	2	y_4

3. Apply ORTHO to x_0, x_1, and x_2 of Exercise 2 and then apply ABDO. Note the similarity of the results.

Unit 12

1. Given the accompanying ABDO matrix and given that the residual mean square is 6.48 with 15 degrees of freedom, find a 95% confidence interval for $\beta_0 + 3\beta_1 - 2\beta_3$.

x_0	x_1	x_2	y
18	4	8	13.24
	16	−9	12.37
		19	12.46

2. Given the accompanying ABDO matrix and given that the residual mean square is 10.64 with 17 degrees of freedom, find a 95% confidence interval for $\beta_0 + 1.80\beta_1 - 1.25\beta_2$.

x_0	x_1	x_2	y
20	14	-3	50.8
	22	-22	36.4
		0	0

3. Refer to Exercise 2. Attempt to find a 95% confidence interval for $\beta_0 + 2\beta_1 + 2\beta_2$. Note that ELIM cannot be carried to completion. When this happens, the linear combination of the parameters is said to be *nonestimable*.

4. Suppose the results of six basketball games involving the University of Kentucky, the University of Tennessee, and Vanderbilt University are as follows:

Kentucky	105	Vanderbilt	92
Kentucky	95	Tennessee	98
Tennessee	53	Vanderbilt	45
Vanderbilt	80	Kentucky	85
Tennessee	60	Kentucky	70
Vanderbilt	82	Tennessee	81

Assuming you have no other information, predict the outcome of the upcoming tournament game between Kentucky and Tennessee. (Hint: Let the response variable be the difference between the second column and first column. Let the parameters be β_K, β_T, and β_V. Assume that a difference in scores is an observation on the corresponding difference in the β's. Find the **X** matrix by using the method given in Exercise 2 of Unit 3.)

Unit 13

1. For a given set of data

$$\mathbf{X} = \begin{bmatrix} 1 & 4 & 3 & 3 \\ 1 & 1 & 1 & 2 \\ 1 & 2 & 8 & 1 \\ 1 & 2 & 4 & 4 \\ 1 & 4 & 2 & 7 \\ 1 & 5 & 1 & 3 \\ 1 & 3 & 5 & 6 \\ 1 & 7 & 1 & 9 \end{bmatrix} \quad \text{and} \quad \boldsymbol{\beta} = \begin{bmatrix} \beta_0 \\ \beta_1 \\ \beta_2 \\ \beta_3 \end{bmatrix}$$

a. Suppose $\beta_2 = 0$. Find a new **X** matrix and a β vector to reflect this result. Find a third **X** matrix to test the hypothesis that $\beta_2 = 0$ by the analysis of variance technique.
b. Suppose $\beta_1 = \beta_2 = \beta_3$. Find a new **X** matrix and a β vector to reflect this result. Find a third **X** matrix to test the hypothesis that $\beta_1 = \beta_2 = \beta_3$, using the analysis of variance technique.

2. With the accompanying data test the hypothesis that the slope of the linear relationship of y with x is the same for each group.

Group 1		Group 2		Group 3	
x	y	x	y	x	y
1	16.9	1	13.6	1	20.1
2	17.8	2	14.5	2	22.8
3	17.3	3	16.4	3	23.0
4	19.4	4	17.9	4	26.9
5	20.2	5	20.0	5	27.8

Unit 15

1. A clinical trial to assess effectiveness of a new drug for reducing high blood pressure produced the accompanying data. The placebo is a capsule that has no active ingredient but is indistinguishable from the other drugs by the patient. The standard is a currently used drug. The numbers represent the reduction from a baseline after one week of therapy. If the new drug is better than the placebo and as good as the standard, it has potential. Perform a statistical analysis and state your conclusions.

Placebo	Standard	New Drug
5	20	19
−2	19	20
4	25	21
4	23	18
1	24	19
	23	21
	19	20

2. In what situations would you use the general means model and in what situations would you use the equal means model and the general means model together?

Unit 17

1. A study was conducted to examine the effects of personal appearance on ratings for employment. A photograph of an attractive person and one of an unattractive person were selected (both people were of the same sex).

Three letters of recommendation were written, one giving a poor recommendation, one giving a fair recommendation, and one giving an excellent recommendation. The combination of the two photographs and the three letters represents six hypothetical job applicants. Thirty-six people were randomly divided into six groups. Members of each group rated one of the hypothetical applicants. The results are shown in the accompanying table. Analyze the data and write your conclusions.

	Appearance	
Recommendation	Unattractive	Attractive
Poor	16, 11, 9, 10, 7.5, 13	12.5, 13, 11, 9, 9, 9.5
Fair	13, 21, 22, 19, 19.5, 17	14, 12, 11, 13, 15, 15
Excellent	23, 22, 25.5, 28, 26, 27	22.5, 17, 22, 17, 22, 17

2. In a clinical trial to assess the effects on the blood pressure of combining two drugs in varying amounts, the accompanying data were collected. Each number represents a posttherapy blood pressure measurement on an individual having pretherapy blood pressure between 120 and 130. Analyze the data and write your conclusions.

	Drug B		
Drug A	1 mg/kg	2 mg/kg	4 mg/kg
10 mg/kg	125, 120, 118	116, 119, 121	111, 114, 116
20 mg/kg	119, 118, 120	120, 117, 115	110, 112, 115
40 mg/kg	116, 115, 112	108, 105, 102	95, 96, 100
80 mg/kg	102, 100, 95	98, 95, 94	95, 90, 89

Unit 18

1. The accompanying table of toy prices at various Lexington stores appeared in the *Lexington Herald* on October 26, 1978. Since store markups are usually done on a percentage increase, it is appropriate to use the logarithm of price as the response variable.
   a. Do the stores differ in their prices?
   b. Is the average price at chains with large mail-order operations (Wards, Sears) different from the average price at discount stores (Gold Circle, Wiggs, Hills, K-Mart)?
   c. Are Shillito's (a department store) prices different from the average at the toy stores (Thornbury's, Playthings)?
   d. Is the average price in the stores mentioned in part b different from the average price in the stores mentioned in part c?
   e. Give general conclusions about the classes of stores where toys are cheaper. Include any qualifications that need to be made because of the NA's.

Merchandise	Shillito's	Wards*	Sears	Thornbury's	Gold Circle	CSC	Wiggs	Playthings	Hills	K-Mart
Battlestar Galactica (Parker, 4–14)	$7.99	NA†	NA	$6.89	$5.49	$6.47	$4.99	NA	$3.97	NA
Shoot-out in Space (Tomy, 6–up)	24.99	$24.99	$19.99	NA	19.99	22.96	NA	NA	NA	NA
Merlin (Parker, 7–up)	24.99	24.99	24.97	24.29	24.99	24.97	29.99	$39.00	24.99	NA
Star Wars Dolls (Kenner, 3–up)	NA	NA	NA	10.97	9.99	9.97	8.99	NA	NA	NA
Superman Doll (Mego, 3–up)	12.99	NA	11.69	11.69	NA	NA	NA	NA	NA	$9.00
Microtron (Mego, 3–up)	10.99	NA	NA	NA	NA	NA	8.99	NA	NA	NA
Hungry Hippo (Hasbro, 5–up)	10.99	11.49	7.47	11.69	NA	9.97	7.99	NA	NA	7.50
Humpty-Dumpty Pillow (Fisher Price, 2 mo–2 yr)	NA	NA	4.99	5.29	4.99	4.97	4.99	5.50	4.97	NA
My Baby Beth (Fisher Price, 4–up)	12.99	NA	13.47	13.99	12.99	NA	16.99	16.95	NA	NA
Star Wars X-Wing Fighter (Kenner, 5–up)	NA	14.99	8.97	11.79	NA	NA	NA	NA	NA	10.50
Electronic Football (Mattel, 8–up)	29.99	NA	20.87	25.00	29.99	23.97	NA	39.00	NA	25.00
Looney Bin (Coleco, 6–up)	NA	NA	NA	NA	10.99	NA	NA	NA	NA	NA
Fat Chance (Milton Bradley, 5–up)	NA	6.49	NA	7.99	5.99	6.47	6.99	NA	6.78	6.50

Source: The information for the chart was compiled by the Lifestyle staff from October 18 to October 25. Data courtesy of the *Lexington Herald*.
*At Wards, prices quoted are based on a master list of toys prior to items being put on shelves.
†NA means not available at the time of the survey.

Unit 19

1. Thirty rats are randomly assigned to three groups of ten each. Each group is fed a different ration. The objective of the experiment is to compare rations with regard to their ability to satisfy appetites. For the jth group assume the model is the straight line

$$y_{j\ell} = \beta_{0j} + \beta_{1j} x_{j\ell} + \varepsilon_{j\ell}$$

where $x_{j\ell}$ is the amount of ration (grams) consumed during the period from 8:00 A.M. to 11:00 A.M. and $y_{j\ell}$ is the amount of ration (grams) consumed during the period from 1:00 P.M. to 4:00 P.M. We would speculate that there is a negative relationship between y and x since a rat that eats heartily in the morning may be less hungry in the afternoon. Rations that are particularly satisfying would have the steeper slopes. The data are shown in the accompanying table.

Group 1		Group 2		Group 3	
Morning Consumption	Afternoon Consumption	Morning Consumption	Afternoon Consumption	Morning Consumption	Afternoon Consumption
7.5	8.0	4.5	9.0	3.0	8.0
7.4	10.5	7.0	10.5	9.7	6.5
7.5	10.4	7.1	8.0	5.9	9.9
6.1	10.9	6.6	6.5	6.7	6.7
6.2	10.1	9.4	9.9	9.6	5.7
7.5	9.8	6.8	6.7	4.7	9.0
11.1	8.9	6.0	8.7	3.8	8.5
8.4	9.3	1.4	9.5	9.3	2.9
7.5	8.7	7.0	9.0	9.3	7.6
8.7	9.5	8.4	10.5	6.5	8.5

   a. Estimate all the slopes and intercepts.
   b. Test to see if the slopes are equal.
   c. If you decide the slopes are equal, estimate the slope and calculate the adjusted treatment means.

Unit 20

1. It is desirable for tablets to dissolve quickly. A pharmaceutical company wants to identify factors that explain the variability among the 30-second dissolution percentages in their tablets. Possible explanatory variables are the mixer, method of making the paste, the tablet press, and the amounts of various mediums used. Identify sets of important factors affecting dissolution percentage. Do you think there are other factors not represented in the data that affect dissolution percentage? The data are shown in the accompanying table.

| | Paste | Tablet | Mediums | | | | | | Dissolution |
Mixer	Method	Press	1	2	3	4	5	6	Percentage
2	3	2	27	26	26	26	28	27	738
2	3	2	29	29	29	28	30	29	826
1	1	4	31	32	32	32	31	32	912
1	1	3	33	33	33	33	33	32	825
1	1	4	33	33	33	33	33	33	587
1	1	3	33	33	33	33	33	33	783
1	1	2	33	33	33	33	33	33	757
2	3	3	33	33	33	33	34	33	816
1	1	3	33	33	33	33	34	33	509
1	2	4	33	33	33	33	34	33	708
1	2	2	34	33	33	33	34	33	408
1	2	4	34	34	33	33	34	33	660
1	2	3	34	34	33	33	34	33	409
1	2	2	34	34	33	34	34	33	400
1	2	4	34	34	33	34	34	33	756
2	3	3	34	35	33	34	34	33	808
1	2	2	34	35	33	34	34	33	491
2	3	2	34	35	33	34	34	33	858
2	3	3	34	34	34	34	34	33	757
1	2	2	35	34	34	34	34	33	763
1	2	4	35	34	34	34	34	35	457
1	2	4	35	34	34	34	35	35	516
1	2	4	35	34	34	34	35	35	561
1	2	2	35	34	34	34	35	35	763

2. Acetylene data are shown in the accompanying table. The objective is to find a prediction equation for y in terms of x_1, x_2, and x_3. You may use functions of these basic x variables as additional x variables.

x_1, Reactor Temperature (°C)	x_2, Ratio of H_2 to n-Heptane (mole ratio)	x_3, Contact Time (sec)	y, Conversion of n-Heptane to Acetylene (%)
1300	7.5	0.0120	49.0
1300	9.0	0.0120	50.2
1300	11.0	0.0115	50.5
1300	13.5	0.0130	48.5
1300	17.0	0.0135	47.5
1300	23.0	0.0120	44.5

x_1, Reactor Temperature (°C)	x_2, Ratio of H_2 to n-Heptane (mole ratio)	x_3, Contact Time (sec)	y, Conversion of n-Heptane to Acetylene (%)
1200	5.3	0.0400	28.0
1200	7.5	0.0380	31.5
1200	11.0	0.0320	34.5
1200	13.5	0.0260	35.0
1200	17.0	0.0340	38.0
1200	23.0	0.0410	38.5
1100	5.3	0.0840	15.0
1100	7.5	0.0980	17.0
1100	11.0	0.0920	20.5
1100	17.0	0.0860	29.5

Source: Donald W. Marquardt and Ronald D. Snee. "Ridge Regression in Practice," *American Statistician* 29 (1975): 3–20.

Unit 21

1. Using the data and the selected model from Exercise 2 of Unit 20, predict y when $x_1 = 1100$, $x_2 = 8$, and $x_3 = 0.0120$. Assess this predictor. Use, as prior information, the notion that the response for low values of both x_1 and x_3 should be equal to the response for high values of both x_1 and x_3.

Unit 22

1. Nine subjects with angina pectoris were randomly divided into two groups. One group received a drug and the other group received a placebo (control). The subjects were observed weekly for five weeks. In some instances the appointments were not kept and hence there were some missing data. Several responses were measured that are related to the safety and effectiveness of the drug. The data on systolic blood pressure are presented here. What do you think the objective of this trial is? Analyze the data with respect to this objective. State the assumptions underlying your analysis.

		Week				
	Subject	1	2	3	4	5
Drug	1	135	140	150	—	140
	2	130	120	130	135	120
	3	115	120	130	130	120
	4	—	110	120	130	—
	5	130	130	130	125	130

	Subject	\multicolumn{5}{c}{Week}
		1	2	3	4	5
Control	6	160	165	180	170	150
	7	160	170	165	—	130
	8	150	170	170	170	—
	9	—	190	170	180	180

Note: — designates a missing value.

Unit 23

1. The concentration of a drug in the plasma of a rat is measured at several times after an intramuscular injection of the drug. The data are given in the accompanying table.

Hours	Concentration	Hours	Concentration
0.25	6.82	3.00	7.43
0.50	9.96	3.50	6.42
0.75	11.32	4.00	5.97
1.00	11.16	4.50	4.23
1.25	11.48	5.00	3.81
1.50	11.40	6.00	3.25
1.75	10.87	7.00	1.81
2.00	10.32	8.00	1.19
2.50	9.40		

   a. Using a computer program, fit the drug assimilation model in the text to the data.
   b. Find the estimates of the parameters and their standard errors.
   c. Plot the residuals with the predicted values.

2. A model that results from assuming that the body tissue is a compartment that interchanges with the plasma is

$$y = R \cdot KA \left(\frac{K21-A}{(KA-A)(B-A)} \exp(-A \cdot T) + \frac{K21-B}{(KA-B)(A-B)} \exp(-B \cdot T) + \frac{K21-KA}{(KA-A)(KA-B)} \exp(KA \cdot T) \right) + \varepsilon$$

   a. Fit this model to the data of Exercise 1.
   b. Find the estimates of the parameters and their standard errors.
   c. Plot the residuals with the predicted values.

3. Which of the models from Exercises 1 and 2 seems more appropriate? Explain.

Appendix B

Annotated Computer Output

In this unit a brief description of three widely distributed statistical systems is given. The information is from the manuals of the respective systems. The use of these systems is demonstrated with data sets of earlier units. There are other excellent systems available. These particular systems were selected for presentation because of the authors' experiences and because at least one of them is available for virtually any minicomputer or large computer.

The Biomedical Computer Programs P-Series (BMDP) were developed by the Health Sciences Computing Facility, University of California, Los Angeles, CA 90024. The series contains 33 programs including data description, linear regression, nonlinear regression, and analysis of variance and covariance. These programs have been adapted for use with a large variety of computers.

Minitab is a very easy to use, flexible, and powerful statistical computing system. It has been checked on a wide variety of computers. It can be used on almost any computer that has sufficient memory to store the program and worksheet. It runs interactively or in batch mode. For information, write Minitab Project, Statistics Department, 215 Pond Laboratory, The Pennsylvania State University, University Park, Pennsylvania 16802.

SAS is a computer system for data analysis. In one easy-to-use system it provides all the tools needed for data analysis: information storage and retrieval, data modification and programming, report writing, statistical analysis, and file handling. SAS runs only on IBM 360/370 computers and plug-compatible machines. For information, write to SAS Institute, Inc., Box 8000, Cary, NC 27511.

The following table is a guide to the annotated computer output in this section.

Data Set	Units	System	Page
Arsenic	4, 5, 6	BMDP	346
Firefly	10, 12, 13	SAS	351
Electricity load	13	BMDP	363
Fat digestibility	17	SAS	365
Swamp pH	18	Minitab	368
Soybean physiological	19	Minitab	370

Appendix B

```
BMDP9R - ALL POSSIBLE SUBSETS REGRESSION
HEALTH SCIENCES COMPUTING FACILITY
UNIVERSITY OF CALIFORNIA, LOS ANGELES 90024
PROGRAM REVISED NOVEMBER 1979
MANUAL REVISED -- 1979
COPYRIGHT (C) 1979 REGENTS OF UNIVERSITY OF CALIFORNIA
```
⎯ The regression program in the BMDP series.

```
 -- IF THERE ARE FEWER THAN THREE INDEPENDENT VAR-
    IABLES, THEN METHOD=NONE. WILL BE USED.
 -- IF STATISTICS. IS STATED IN THE PLOT PARAGRAPH,
    THEN STATISTICS AS IN BMDP6D WILL ACCOMPANY EACH
    PLOT.
 -- TO LIMIT THE NUMBER OF VARIABLES IN THE REPORTED
    SUBSETS, IN THE PRINT PARAGRAPH STATE MAXVAR=N.
    WHERE N IS THE MAXIMUM NUMBER OF VARIABLES THAT
    YOU DESIRE. A SUBSET WITH GREATER THAN N VAR-
    IABLES WILL NOT BE REPORTED UNLESS IT IS ONE OF
    THE BEST SUBSETS BY THE CP OR ADJUSTED R-SQUARED
    CRITERIA.
 -- TO OBTAIN THE COVARIANCE MATRIX OF THE REGRESSION
    COEFFICIENTS, INCLUDE CREG IN THE MATRIX STATEMENT
    OF THE PRINT PARAGRAPH, E.G.,
           MATRIX=CORR,RESID,CREG.
 -- IF RESIDUALS ARE COMPUTED OR IF YOU STATED
    HISTOGRAM. IN THE PLOT PARAGRAPH, A HISTOGRAM OF
    THE STANDARDIZED (STUDENTIZED) RESIDUALS WILL BE
    MADE.
```
⎯ This information pertains to a many-variable regression model and is not pertinent for the present example.

```
 PROGRAM CONTROL INFORMATION

/PROBLEM  TITLE='ARSENIC DATA'.        ⎯ Title to be printed at the top of each page.
/INPUT    VARIABLES ARE 2.
          FORMAT IS '(F4.0,F4.0)'.     ⎯ Standard FORTRAN format.
/VARIABLE NAMES ARE CONCEN, MILES.     ⎯ Names the variables.
/REGRESS  INDEPENDENT IS MILES.
          DEPENDENT IS CONCEN.
          METHOD=NONE.                 ⎯ No variable selection is to be done.
/PRINT    MATRICES ARE RESIDUAL.
/PLOT     YVAR ARE RESIDUAL, CONCEN.
          XVAR ARE PREDICTD, MILES.    ⎯ Plot 1.
/END                                   ⎯ Plot 2.
```
⎯ This is input to the program

```
 PROBLEM TITLE . . . . .
 ARSENIC DATA

 NUMBER OF VARIABLES TO READ IN. . . . . . . .        2
 NUMBER OF VARIABLES ADDED BY TRANSFORMATIONS. .      0
 TOTAL NUMBER OF VARIABLES . . . . . . . . . .        2
 NUMBER OF CASES TO READ IN. . . . . . . . . .  1000000
 CASE LABELING VARIABLES . . . . . . . . . . .

 LIMITS AND MISSING VALUE CHECKED BEFORE TRANSFORMATIONS
 BLANKS ARE. . . . . . . . . . . . . . . . . .    ZEROS
 INPUT UNIT NUMBER . . . . . . . . . . . . . .        5
 REWIND INPUT UNIT PRIOR TO READING. . DATA. . .     NO
 NUMBER OF WORDS OF DYNAMIC STORAGE. . . . . .    56320
 INPUT FORMAT. . . . .
 (F4.0,F4.0)

 VARIABLES TO BE USED
   1 CONCEN     2 MILES

 INDEPENDENT VARIABLES ARE
     2 MILES

 DEPENDENT VARIABLE. . . . . . . . . . . . . .    1 CONCEN
 NUMBER OF 'BEST' REGRESSIONS. . . . . . . . .        5
 SELECTION CRITERION . . . . . . . . . . . . .     NONE
 WEIGHT VARIABLE . . . . . . . . . . . . . . .
 PRECISION . . . . . . . . . . . . . . . . . .   DOUBLE
 TOLERANCE FOR MATRIX INVERSION. . . . . . . . 0.0001000

 PRINT CORRELATION MATRIX. . . . . . . . . . .       NO
 PRINT COVARIANCE MATRIX . . . . . . . . . . .       NO
 PRINT RESIDUALS . . . . . . . . . . . . . . .      YES
 PRINT COVARIANCE MATRIX FOR REGRESSION COEFS. .     NO
 PRINT CORRELATION MATRIX FOR REGRESSION COEFS .     NO
 MAX. NO. OF VARS. IN ANY REPORTED SUBSET  . . .      1

 NORMAL PROB. PLOT OF STANDARDIZED RESIDUALS . .     NO
 NUMBER OF PAIRS OF VARIABLES TO BE PLOTTED. . .      2
```
⎯ This confirms input and gives default values for various control parameters.

```
DATA AFTER TRANSFORMATIONS FOR FIRST      5 CASES
CASES WITH ZERO WEIGHTS AND MISSING DATA NOT INCLUDED.

CASE    CASE      CASE
LABEL   NUMBER    WEIGHT    2 MILES    1 CONCEN
        1         1.00000   2.00000    3.19000
        2         1.00000   4.00000    3.26000
        3         1.00000   8.00000    1.82000
        4         1.00000   10.00000   1.02000
        5         1.00000   12.00000   1.85000

NUMBER OF CASES READ. . . . . . . . . .    10
```

UNIVARIATE SUMMARY STATISTICS

VARIABLE	MEAN	STANDARD DEVIATION	COEFFICIENT OF VARIATION	SMALLEST VALUE	LARGEST VALUE	SMALLEST STANDARD SCORE	LARGEST STANDARD SCORE	SKEWNESS	KURTOSIS
2 MILES	16.10000	11.18978	0.695017	2.00000	36.00000	-1.26	1.78	0.40	-1.32
1 CONCEN	1.62800	1.31231	0.621813	0.30000	3.26000	-1.31	1.61	0.40	-1.31

Enough data is printed so that the input can be verified.

VALUES FOR KURTOSIS GREATER THAN ZERO INDICATE DISTRIBUTIONS
WITH HEAVIER TAILS THAN THE NORMAL DISTRIBUTION.

```
STATISTICS FOR 'BEST' SUBSET
SQUARED MULTIPLE CORRELATION         0.74624
MULTIPLE CORRELATION                 0.86385
ADJUSTED SQUARED MULT. CORR.         0.71452
RESIDUAL MEAN SQUARE                 3.292548
STANDARD ERROR OF EST.               0.540877
F-STATISTIC                          23.53
NUMERATOR DEGREES OF FREEDOM         1
DENOMINATOR DEGREES OF FREEDOM       8
SIGNIFICANCE                         0.0013
```

Summary statistics for the model.

VARIABLE NO. NAME	REGRESSION COEFFICIENT	STANDARD ERROR	STAND. COEF.	T-STAT.	2TAIL SIG.	TOLERANCE	CONTRIBUTION TO R-SQUARED
INTERCEPT	2.88623	0.310720	2.851	9.29	0.000		
2 MILES	-3.3781507	0.3161122	-3.864	-4.85	0.001	1.000000	0.746243

Standard type of output.

THE CONTRIBUTION TO R-SQUARED FOR EACH VARIABLE IS THE AMOUNT
BY WHICH R-SQUARED WOULD BE REDUCED IF THAT VARIABLE WERE
REMOVED FROM THE REGRESSION EQUATION.

IN THE TABLE BELOW, THE STANDARDIZED RESIDUAL IS THE RESIDUAL
DIVIDED BY ITS STANDARD ERROR. THE COLUMN LABEL DELETED
(PRESS) RESIDUAL CONTAINS THE RESIDUAL FOR EACH CASE FROM
PREDICTING THAT CASE FROM THE OTHER CASES, I.E, THE
RESIDUAL FOR THE CASE AFTER REMOVING THE EFFECT OF
THAT CASE FROM THE REGRESSION COEFFICIENTS. SIMILARLY, THE
ADJUSTED (PRESS) PREDICTED VALUE IS THE PREDICTED VALUE FOR
THE CASE AFTER REMOVING THE EFFECT OF THAT CASE FROM THE
REGRESSION COEFFICIENTS. FOR A DISCUSSION OF COOK'S
DISTANCE, SEE R. D. COOK, TECHNOMETRICS, FEB., 1977.

CASE LABEL	CASE NO.	OBSERVED CONCEN	PREDICTED VALUE	STANDARD ERROR OF PRED.VAL.	RESIDUAL	CASE WEIGHT	WEIGHTED RESIDUAL	STAND-ARDIZED RESIDUAL	DELETED (PRESS) RESIDUAL	ADJUSTED (PRESS) PRED.VAL.	MAHALA-NOBIS DISTANCE	COOK'S DISTANCE
	1	3.1930	2.7299	0.2844	0.4601	1.000	0.4601	1.00	0.6358	2.5542	1.59	0.19
	2	3.2600	2.5736	0.2594	0.6864	1.000	0.6864	1.45	0.8913	2.3687	1.17	0.31
	3	1.8200	2.2613	0.2151	-0.4413	1.000	-0.4413	-0.89	-0.5239	2.3439	0.52	0.07
	4	1.0200	2.1047	0.1973	-1.0847	1.000	-1.0847	-2.15	-1.2511	2.2711	0.30	0.36
	5	1.8500	1.9484	0.1834	-0.0984	1.000	-0.0984	-0.19	-0.1112	1.9612	0.13	0.00
	6	2.0500	1.7140	0.1720	0.3360	1.000	0.3360	0.66	0.3738	1.6762	0.01	0.02
	7	1.3400	1.2451	0.1884	0.0949	1.000	0.0949	0.19	0.1080	1.2320	0.19	0.00
	8	0.7900	1.0888	0.2040	-0.2988	1.000	-0.2988	-0.60	-0.3483	1.1383	0.38	0.03
	9	0.6600	0.5417	0.2818	0.1183	1.000	0.1183	0.26	0.1624	0.4976	1.54	0.01
	10	0.3000	0.0728	0.3634	0.2272	1.000	0.2272	0.57	0.4142	-0.1142	3.16	0.13

← Press residuals as discussed in Unit 20.

Annotated Computer Output

```
SUMMARY STATISTICS FOR RESIDUALS

(CASES WITH POSITIVE WEIGHT)
AVERAGE RESIDUAL                        0.0000
RESIDUAL MEAN SQUARE                    0.29254818
AVERAGE DELETED RESIDUAL                0.0351
AVE. SQUARED DELETED RESIDUAL
   (PREDICTION MEAN SQUARE)             0.35215531
SERIAL CORRELATION                      0.2538
DURBIN-WATSON STATISTIC                 1.4088
------------------------------------------------
HISTOGRAM OF STANDARDIZED (STUDENTIZED) RESIDUALS
EACH BIN OF THE HISTOGRAM IS LABELED WITH ITS LOWER LIMIT.
NOTE THAT IF THE COUNT FOR A BIN EXCEEDS 100, ONLY
100 ASTERISKS WILL BE PRINTED.

         -2.2    1 *
         -2.0    0
         -1.8    0
         -1.6    0               For large data sets this should be bell-shaped.
         -1.4    0
         -1.2    0
         -1.0    1 *
         -0.8    0
         -0.6    1 *
         -0.4    0
         -0.2    1 *
          0.0    1 *
          0.2    1 *
          0.4    1 *
          0.6    1 *
          0.8    1 *
          1.0    0
          1.2    0
          1.4    1 *

------------------------------------------------
    3.16 IS THE MAXIMUM VALUE OF MAHALANOBIS DISTANCE AMONG CASES WITH
POSITIVE CASE WEIGHT. THIS OCCURRED FOR CASE NUMBER     10, CASE LABEL =

------------------------------------------------
   -2.15 IS THE LARGEST STANDARDIZED RESIDUAL (IN ABSOLUTE VALUE) AMONG
CASES WITH POSITIVE CASE WEIGHT. THIS OCCURRED FOR CASE NUMBER    4, CASE LABEL =

------------------------------------------------
    0.36 IS THE MAXIMUM VALUE OF COOK'S DISTANCE AMONG CASES
WITH POSITIVE WEIGHT. THIS OCCURRED FOR CASE NUMBER    4, CASE LABEL =
IF THIS CASE WERE OMITTED, THE REGRESSION COEFFICIENTS WOULD
MOVE FROM THE VALUES REPORTED ABOVE TO THE EDGE OF A   43.27
PERCENT CONFIDENCE ELLIPSOID.

------------------------------------------------




COMPARISON OF ESTIMATES OF REGRESSION COEFFICIENTS
(RELATIVE DIFFERENCE IS DIFFERENCE DIVIDED BY ORDINARY COEF.
STANDARD ERROR IS THAT OF ORDINARY COEFFICIENT.)

                             OMITTING                      DIFFERENCE
                ORDINARY     CASE WITH                     DIVIDED BY
                   LEAST     LARGEST COOK    RELATIVE      STANDARD
                 SQUARES     DISTANCE        DIFFERENCE    ERROR

   INTERCEPT    2.886225     3.120378        -0.0811       -0.7536
   2 MILES     -0.078151    -0.084923        -0.0867        0.4203

------------------------------------------------
NUMERICAL CONSISTENCY CHECK

RESIDUAL MEAN SQUARES ARE COMPUTED FROM BOTH COVARIANCE MATRIX AND RESIDUALS, AND
RELATIVE DIFFERENCE (DIFFERENCE DIVIDED BY SMALLER OF TWO ESTIMATES) IS COMPUTED.

       RESIDUAL MEAN SQUARES COMPUTED FROM

       COVARIANCE MATRIX     RESIDUALS      RELATIVE DIFFERENCE

          0.292548D 00       0.292548D 00      -0.37950D-15


IN THE BIVARIATE PLOTS WHICH FOLLOW, A = 10 CASES, B = 11 CASES, ..., AND * = 20 OR MORE CASES.
AN * IN THE BORDER OF ANY PLOT INDICATES A MISSING VALUE OR
A VALUE OUT OF RANGE.
```

PAGE 7   ARSENIC DATA

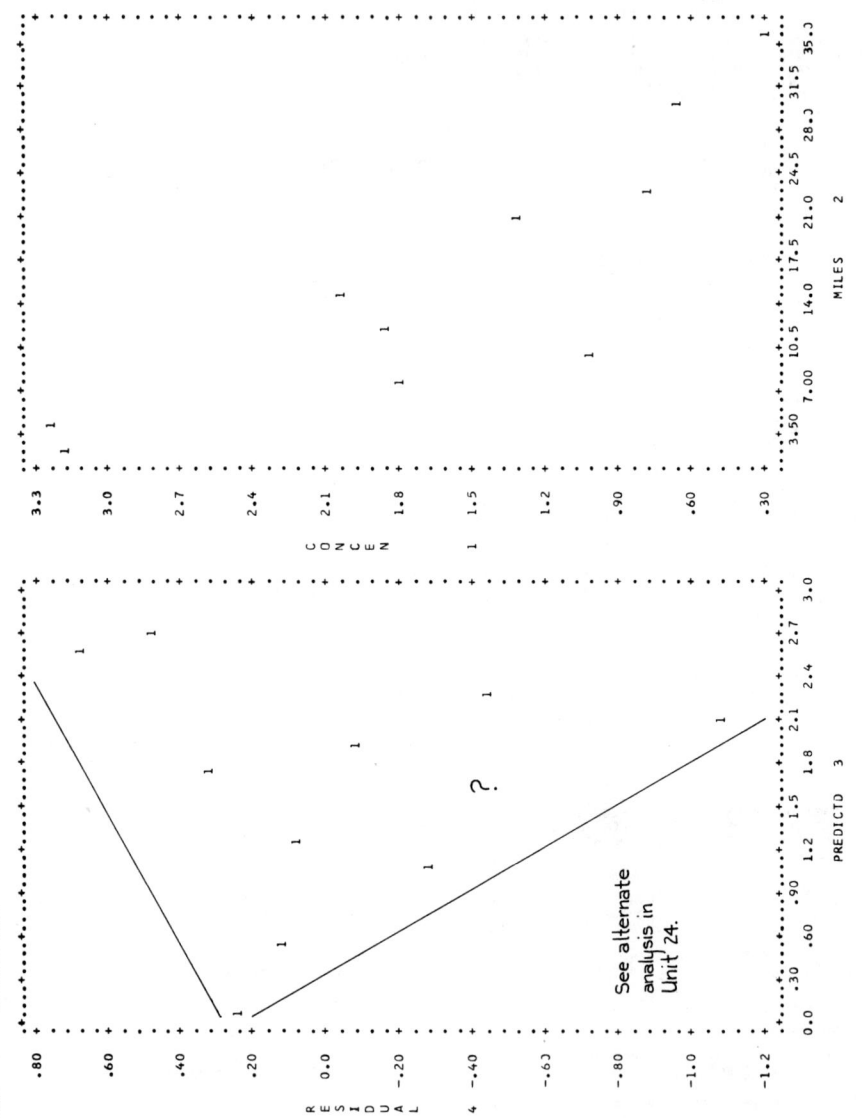

See alternate analysis in Unit 24.

350

STATISTICAL ANALYSIS SYSTEM

NOTE: SAS OPTIONS SPECIFIED ARE:
SORT=4

```
1       TITLE FIREFLY DATA;        ← Prints a heading on each page.
2       DATA FIREFLY;
3       INPUT FTIME LIGHT TEMP;    ← Causes the data cards to be printed.
4       LIST;
5       CARDS;
RULE:   12345678 10123456 20123456 30123456 40123456 50123456 60123456 70123456 80
6       45 26 21.1  ⎫
7       40 35 23.9  ⎪
8       58 40 17.8  ⎪
9       50 41 22.0  ⎪
10      31 45 22.3  ⎪
11      52 55 23.3  ⎬  Data cards in free format.
12      54 55 20.5  ⎪
13      38 56 25.5  ⎪
14      40 70 21.7  ⎪
15      28 75 26.7  ⎪
16      38 79 25.0  ⎪
17      36 87 24.4  ⎪
18      36 100 22.3 ⎪
19      46 100 25.5 ⎪
20      40 113 26.7 ⎪
21      31 130 25.5 ⎪
22      40 140 26.7 ⎭
```

NOTE: DATA SET WORK.FIREFLY HAS 17 OBSERVATIONS AND 3 VARIABLES. 465 OBS/TRK.
NOTE: THE DATA STATEMENT USED 0.17 SECONDS AND 170K.

```
23      PROC PRINT;
24          VAR LIGHT TEMP FTIME;    ← Causes selected variables to be neatly printed.
```

NOTE: THE PROCEDURE PRINT USED 0.27 SECONDS AND 172K AND PRINTED PAGE 1.

```
25      PROC PLOT;
26          PLOT FTIME*LIGHT FTIME*TEMP LIGHT*TEMP;  ← Produces graphs of all pairs of variables.
```

NOTE: THE PROCEDURE PLOT USED 0.54 SECONDS AND 174K AND PRINTED PAGES 2 TO 4.

```
27      PROC REG;            ⎫ Regression
28          MODEL FTIME=LIGHT TEMP / P SS1 SS2 SEQB;   ← Calculates predicted values and residuals.
                                                              ↑      ↑     ↑
29          OUTPUT OUT=NEW1 PREDICTED=YHAT1 RESIDUAL=RESID1;     │      │     └ Prints sequential coefficients.
30          MODEL FTIME= TEMP LIGHT/SS1 SS2 SEQB PARTIAL;        │      └ Prints partial sums of squares.
31          MODEL FTIME=TEMP / P SS1 SS2 SEQB;                   └ Prints sequential sums of squares.
32          OUTPUT OUT=NEW2 PREDICTED=YHAT2 RESIDUAL=RESID2;
```
Creates and saves a data set NEW1.

NOTE: DATA SET WORK.NEW1 HAS 17 OBSERVATIONS AND 5 VARIABLES. 296 OBS/TRK.
NOTE: DATA SET WORK.NEW2 HAS 17 OBSERVATIONS AND 5 VARIABLES. 296 OBS/TRK.
NOTE: THE PROCEDURE REG USED 0.81 SECONDS AND 174K AND PRINTED PAGES 5 TO 10.

```
33      PROC PLOT DATA=NEW1;
```

SAS prints out the input and gives the page number of related output.

Creates a data set FIREFLY containing variables FTIME, LIGHT, and TEMP.

351

S T A T I S T I C A L   A N A L Y S I S   S Y S T E M

34.   PLOT RESID1*YHAT1/VREF=0; ← Plots the residuals with the predicted values from the first model and puts a reference line at zero.

NOTE: THE PROCEDURE PLOT USED 0.34 SECONDS AND 174K AND PRINTED PAGE 11.

35   PROC PLOT DATA=NEW2;
36   PLOT RESID2*TEMP/VREF=0;

NOTE: THE PROCEDURE PLOT USED 0.32 SECONDS AND 174K AND PRINTED PAGE 12.

NOTE: SAS USED 174K MEMORY.

NOTE: SAS INSTITUTE INC.
      SAS CIRCLE
      BOX 8000
      CARY, N.C. 27511

FIREFLY DATA

OBS	LIGHT	TEMP	FTIME
1	26	21.1	45
2	35	23.9	40
3	40	17.8	58
4	41	22.0	50
5	45	22.3	31
6	55	23.3	52
7	55	23.5	54
8	56	25.5	38
9	70	21.7	40
10	75	26.7	28
11	79	25.0	38
12	87	24.4	36
13	100	22.3	36
14	100	25.5	46
15	110	26.7	40
16	130	25.5	31
17	143	26.7	43

Data, neatly formatted as requested by input lines 23 and 24.

PLOT OF FTIME*LIGHT   FIREFLY DATA
                             LEGEND: A = 1 OBS, B = 2 OBS, ETC.

First plot as requested by lines 25 and 2b.

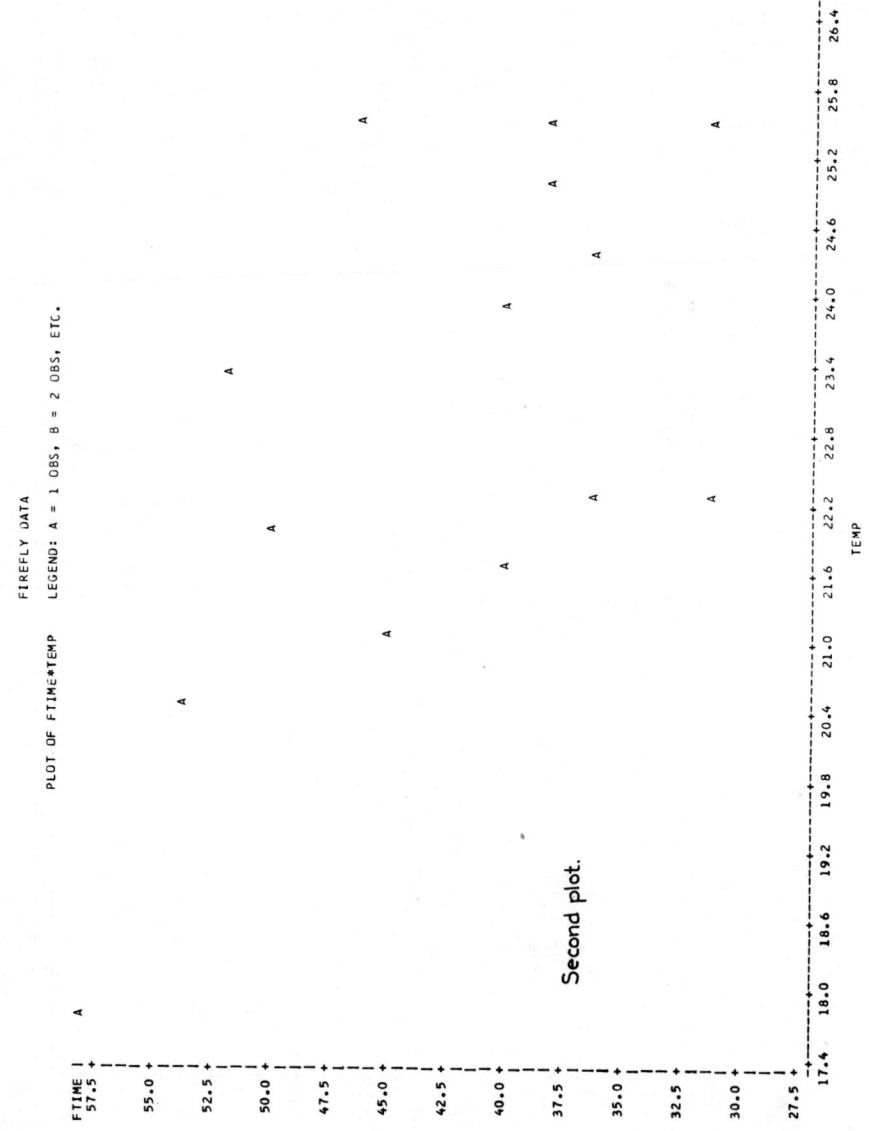

Second plot.

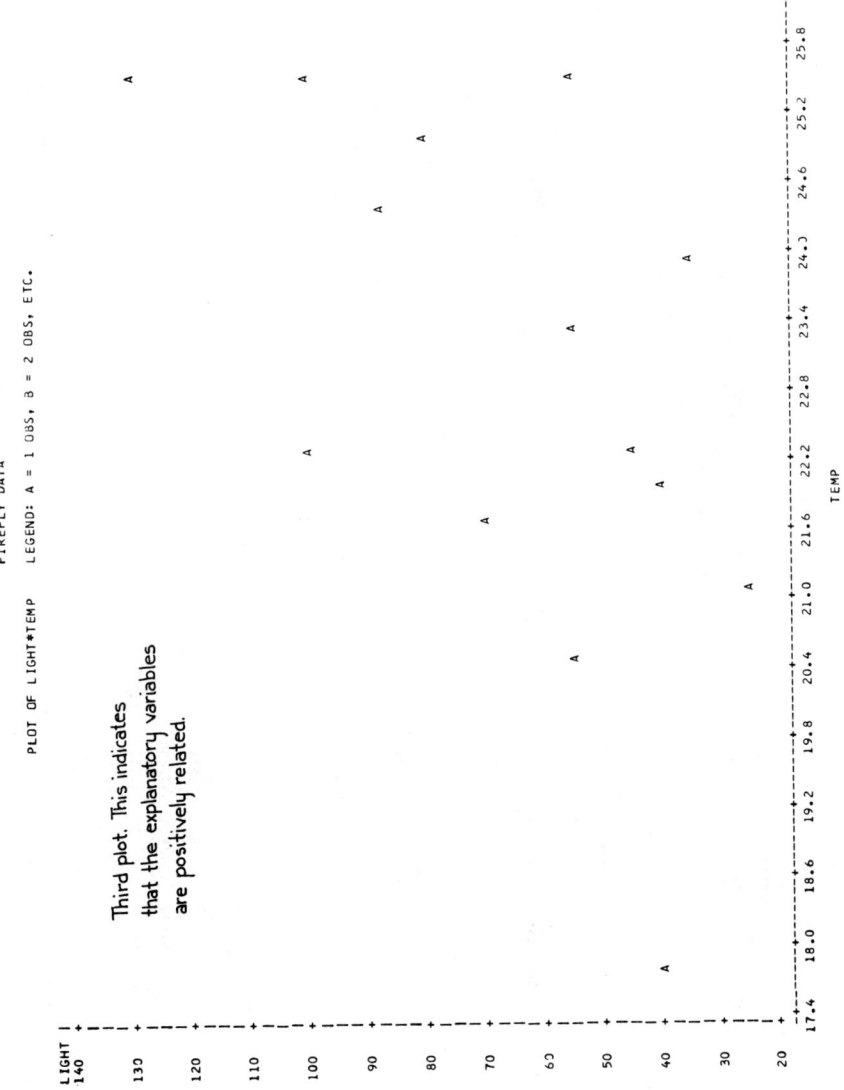

Third plot. This indicates that the explanatory variables are positively related.

FIREFLY DATA

SEQUENTIAL PARAMETER ESTIMATES

```
INTERCEP  41.3529
LIGHT     48.8962  -.103083                    ← Regression coefficients with just INTERCEP in the model.
TEMP      91.4745   6.9E-04  -2.12753          ← Regression coefficients with INTERCEP and LIGHT in the model.
                                               ← Regression coefficients with INTERCEP, LIGHT, and TEMP in the model.
```

Note that the coefficient of LIGHT changes sign when TEMP is included.

DEP VARIABLE: FTIME

SOURCE	DF	SUM OF SQUARES	MEAN SQUARE	F VALUE	PROB>F
MODEL	2	456.739	228.370	4.680	0.0182
ERROR	14	683.143	48.795924		
C TOTAL	16	1139.882			

ROOT MSE	6.985408	R-SQUARE	0.4007
DEP MEAN	41.352941	ADJ R-SQ	0.3151
C.V.	16.89217		

VARIABLE	DF	PARAMETER ESTIMATE	STANDARD ERROR	T FOR H0: PARAMETER=0	PROB > |T|	TYPE I SS	TYPE II SS
INTERCEP	1	91.474451	18.825133	4.859	0.0003	29371.118	1152.148
LIGHT	1	0.000691989	0.068339	0.010	0.9921	194.421	0.005003143
TEMP	1	-2.127529	0.917599	-2.319	0.0361	262.318	262.318

Note that the diagonal elements, 41.3529, -.103083, and -2.12753, are the sequential coefficients.

Partial sums of squares.
Sequential sums of squares.

OBS	ACTUAL	PREDICT VALUE	RESIDUAL
1	45.000	46.602	-1.602
2	40.000	40.651	-0.653720
3	58.000	53.632	4.368
4	50.000	44.697	5.333
5	31.000	44.062	-13.062
6	52.000	41.941	10.059
7	54.000	47.898	6.102
8	38.000	37.261	0.738795
9	40.000	45.356	-5.356
10	28.000	34.721	-6.721
11	38.000	38.341	-0.340885
12	36.000	39.623	-3.623
13	36.000	44.100	-8.100
14	46.000	37.292	8.708
15	40.000	34.746	5.254
16	31.000	37.312	-6.312
17	40.000	34.766	5.234

SUM OF RESIDUALS   4.72511E-13
SUM OF SQUARED RESIDUALS 683.1429

```
SEQUENTIAL PARAMETER ESTIMATES
INTERCEP   41.3529
TEMP       91.3816  -2.12144
LIGHT      91.4745  -2.12753  6.9E-04
```

Similar output, but with the order of LIGHT and TEMP reversed.

DEP VARIABLE: FTIME

SOURCE	DF	SUM OF SQUARES	MEAN SQUARE	F VALUE	PROB>F
MODEL	2	456.739	228.370	4.680	0.0182
ERROR	14	683.143	48.795924		
C TOTAL	16	1139.882			

ROOT MSE	6.985408	R-SQUARE	0.4007	
DEP MEAN	41.352941	ADJ R-SQ	0.3151	
C.V.	16.89217			

VARIABLE	DF	PARAMETER ESTIMATE	STANDARD ERROR	T FOR H0: PARAMETER=0	PROB > |T|	TYPE I SS	TYPE II SS
INTERCEP	1	91.474451	18.825103	4.859	0.0003	29071.118	1152.148
TEMP	1	-2.127529	0.917599	-2.319	0.0361	456.734	262.318
LIGHT	1	0.0006919891	0.068339	0.010	0.9921	0.005003143	0.005003143

FIREFLY DATA

PARTIAL REGRESSION RESIDUAL PLOTS

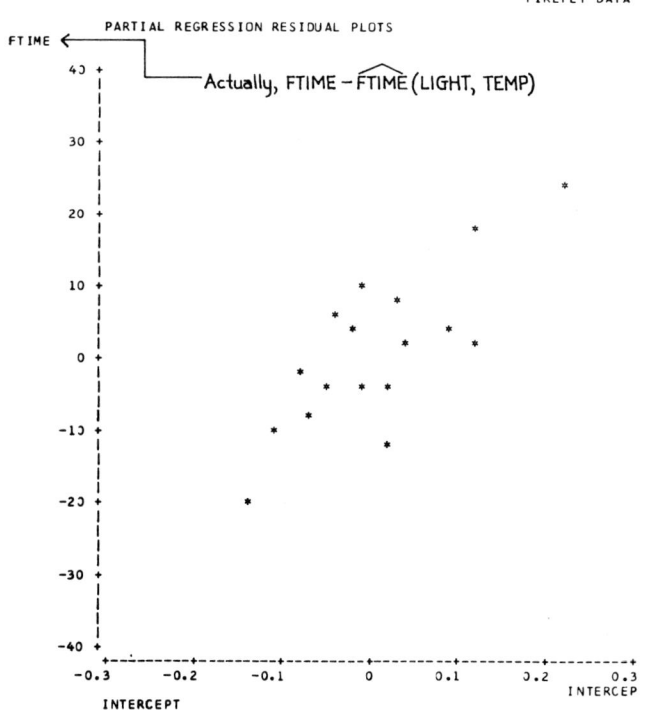

Actually, FTIME − $\widehat{\text{FTIME}}$ (LIGHT, TEMP)

INTERCEPT − $\widehat{\text{INTERCEPT}}$ (LIGHT, TEMP)

This plot and the next two were produced in response to the PARTIAL option on input line 30.

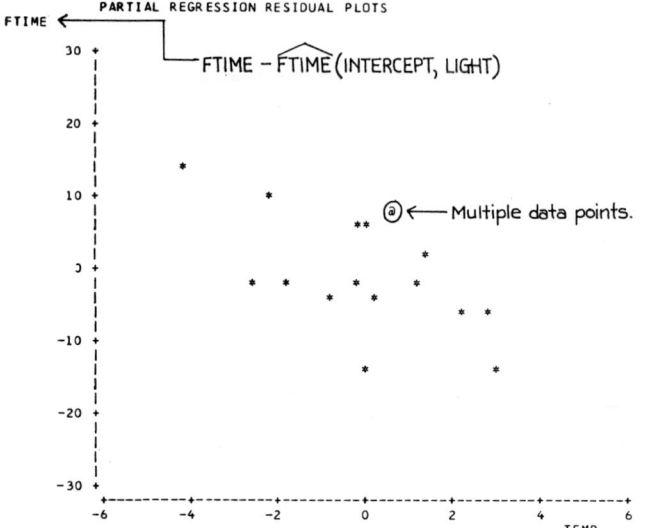

PARTIAL REGRESSION RESIDUAL PLOTS

These adjusted variables are such that a simple through-the-origin regression gives the same regression coefficient as the partial coefficient of the full model.
They are useful for identifying influential observations.

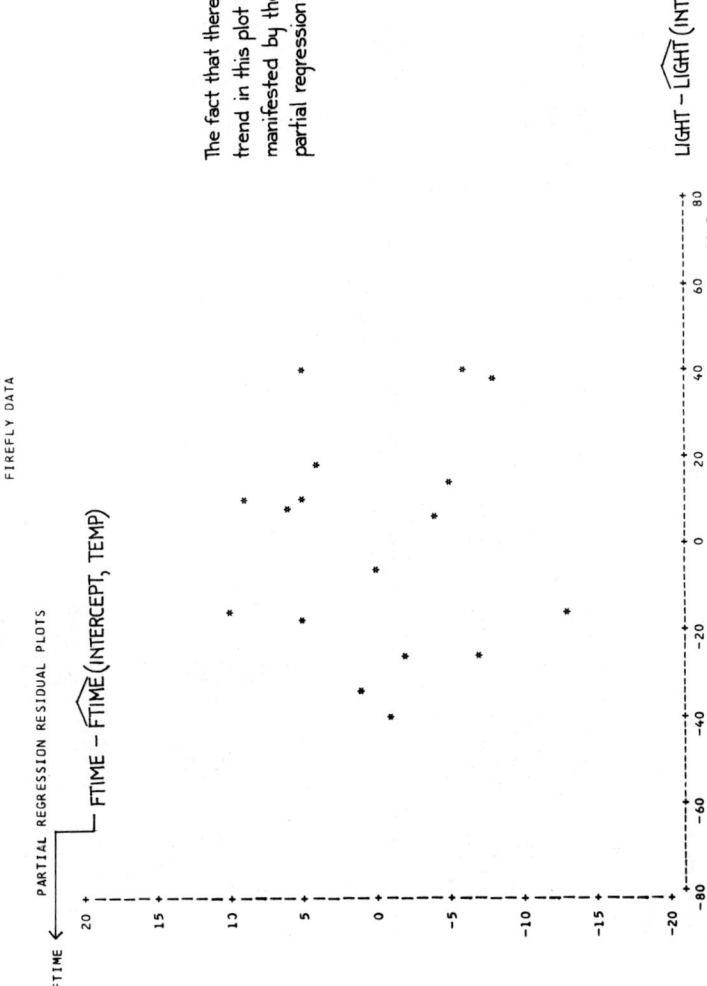

FIREFLY DATA

SEQUENTIAL PARAMETER ESTIMATES

INTERCEP   41.3529
TEMP   91.3816   -2.12144

DEP VARIABLE: FTIME

SOURCE	DF	SUM OF SQUARES	MEAN SQUARE	F VALUE	PROB>F
MODEL	1	456.734	456.734	10.029	0.0017
ERROR	15	683.148	45.543196		
C TOTAL	16	1139.882			

		R-SQUARE	
ROOT MSE	6.748570	R-SQUARE	0.4007
DEP MEAN	41.352941	ADJ R-SQ	0.3607
C.V.	16.31944		

VARIABLE	DF	PARAMETER ESTIMATE	STANDARD ERROR	T FOR H0: PARAMETER=0	PROB > |T|	TYPE I SS	TYPE II SS
INTERCEP	1	91.381582	15.882436	5.754	0.0001	29071.118	1507.671
TEMP	1	-2.121444	0.669902	-3.167	0.0064	456.734	456.734

OBS	ACTUAL	PREDICT VALUE	RESIDUAL
1	45.000	46.619	-1.619
2	40.000	40.679	-.679071
3	58.000	53.620	4.380
4	50.000	44.710	5.290
5	31.000	44.073	-13.073
6	52.000	41.952	10.048
7	54.000	47.892	6.108
8	38.000	37.285	0.715240
9	40.000	45.346	-5.346
10	28.000	34.739	-6.739
11	38.000	38.345	-.345482
12	36.000	39.618	-3.618
13	36.000	44.073	-8.073
14	46.000	37.285	8.715
15	40.000	34.739	5.261
16	31.000	37.285	-6.285
17	40.000	34.739	5.261

SUM OF RESIDUALS   3.65930E-13
SUM OF SQUARED RESIDUALS   683.1479

The remainder of the output is to give more detail on the submodel with INTERCEPT and TEMP. It was requested on input lines 31-36.

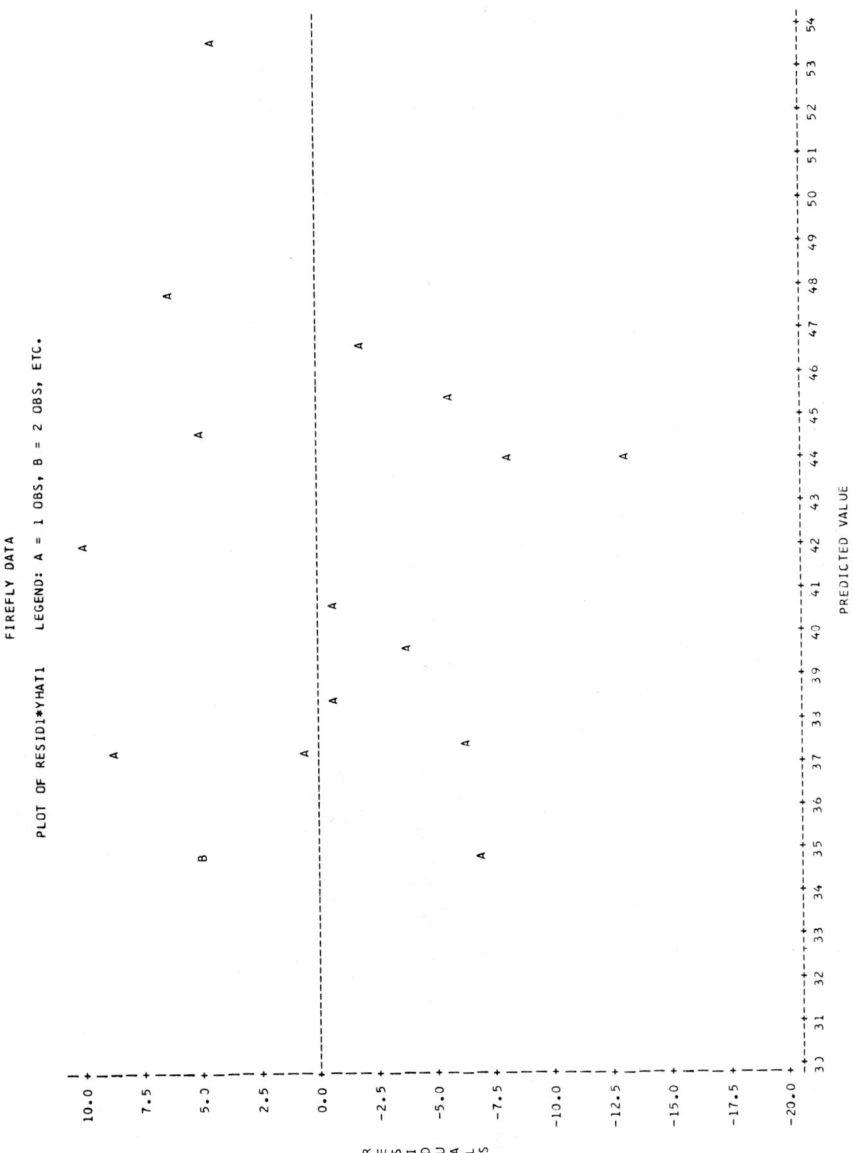

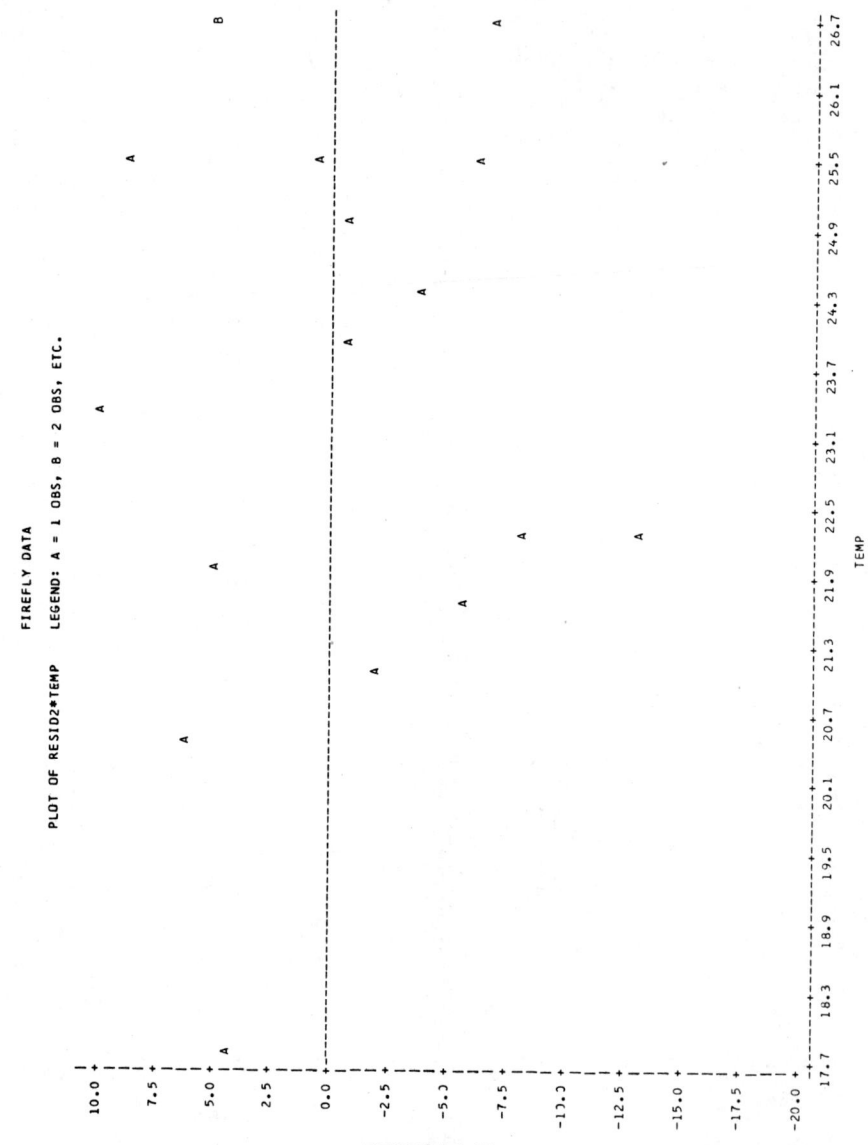

Annotated Computer Output

```
BMDP1V - ONE-WAY ANALYSIS OF VARIANCE AND COVARIANCE
HEALTH SCIENCES COMPUTING FACILITY
UNIVERSITY OF CALIFORNIA, LOS ANGELES 90024
PROGRAM REVISED NOVEMBER 1979
MANUAL REVISED -- 1979
COPYRIGHT (C) 1979 REGENTS OF UNIVERSITY OF CALIFORNIA



    PROGRAM CONTROL INFORMATION

/PROBLEM   TITLE IS 'ELECTRICAL LOAD DATA'.
/INPUT     VARIABLES ARE 3.
           FORMAT IS '(F1.0,F3.0,F5.0)'.
/VARIABLE  NAMES ARE CLASS, TEMP, ELOAD.
           GROUP IS CLASS.
/DESIGN    DEPENDENT IS ELOAD.
           INDEPENDENT IS TEMP.
/END
```
} Input to the program.

```
    PROBLEM TITLE . . . . . .
 ELECTRICAL LOAD DATA

    NUMBER OF VARIABLES TO READ IN. . . . . . . .       3
    NUMBER OF VARIABLES ADDED BY TRANSFORMATIONS. .     0
    TOTAL NUMBER OF VARIABLES . . . . . . . . . .       3
    NUMBER OF CASES TO READ IN. . . . . . . . . . 1000000
    CASE LABELING VARIABLES . . . . . . . . . .
    LIMITS AND MISSING VALUE CHECKED BEFORE TRANSFORMATIONS
    BLANKS ARE. . . . . . . . . . . . . . . . .    ZEROS
    INPUT UNIT NUMBER . . . . . . . . . . . . . .       5
    REWIND INPUT UNIT PRIOR TO READING. . DATA. . .    NO
    NUMBER OF WORDS OF DYNAMIC STORAGE. . . . . .   54784
    INPUT FORMAT. . . . .
 (F1.0,F3.0,F5.0)


    VARIABLES TO BE USED
       1 CLASS         2 TEMP         3 ELOAD

    GROUPING VARIABLE IS. . . . . . . . . . . . . .CLASS

    NUMBER OF CASES READ. . . . . . . . . . . .        31
    NUMBER OF GROUPS FOUND. . . . . . . . . . .         3

                      BEFORE TRANSFORMATION                                 INTERVAL RANGE
 VARIABLE      MINIMUM     MAXIMUM    MISSING    CATEGORY  CATEGORY     GREATER     LESS THAN
 NO. NAME      LIMIT       LIMIT      CODE       CODE      NAME         THAN        OR EQUAL TO

  1   CLASS                                      1.00000      1
                                                 2.00000      2
                                                 3.00000      3


 NUMBER OF CASES PER GROUP
 -----------------------
          1      21.
          2       5.
          3       5.
 TOTAL           31.
```

```
*********************************************************************************
*********************************************************************************

ESTIMATES OF MEANS
-----------------

                        1           2           3         TOTAL
                        1           2           3           4
         TEMP    2    75.6666     76.2000     77.6000     76.0644
         ELOAD   3  1117.9333    971.9998    859.3999   1052.6760
```
Level names were not supplied, so the groups are numbered 1, 2, and 3.

```
DEPENDENT VARIABLE IS  ELOAD
*********************************************************************************

         COVARIATE     REG.COEFF.     STD.ERR.       T-VALUE
           TEMP         12.75555       1.50924       8.45163


         GROUP    N        GRP.MEAN     ADJ.GRP.MEAN     STD.ERR.
           1     21.     1117.90332      1122.97778       7.21553
           2      5.      971.99976       970.27026      14.73757
           3      5.      859.39990       839.81274      14.91728
                                              ↑
```
——— See Unit 19.

```
ANALYSIS OF VARIANCE

SOURCE OF VARIANCE                    D.F.    SUM OF SQ.      MEAN SQ.     F-VALUE      TAIL AREA PROBABILITY

EQUALITY OF ADJ. CELL MEANS             2    353750.6875    176875.3125    162.9030          0.0000
ZERO SLOPE                              1     77556.6250     77556.6250     71.4330          0.0000
ERROR                                  27     29315.8125      1085.7708

EQUALITY OF SLOPES                      2      2667.9336      1333.9668      1.2515          0.3034
ERROR                                  25     26647.8789      1065.9150
                                                  ↑
```
——— These are all partial sums of squares.

```
SLOPE WITHIN EACH GROUP.
-----------------------

                        1           2           3
                        1           2           3
         TEMP    2     13.3093     13.5714      4.1205


T-TEST MATRIX FOR ADJUSTED GROUP MEANS ON    27 DEGREES OF FREEDOM
-----------------------------------------------------------------

                          1           2           3
                          1           2           3
            1    1      0.0
            2    2     -9.3020      0.0
            3    3    -17.0023     -6.2280      0.0
```
All possible differences among adjusted group means.

```
PROBABILITIES FOR THE T-VALUES ABOVE
------------------------------------

                          1           2           3
                          1           2           3
            1    1      1.0000
            2    2      0.0000      1.0000
            3    3      0.0         0.0000      1.0000
```

Annotated Computer Output

```
1       TITLE FAT DIGESTIBILITY DATA;              ← Title to be put at the top of each page.
2       DATA FAT_DIG;                              ← This designates FAT as a nonnumeric variable.
3       INPUT PERIOD FAT $ LECITHIN Y;
4            IF FAT='T' AND LECITHIN=0 THEN TREAT=1;  ⎫
5       ELSE IF FAT='C' AND LECITHIN=0 THEN TREAT=2;  ⎬ These create a variable TREAT
6       ELSE IF FAT='T' AND LECITHIN=1 THEN TREAT=3;  ⎭ from FAT and LECITHIN.
7       ELSE                           TREAT=4;
8       LIST;
9       CARDS;
RULE:   1234567 101234567 201234567 301234567 401234567 501234567 601234567 701234567 80
10      1 T 0 64.6   ⎫
11      2 T 0 52.4   ⎪
12      3 T 0 53.8   ⎪
13      1 C 0 66.0   ⎪
14      2 C 0 60.1   ⎪
15      3 C 0 64.4   ⎬ List of input cards.
16      1 T 1 85.0   ⎪
17      2 T 1 68.9   ⎪
18      3 T 1 77.5   ⎪
19      1 C 1 96.0   ⎪
20      2 C 1 90.4   ⎪
21      3 C 1 98.2   ⎭
NOTE: DATA SET WORK.FAT_DIG HAS 12 OBSERVATIONS AND 5 VARIABLES. 296 OBS/TRK.
NOTE: THE DATA STATEMENT USED 0.37 SECONDS AND 128K.
22      PROC PRINT;
NOTE: THE PROCEDURE PRINT USED 0.25 SECONDS AND 130K AND PRINTED PAGE 1.
23      PROC GLM;                                  ← General linear model procedure.
24      CLASSES PERIOD TREAT;                      ← This causes PERIOD and TREAT to be expanded
25      MODEL Y=PERIOD TREAT;                         into indicator variables.
26      OUTPUT OUT=NEW RESIDUAL=RESID PREDICTED=YHAT;
27      ESTIMATE 'W VS WO LECITHIN'                ← A contrast to compare with LECITHIN to without LECITHIN.
28           TREAT .5 .5 -.5 -.5;
29      ESTIMATE 'FAT DIFF WO LECITHIN'            ← A contrast to compare fats when there is no LECITHIN.
30           TREAT 1 -1 0 0;
31      ESTIMATE 'FAT DIFF W LECITHIN'             ← A contrast to compare fats when there is LECITHIN.
32           TREAT 0 0 1 -1;
NOTE: DATA SET WORK.NEW HAS 12 OBSERVATIONS AND 7 VARIABLES. 217 OBS/TRK.
NOTE: THE PROCEDURE GLM USED 0.60 SECONDS AND 176K AND PRINTED PAGES 2 TO 3.
33      PROC PLOT;
34          PLOT RESID*YHAT/VREF=0;                ← This causes a reference line to be drawn at zero.
NOTE: THE PROCEDURE PLOT USED 0.36 SECONDS AND 130K AND PRINTED PAGE 4.
```

FAT DIGESTIBILITY DATA

OBS	PERIOD	FAT	LECITHIN	Y	TREAT
1	1	T	0	64.6	1
2	2	T	0	52.4	1
3	3	T	0	53.8	1
4	1	C	0	66.0	2
5	2	C	0	60.1	2
6	3	C	0	64.4	2
7	1	T	1	85.0	3
8	2	T	1	68.9	3
9	3	T	1	77.5	3
10	1	C	1	96.0	4
11	2	C	1	90.4	4
12	3	C	1	98.2	4

When no variables are specified with PROC PRINT, all variables are printed.

FAT DIGESTIBILITY DATA

GENERAL LINEAR MODELS PROCEDURE

CLASS LEVEL INFORMATION

CLASS	LEVELS	VALUES
PERIOD	3	1 2 3
TREAT	4	1 2 3 4

NUMBER OF OBSERVATIONS IN DATA SET = 12

FAT DIGESTIBILITY DATA
GENERAL LINEAR MODELS PROCEDURE

DEPENDENT VARIABLE: Y

SOURCE	DF	SUM OF SQUARES	MEAN SQUARE	F VALUE	PR > F	R-SQUARE	C.V.
MODEL	5	2729.54083333	545.90816667	46.06	0.0001	0.974610	4.7089
ERROR	6	71.10833333	11.85138889			STD DEV	Y MEAN
CORRECTED TOTAL	11	2800.64916667				3.44258462	73.10833333

SOURCE	DF	TYPE I SS	F VALUE	PR > F		DF	TYPE IV SS	F VALUE	PR > F
PERIOD	2	198.81166667	8.39	0.0183		2	198.81166667	8.39	0.0183
TREAT	3	2530.72916667	71.18	0.0001		3	2530.72916667	71.18	0.0001

Sequential sums of squares ⬏   ⬐ Partial sums of squares

PARAMETER	ESTIMATE	T FOR H0: PARAMETER=0	PR > |T|	STD ERROR OF ESTIMATE
W VS WO LECITHIN	-25.78333333	-12.97	0.0001	1.98757716
FAT DIFF WO LECITHIN	-6.56666667	-2.34	0.0581	2.81085857
FAT DIFF W LECITHIN	-17.73333333	-6.31	0.0007	2.81085857

} Requested information on contrasts.

Since the data is balanced, the sequential sums of squares and partial sums of squares are equal.

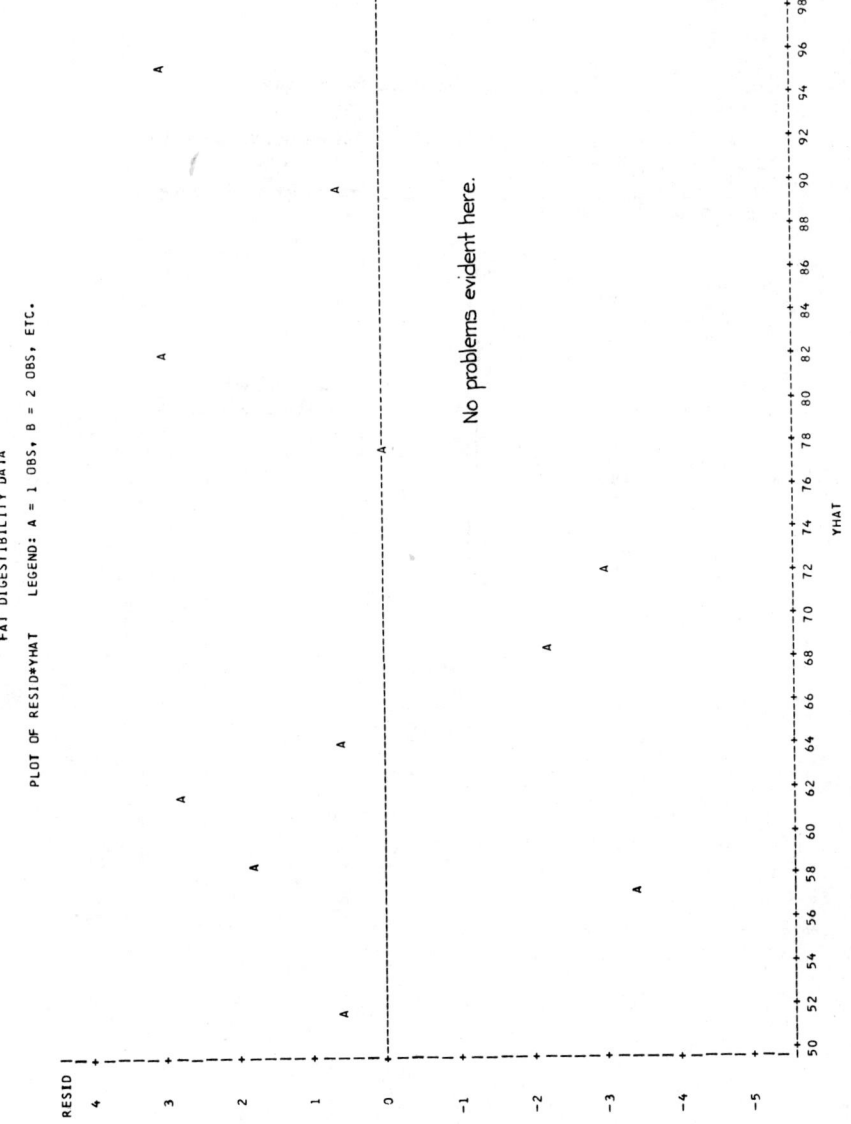

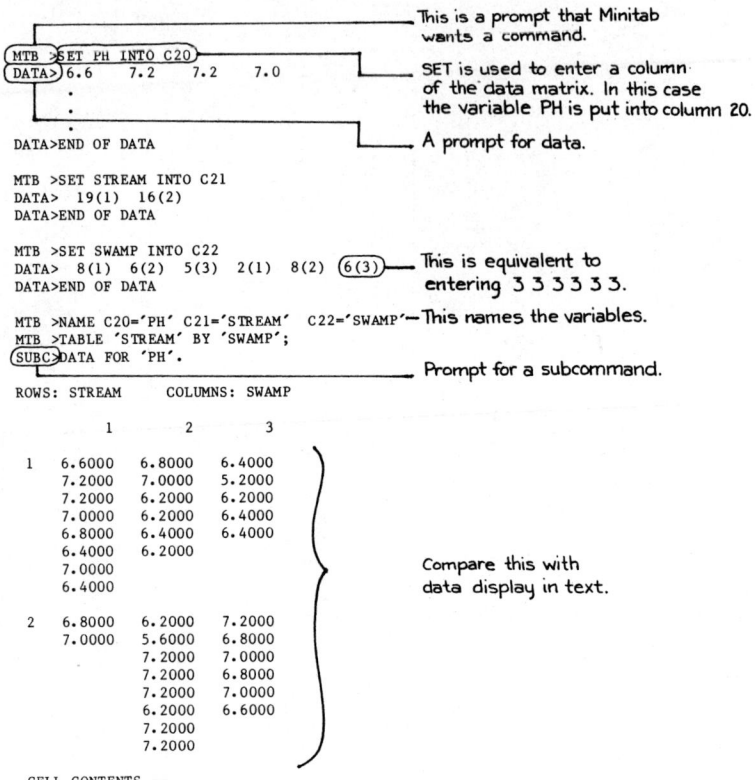

```
MTB >SET PH INTO C20
DATA> 6.6    7.2    7.2    7.0
      .
      .
      .
DATA>END OF DATA

MTB >SET STREAM INTO C21
DATA>  19(1)   16(2)
DATA>END OF DATA

MTB >SET SWAMP INTO C22
DATA>  8(1)   6(2)   5(3)   2(1)   8(2)   6(3)
DATA>END OF DATA

MTB >NAME C20='PH' C21='STREAM'  C22='SWAMP'
MTB >TABLE 'STREAM' BY 'SWAMP';
SUBC>DATA FOR 'PH'.

  ROWS: STREAM      COLUMNS: SWAMP

              1        2        3

    1    6.6000   6.8000   6.4000
         7.2000   7.0000   5.2000
         7.2000   6.2000   6.2000
         7.0000   6.2000   6.4000
         6.8000   6.4000   6.4000
         6.4000   6.2000
         7.0000
         6.4000

    2    6.8000   6.2000   7.2000
         7.0000   5.6000   6.8000
                  7.2000   7.0000
                  7.2000   6.8000
                  7.2000   7.0000
                  6.2000   6.6000
                  7.2000
                  7.2000

  CELL CONTENTS --
              PH:DATA
```

Annotated Computer Output

```
MTB >TABLE 'STREAM' BY 'SWAMP';   } The TABLE command used
SUBC>STATISTICS FOR 'PH'.           with a different subcommand.

ROWS: STREAM      COLUMNS: SWAMP

            1         2         3       ALL

    1       8         6         5        19
         6.8250    6.4667    6.1200    6.5263
         0.3284    0.3503    0.5216    0.4724

    2       2         8         6        16
         6.9000    6.7500    6.9000    6.8250
         0.1414    0.6481    0.2098    0.4669

  ALL      10        14        11        35
         6.8400    6.6286    6.5455    6.6628
         0.2952    0.5427    0.5448    0.4869

     CELL CONTENTS --
                    COUNT
               PH:MEAN
                    STD DEV


MTB >LET C25 = 10*'SWAMP' + 2*'STREAM'      A new variable is defined
MTB >LPLOT 'PH' VS C25, USING CODES FOR 'STREAM'   that has a distinct value
                                                    for each combination of
       PH                                           SWAMP and STREAM.
    7.50+  NORTH          MESIC        SHRUB
       -
       -    2              5             B       A = near stream
    7.00+  2  B            A             2       B = away stream
       -   A  B            A             2       Numbers = number of data
       -   A                             B                 points present.
    6.50+
       -    2              A             3
       -                   3  2          A
    6.00+
       -
       -
       -                   B
    5.50+
       -
       -                                 A
       -
    5.00+
         +---------+---------+---------+---------+---------+C25
        12.0      18.0      24.0      30.0      36.0      42.0
```

```
MTB >INDICATOR VARIABLES FOR 'STREAM', PUT INTO C1-C2
MTB >INDICATOR VARIABLES FOR 'SWAMP', PUT INTO C3-C5
MTB >LET C6 = C1*C3  ⎫   ← The first two indicator variables for the
MTB >LET C7 = C1*C4  ⎭      general means model. The remainder would
MTB >NAME C30='RESIDS' C31='PREDS'                          be redundant.
MTB >REGRESS 'PH' ON 5 PRED. C1 C3-C4 C6-C7 , STORE 'RESIDS' 'PREDS'

THE REGRESSION EQUATION IS
Y =    6.90 - 0.780 X1 -0.0000 X2
     - 0.150 X3 + 0.705 X4 + 0.497 X5

                                  ST. DEV.    T-RATIO =
         COLUMN     COEFFICIENT   OF COEF.    COEF/S.D.
          --          6.9000       0.1800       38.34
   X1     C1         -0.7800       0.2670       -2.92
   X2     C3         -0.0000       0.3600       -0.00
   X3     C4         -0.1500       0.2381       -0.63
   X4     C6          0.7050       0.4390        1.61
   X5     C7          0.4967       0.3577        1.39

THE ST. DEV. OF Y ABOUT REGRESSION LINE IS
S = 0.4409
WITH ( 35- 6) = 29 DEGREES OF FREEDOM

R-SQUARED = 30.1 PERCENT
R-SQUARED = 18.0 PERCENT, ADJUSTED FOR D.F.

ANALYSIS OF VARIANCE

  DUE TO      DF        SS       MS=SS/DF
REGRESSION     5      2.4254      0.4851
RESIDUAL      29      5.6363      0.1944
TOTAL         34      8.0617

FURTHER ANALYSIS OF VARIANCE
SS EXPLAINED BY EACH VARIABLE WHEN ENTERED IN THE ORDER GIVEN

  DUE TO      DF        SS                Sequential sum of squares.
REGRESSION     5      2.4254
   C1          1      0.7750  ← R(L|M)
   C3          1      1.0105 ⎫
   C4          1      0.0322 ⎭  1.0427 R(C|M,L)
   C6          1      0.2330 ⎫
   C7          1      0.3747 ⎭  .6077 R(I|M,L,C)

          X1       Y      PRED. Y    ST.DEV.
  ROW     C1      PH       VALUE     PRED. Y   RESIDUAL   ST.RES.
   16    1.00   5.2000    6.1200     0.1972    -0.9200    -2.33R
   20    0.00   6.8000    6.9000     0.3117    -0.1000    -0.32 X
   21    0.00   7.0000    6.9000     0.3117     0.1000     0.32 X
   23    0.00   5.6000    6.7500     0.1559    -1.1500    -2.79R

R DENOTES AN OBS. WITH A LARGE ST. RES.
X DENOTES AN OBS. WHOSE X VALUE GIVES IT LARGE INFLUENCE.
```

Annotated Computer Output

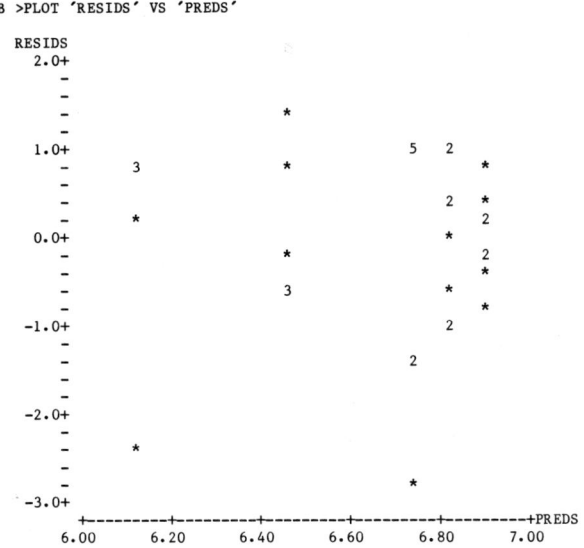

Instead of plotting ordinary residuals, Minitab plots standardized residuals defined by $\dfrac{y_i - \hat{y}_i}{\sqrt{1-v_c}\, s}$, where c is the ith row of X and $s = \sqrt{\text{residual mean square}}$.

```
MTB >REGRESS 'PH' ON 5 PRED. C3-C4 C1 C6-C7     Reorder the variables.

THE REGRESSION EQUATION IS
Y =    6.90 -0.0000 X1 - 0.150 X2
     - 0.780 X3 + 0.705 X4 + 0.497 X5

                                ST. DEV.      T-RATIO =
        COLUMN    COEFFICIENT   OF COEF.      COEF/S.D.
        --        6.9000        0.1800        38.34
   X1   C3       -0.0000        0.3600        -0.00
   X2   C4       -0.1500        0.2381        -0.63
   X3   C1       -0.7800        0.2670        -2.92
   X4   C6        0.7050        0.4390         1.61
   X5   C7        0.4967        0.3577         1.39

THE ST. DEV. OF Y ABOUT REGRESSION LINE IS
S = 0.4409
WITH ( 35- 6) = 29 DEGREES OF FREEDOM

R-SQUARED = 30.1 PERCENT
R-SQUARED = 18.0 PERCENT, ADJUSTED FOR D.F.


ANALYSIS OF VARIANCE

   DUE TO      DF        SS       MS=SS/DF
   REGRESSION   5     2.4254      0.4851
   RESIDUAL    29     5.6363      0.1944
   TOTAL       34     8.0617

FURTHER ANALYSIS OF VARIANCE
SS EXPLAINED BY EACH VARIABLE WHEN ENTERED IN THE ORDER GIVEN

   DUE TO      DF        SS
   REGRESSION   5     2.4254
   C3           1     0.4392  ⎱ .4817 = R(C|M)
   C4           1     0.0425  ⎰
   C1           1     1.3359       R(L|C,M)
   C6           1     0.2330  ⎱ .6077 = R(I|M,L,C)
   C7           1     0.3747  ⎰

          X1          Y      PRED. Y    ST.DEV.
  ROW     C3         PH       VALUE     PRED. Y    RESIDUAL   ST.RES.
   16    0.00      5.2000    6.1200     0.1972     -0.9200    -2.33R
   20    1.00      6.8000    6.9000     0.3117     -0.1000    -0.32 X
   21    1.00      7.0000    6.9000     0.3117      0.1000     0.32 X
   23    0.00      5.6000    6.7500     0.1559     -1.1500    -2.79R

R DENOTES AN OBS. WITH A LARGE ST. RES.
X DENOTES AN OBS. WHOSE X VALUE GIVES IT LARGE INFLUENCE.
```

MTB >NAMES C1='YIELD' C2='HEIGHT' C3='GROUP' ←——— Names the variables.

```
MTB >READ C1-C3
DATA>    12.2   48   1
DATA>    12.4   52   1
DATA>    11.9   42   1
DATA>    11.3   35   1
           .     .   .
           .     .   .
           .     .   .
DATA>END OF DATA
```

MTB >ONEWAY ANALYSIS OF VARIANCE ON 'YIELD' BY 'GROUP'

ANALYSIS OF VARIANCE

DUE TO	DF	SS	MS=SS/DF	F-RATIO	
FACTOR	2	308.187	154.093	535.52	Analysis of variance
ERROR	42	12.085	0.288		ignoring covariable.
TOTAL	44	320.272			

LEVEL	N	MEAN	ST. DEV.
1	15	12.260	0.533
2	15	15.860	0.505
3	15	9.467	0.569

POOLED ST. DEV. =   0.536

INDIVIDUAL 95 PERCENT C. I. FOR LEVEL MEANS
(BASED ON POOLED STANDARD DEVIATION)

```
            +---------+---------+---------+---------+
1                              I*I*I
2                                              I*I*I
3       I*I*I
            +---------+---------+---------+---------+
           9.0      10.5      12.0      13.5      15.0      16.5
```

```
MTB >INDICATORS FOR 'GROUP', PUT INTO C11-C13  ←── Creates indicator variables.
MTB >PRINT C1-C3 C11-C13
```

COLUMN	YIELD	HEIGHT	GROUP	C11	C12	C13
COUNT	45	45	45	45	45	45
ROW						
1	12.20	48.	1.	1.	0.	0.
2	12.40	52.	1.	1.	0.	0.
3	11.90	42.	1.	1.	0.	0.
4	11.30	35.	1.	1.	0.	0.
5	11.80	40.	1.	1.	0.	0.
6	12.10	48.	1.	1.	0.	0.
7	13.10	60.	1.	1.	0.	0.
8	12.70	61.	1.	1.	0.	0.
9	12.40	50.	1.	1.	0.	0.
10	11.40	33.	1.	1.	0.	0.
11	12.30	48.	1.	1.	0.	0.
12	12.20	51.	1.	1.	0.	0.
13	12.60	56.	1.	1.	0.	0.
14	13.20	65.	1.	1.	0.	0.
15	12.30	51.	1.	1.	0.	0.
16	16.60	63.	2.	0.	1.	0.
17	15.80	50.	2.	0.	1.	0.
18	16.50	63.	2.	0.	1.	0.
19	15.00	33.	2.	0.	1.	0.
20	15.40	38.	2.	0.	1.	0.
21	15.60	45.	2.	0.	1.	0.
22	15.80	50.	2.	0.	1.	0.
23	15.80	48.	2.	0.	1.	0.
24	16.00	50.	2.	0.	1.	0.
25	15.80	49.	2.	0.	1.	0.
26	15.00	35.	2.	0.	1.	0.
27	16.20	50.	2.	0.	1.	0.
28	16.70	62.	2.	0.	1.	0.
29	15.80	49.	2.	0.	1.	0.
30	15.90	52.	2.	0.	1.	0.
31	9.50	52.	3.	0.	0.	1.
32	9.50	54.	3.	0.	0.	1.
33	9.60	58.	3.	0.	0.	1.
34	8.80	45.	3.	0.	0.	1.
35	9.50	57.	3.	0.	0.	1.
36	9.80	62.	3.	0.	0.	1.
37	9.10	52.	3.	0.	0.	1.
38	10.30	67.	3.	0.	0.	1.
39	9.50	55.	3.	0.	0.	1.
40	8.50	40.	3.	0.	0.	1.
41	8.60	41.	3.	0.	0.	1.
42	10.40	67.	3.	0.	0.	1.
43	9.40	55.	3.	0.	0.	1.
44	10.20	66.	3.	0.	0.	1.
45	9.30	56.	3.	0.	0.	1.

Annotated Computer Output

```
MTB >LPLOT 'YIELD' VS 'HEIGHT', USE LETTERS FOR 'GROUP'
```

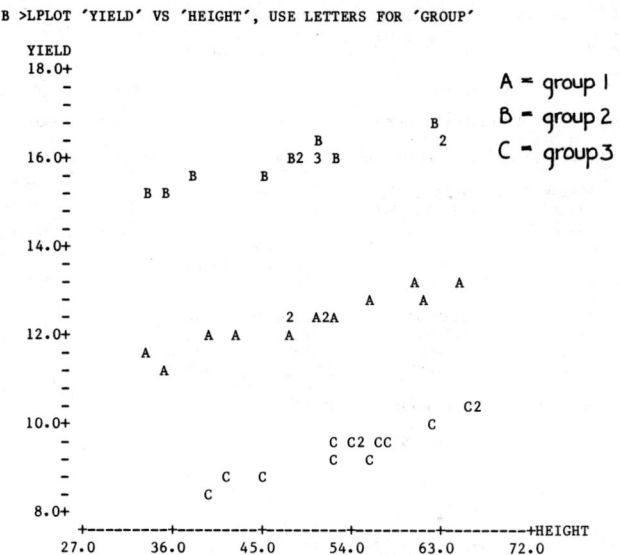

```
MTB >NAME C4='HT-DEV' C5='RESID' C6='YHAT'

MTB >NAME C11='CONTROL' C12 = 'LIGHT' C13 = 'SHADE'

MTB >LET 'HT-DEV' = 'HEIGHT' - AVERAGE('HEIGHT')      ← Creates x-x̄.

MTB >NOCONSTANT IN FOLLOWING MODEL           ← Leave x₀ out.

MTB >REGRESS 'YIELD' ON 4 PRED. 'HT-DEV' C11-C13, STORE 'RESID' 'YHAT'
```
                                                           ↑   ↑— Second item =
                                                                       predicted values.
                                                                    — First item =
                                                                       standardized
                                                                       residuals.

THE REGRESSION EQUATION IS
Y = +0.0584 X1 + 12.4 X2 + 16.0 X3
    + 9.24 X4

```
                                      ST. DEV.    T-RATIO =
              COLUMN    COEFFICIENT    OF COEF.   COEF/S.D.
NOCONSTANT
X1      HT-DEV          0.058366       0.002232     26.16
X2      CONTROL        12.3689         0.0336      368.21
X3      LIGHT          15.9804         0.0337      474.89
X4      SHADE           9.23708        0.03447    267.98
```

THE ST. DEV. OF Y ABOUT REGRESSION LINE IS
S = 0.1291
WITH (45- 4) = 41 DEGREES OF FREEDOM

ANALYSIS OF VARIANCE

```
DUE TO       DF         SS          MS=SS/DF
REGRESSION    4     7383.256        1845.813
RESIDUAL     41        0.683           0.017
TOTAL        45     7383.930
```

FURTHER ANALYSIS OF VARIANCE
SS EXPLAINED BY EACH VARIABLE WHEN ENTERED IN THE ORDER GIVEN

```
DUE TO       DF         SS
REGRESSION    4     7383.256
HT-DEV        1        1.778
CONTROL       1     2271.693
LIGHT         1     3912.925
SHADE         1     1196.862
```

```
           X1          Y      PRED. Y     ST.DEV.
ROW     HT-DEV      YIELD     VALUE       PRED. Y    RESIDUAL    ST.RES.
 27       -1.2     16.2000    15.9104      0.0334     0.2896      2.32R
```

R DENOTES AN OBS. WITH A LARGE ST. RES.

Annotated Computer Output

```
MTB >LPLOT 'RESID' VS 'YHAT', CODES FOR 'GROUP'
```

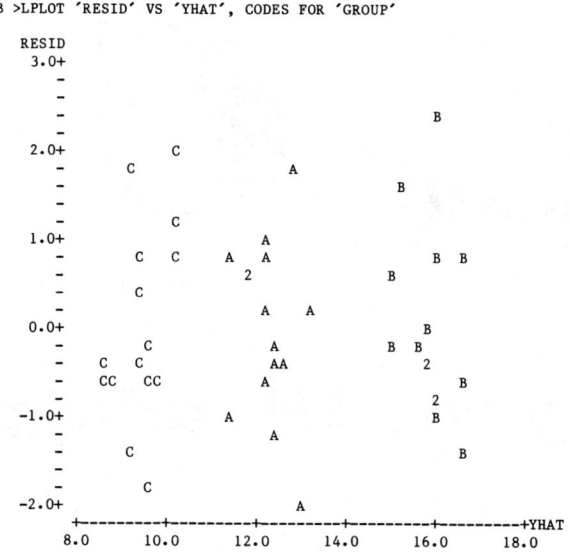

These are standardized residuals.

Appendix C

Suggested Reading

The references given here include articles from the *American Statistician* and the *Journal of Quality Technology* which are particularly helpful for background reading. Several books on regression that emphasize applications are listed. In addition, a few other books and articles are mentioned, with notes indicating a specialized connection with this book. The book by Draper and Smith contains a very large number of references.

Allen, David M. 1977. "Computational Methods for Estimation of Linear Combinations of Parameters." In *Statistical Computing Section Proceedings of the American Statistical Association*. (This article is relevant to Unit 12.)

Anderson, R. L. and T. A. Bancroft 1952. *Statistical Theory in Research*, New York: McGraw-Hill. (This book contains a discussion of the abbreviated Doolittle algorithm.)

Armitage, P. 1973. *Statistical Methods in Medical Research*. 2nd ed. New York: Wiley. (The material on split-unit designs is relevant to Unit 22.)

Brogan, Donna R. and Michael H. Kutner 1980. "Comparative Analysis of Pretest-Posttest Research Designs." *American Statistician* 34 (no. 4): 229–232.

Carmer, Elliot M. 1972. "Significance Tests and Tests of Models in Regression." *American Statistician* 26 (no. 4): 26–30.

Chatterjee, S., and Bertram Price. 1977. *Regression Analysis by Example*. New York: Wiley.

Daniel, Cuthbert, and Fred S. Wood, with John W. Gorman. 1980. *Fitting Equations to Data: Computer Analysis of Multifactor Data*. 2nd ed. New York: Wiley.

Darlington, Richard B. 1969. "Deriving Least-squares Weights Without Calculus." *American Statistician* 23 (no. 5): 41–42.

Draper, Norman, and Harry Smith. 1981. *Applied Regression Analysis*. 2nd ed. New York: Wiley.

Dutka, Alan F. and Fred J. Ewens. 1971. "A Model for Improving the Accuracy of Polynomial Regression Analysis." *Journal of Quality Technology* 3 (no. 4): 149–155.

Freund, Richard A. 1974. "Contrast Analysis of Experiments." *Journal of Quality Technology* 6 (no. 1): 2–21.

Gallant, A. R. 1975. "Nonlinear Regression." *American Statistician* 29 (no. 2): 73–81.

Goodnight, James H. 1979. "A Tutorial on the Sweep Operator." *American Statistician* 33 (no. 3): 149–158.

Gujarati, Damodar. 1970. "Use of Dummy Variables in Testing for Equality between Sets of Coefficients in Linear Regression: A Note." *American Statistician* 24 (no. 1): 50–52.

———1970. "Use of Dummy Variables in Testing for Equality between Sets of Coefficients in Linear Regression: A Generalization." *American Statistician* 24 (no. 5): 18–22.

Gunst, Richard F., and Robert L. Mason. 1980. *Regression Analysis and Its Application*. New York: Dekker.

Hahn, Gerald J. 1977. "Fitting Regression with No Intercept Term." *Journal of Quality Technology* 9 (no. 2): 56–61.

———1977. "The Hazards of Extrapolation in Regression Analysis." *Journal of Quality Technology* 9 (no. 4): 159–165.

Hinchen, John D. 1970. "Multiple Regression with Unbalanced Data." *Journal of Quality Technology* 2 (no. 1): 22–29.

Jaech, J. L. 1966. "Understanding Multiple Regression." *Industrial Quality Control* 23 (no. 6): 260–264. (This article is relevant to Unit 11.)

Kempthorne, Oscar. 1952. *The Design and Analysis of Experiments*. New York: Wiley. (The discussion of the scientific method is relevant to Unit 1.)

Kleinbaum, David G., and Lawrence L. Kupper. 1978. *Applied Regression Analysis and Other Multivariate Methods*. Belmont, Calif.: Duxbury Press.

Li, C. C. 1964. "Two Additional Views of Linear Regression Coefficients." *American Statistician* 18 (no. 4): 27–28.

Mandel, John. 1976. "Models, Transformations of Scale and Weighting." *Journal of Quality Technology* 8 (no. 2): 86–97.

Margolin, Marvin S. 1979. "Perpendicular Projections and Elementary Statistics." *American Statistician* 33 (no. 3): 131–135.

Marquardt, Donald W., and Ronald D. Snee. 1975. "Ridge Regression in Practice." *American Statistician* 29 (no. 1): 3–20.

Mendenhall, William. 1968. *Introduction to Linear Models and the Design and Analysis of Experiments*. Belmont, CA: Wadsworth Publishing Company, Inc. (The material on dummy (indicator) variables is relevant to Units 13–19.)

Mosteller, Frederick, and John W. Tukey. 1977. *Data Analysis and Regression*. Reading, Mass.: Addison-Wesley.

Mullet, Gary M. 1976. "Why Regression Coefficients Have the Wrong Sign." *Journal of Quality Technology* 8 (no. 3): 121–126.

Nelson, Lloyd S. 1976. "Fitting a Least-Squares Straight Line." *Journal of Quality Technology* 8 (no. 2): 115–116.

Schilling, Edward G. 1973. "A Systematic Approach to the Analysis of Means. Part I: Analysis of Treatment Effects." *Journal of Quality Technology* 5 (no. 3): 93–108.

——— 1973. "A Systematic Approach to the Analysis of Means. Part II: Analysis of Contrasts." *Journal of Quality Technology* 5 (no. 4): 147–155.

———1974. "The Relationship of Analysis of Variance to Regression. Part I: Balanced Designs." *Journal of Quality Technology* 6 (no. 2): 74–83.

———1974. "The Relationship of Analysis of Variance to Regression. Part II: Unbalanced Designs." *Journal of Quality Technology* 6 (no. 3): 146–153.

Searle, S. R. 1971. *Linear Models*, New York, NY: John Wiley & Sons. (Discussion of unbalanced 2-way crossed classification is relevant to Unit 18.)

Smith, John H. 1972. "Families of Transformations for Use in Regression Analysis." *American Statistician* 26 (no. 3): 59–61.

Snedecor, George W. and William G. Cochran 1980. *Statistical Methods*, 7th ed. Ames, Iowa: Iowa State University Press. (General reference)

Snee, Ronald D. 1973. "Some Aspects of Nonorthogonal Data Analysis. Part I: Development of Prediction Equations." *Journal of Quality Technology* 5 (no. 2): 67–79.

———1973. "Some Aspects of Nonorthogonal Data Analysis. Part II: Comparison of Means." *Journal of Quality Technology* 5 (no. 3): 109–122.

Swallow, William H., and J. Richard Trout. 1976. "On the Plot of Residuals Versus Observed Y's in Regression." *Journal of Quality Technology* 8 (no. 1):34–36.

Swindel, Benee F. 1974. "Instability of Regression Coefficients Illustrated." *American Statistician* 28 (no. 2): 63–65.

Walls, Robert C., and David L. Weeks. 1969. "A Note on the Variance of a Predicted Value in Regression." *American Statistician* 23 (no. 3): 24–26.

Weisberg, Sanford. 1980. *Applied Linear Regression*. New York: Wiley.

Wooding, W. M. 1969. "The Computation and Use of Residuals in the Analysis of Experimental Data." *Journal of Quality Technology* 1 (no. 3): 175–188.

Wooding, W. M. 1973. "The Split-Plot Design." *Journal of Quality Technology* 5 (no. 1): 16–33.

Appendix D

Answers to Selected Exercises

3.2 $\mathbf{X} = \begin{bmatrix} 1 & 1 \\ 1 & 2 \\ 1 & 2 \\ 1 & 3 \\ 1 & 3 \\ 1 & 4 \end{bmatrix}$ $\mathbf{Y} = \begin{bmatrix} 4.9 \\ 4.7 \\ 3.8 \\ 4.2 \\ 4.1 \\ 3.9 \end{bmatrix}$

3.3 $\mathbf{X} = \begin{bmatrix} 1 & 0 & 0 \\ 1 & 0 & 0 \\ 1 & 0 & 0 \\ 1 & 0 & 0 \\ 0 & 1 & 0 \\ 0 & 1 & 0 \\ 0 & 1 & 0 \\ 0 & 1 & 0 \\ 0 & 0 & 1 \\ 0 & 0 & 1 \\ 0 & 0 & 1 \\ 0 & 0 & 1 \end{bmatrix}$ $\mathbf{Y} = \begin{bmatrix} 18.13 \\ 19.14 \\ 17.37 \\ 18.54 \\ 13.45 \\ 18.36 \\ 14.29 \\ 17.84 \\ 12.21 \\ 13.63 \\ 14.36 \\ 12.98 \end{bmatrix}$

3.4 $\begin{bmatrix} 7 & 4 & 2 \\ 5 & 10 & 10 \end{bmatrix}$; not defined; not defined; $\begin{bmatrix} 10 & 10 & 19 \\ 0 & 10 & 33 \end{bmatrix}$

3.5 $\mathbf{b} = \begin{bmatrix} 2.5494 \\ -0.0050 \\ -0.0196 \end{bmatrix}$

5.1 (a) The mean and slope are 9.8417 and 0.3100.
(b) The intercept and slope are 4.8762 and 0.3100.

6.2 The respective $\hat{y}$'s are

44.68	45.08	52.67	49.70
44.68	29.23	20.39	22.82
44.68	56.96	46.10	43.39
44.68	49.04	50.48	61.31
44.68	64.88	64.52	69.46
44.68	41.12	48.02	45.92
44.68	33.19	22.76	18.54
44.68	29.23	34.07	39.15
44.68	41.12	50.30	44.84
44.68	56.96	57.50	51.67

7.2 The analysis of variance table is

Source	Sum of Squares	d.f.	Mean Square
X_0	212,045.	1	212,045.
X_1	26,521.6	1	26,521.6
X_2	2,211.87	1	2,211.87
Residual	13.0622	11	1.1875

The analysis of variance for the individual observations would be identical except that the residual term would also include the within-location variability.

8.2 The relationships are as follows:
 (a) $y = x^2$
 (b) $y = \ln(x+1)$
 (c) $y = \exp(-x/10)$
 (d) $y = \text{SQRT}(x)$

8.3 $c = 1.0$

10.1 $y(0,1) = 4.9857 - 1.3936 x_1$; $y(0,1,2) = 5.2041 + 0.1704 x_1 + 1.4899 x_2$.

10.2 $b_0 = 1.2696$; $b_1 = 2.8184$; $b_2 = 0.0189$. If the x's are unrelated, then the b's are interpreted as slopes. If x_2 is the square of x_1, then the relationship between y and x_1 may be curved, and there is no single slope.

11.2 ABDO matrix:

$$\begin{array}{ccc|c}
4 & 6 & 6 & y_1 + y_2 + y_3 + y_4 \\
1 & 1.5 & 1.5 & (y_1 + y_2 + y_3 + y_4)/4 \\
\hline
 & 1 & 0 & (-y_1 - y_2 + y_3 + y_4)/2 \\
 & 1 & 0 & (-y_1 - y_2 + y_3 + y_4)/2 \\
\hline
 & & 1 & (-y_1 + y_2 - y_3 + y_4)/2 \\
 & & 1 & (-y_1 + y_2 - y_3 + y_4)/2
\end{array}$$

ABDO after ORTHO:

$$\begin{array}{ccc|c}
4 & 0 & 0 & y_1 + y_2 + y_3 + y_4 \\
1 & 0 & 0 & (y_1 + y_2 + y_3 + y_4)/4 \\
\hline
 & 1 & 0 & (-y_1 - y_2 + y_3 + y_4)/2 \\
 & 1 & 0 & (-y_1 - y_2 + y_3 + y_4)/2 \\
\hline
 & & 1 & (-y_1 + y_2 - y_3 + y_4)/2 \\
 & & 1 & (-y_1 + y_2 - y_3 + y_4)/2
\end{array}$$

12.1 estimate: 2.3048; v_c: 0.5788; 95% confidence interval: -1.8223 to 6.4319

12.2 estimate: 4.3600; v_c: 0.1050; 95% confidence interval: 2.1297 to 6.5903

12.3 $$X = \begin{bmatrix} 1 & -1 & 0 \\ 1 & 0 & -1 \\ 0 & -1 & 1 \\ -1 & 1 & 0 \\ -1 & 0 & 1 \\ 0 & 1 & -1 \end{bmatrix} \quad Y = \begin{bmatrix} 13 \\ -3 \\ 8 \\ -5 \\ -10 \\ 1 \end{bmatrix} \quad \beta = \begin{bmatrix} KY \\ VANDY \\ TN \end{bmatrix}$$

Linear combination:

Title	Estimate	v_c	Standard Error
KY minus TN	4.1667	0.3333	3.6477

13.1 (a) Denote the respective columns of **X** by x_0, x_1, x_2, x_3. If $b_2 = 0$, then

$$\mathbf{X} = [x_0 \quad x_1 \quad x_3]$$

To test the hypothesis by using the analysis of variance technique, use

$$\mathbf{X} = [x_0 \quad x_1 \quad x_3 \quad x_2]$$

(b) The appropriate model if $b_1 = b_2 = b_3$ is

$$\mathbf{X} = [x_0 \quad x_1 + x_2 + x_3]$$

To test the hypothesis, use

$$\mathbf{X} = [x_0 \quad x_1 + x_2 + x_3 \quad x_1 \quad x_2 \quad x_3]$$

13.2 The analysis of variance table is

Source	Sum of Squares	d.f.	Mean Square
Group	5944.94	3	1981.65
X	64.2403	1	64.2403
Group * X	6.7527	2	3.3763
Residual	4.0910	9	0.4546

Slopes are significantly different at the 5% level.

15.1 The linear combinations are

Title	Estimate	v_c	Standard Error
New minus placebo	17.4333	0.3667	1.3710
New minus standard	-2.0238	0.3095	1.2596

The new drug is definitely more effective than the placebo and is not significantly different from the standard drug.

15.2 Use the general means model to estimate means and linear combinations of means. Use the equal-means model and the general means model together to test the hypothesis of equal means with an analysis of variance.

17.1 The analysis of variance is

Source	Sum of Squares	d.f.	Mean Square
Mean	9702.25	1	9702.25
Appear	128.445	1	128.445
Recommend	803.042	2	401.521
Appear * Recommend	51.0972	2	25.5486
Residual	194.667	30	6.4889

The above analysis of variance indicates an interaction between recommendation and appearance. A comparison of attractive versus unattractive for each value of recommendation is given below.

Title	Estimate	v_c	Standard Error
Poor	−0.4167	0.3333	1.4381
Fair	−5.2500	0.3333	1.4781
Excellent	−5.6667	0.3333	1.4954

We can now arrive at the following conclusions. If the recommendation is poor, then appearance has little influence on the rating. However, if the recommendation is better than poor, then an attractive appearance has a positive influence on the rating.

17.2 The analysis of variance is

Source	Sum of Squares	d.f.	Mean Square
Mean	425,321.	1	425,321.
A	2,954.31	3	984.770
B	574.390	2	287.195
A * B	122.278	6	20.3796
Residual	174.667	24	7.2778

There is a significant interaction of the two drugs. A product containing a combination of the drugs might be considered. The cell means are

	$B=1$	$B=2$	$B=3$
$A=10$	121.000	118.667	113.667
$A=20$	119.000	117.333	112.333
$A=40$	114.333	105.000	97.0000
$A=80$	99.0000	95.6667	91.3333

The combination of the larger doses is most effective for reducing blood pressure.

Answers to Selected Exercises

18.1 (a) The analysis of variance table is

Source	Sum of Squares	d.f.	Mean Square
Mean	417.234	1	417.234
Toy	23.9049	11	2.1732
Store	0.6902	9	0.0767
Residual	0.7508	47	0.0160

The stores definitely differ in their prices. Relevant linear combinations are

Title	Estimate	v_c	Standard Error
Mail order and discount	0.0322	0.1452	0.0482
Department and toy	−0.0537	0.2365	0.0615
Mail, discount and department, toy	−0.1990	0.0883	0.0376

(b) no
(c) no
(d) yes
(e) The average of mail-order and discount stores is lower than the average of department and toy stores. Other differences are not apparent. The analysis of variance is

Source	Sum of Squares	d.f.	Mean Square
Mean	417.234	1	417.234
Store	3.1275	9	0.3475
Toy	21.4677	11	1.9516
Residual	0.7508	47	0.0160

The difference between ANOVA tables with different orderings of variables is an indicator of the amount of inbalance. In this case the tables do not differ enough to complicate inference.

19.1 (a) The regression coefficients are

Variable	Estimate	v_c	Standard Error
Ration 1	11.9860	3.4659	2.5309
Ration 2	8.5127	1.0546	1.3961
Ration 3	10.8494	0.9163	1.3013
Ration 1 * morning	−0.3050	0.0555	0.3202
Ration 2 * morning	0.0494	0.0232	0.2069
Ration 3 * morning	−0.5138	0.0174	0.1793

(b) The analysis of variance is

Source	Sum of Squares	d.f.	Mean Square
Ration	2240.50	3	746.833
Morning	9.1196	1	9.1196
Ration * morning	7.8377	2	3.9189
Residual	44.3536	24	1.8481

Slopes are not significantly different at 0.10 level.
(c) The regression coefficients are

Variable	Estimate	v_c	Standard Error
Ration 1	11.7693	0.6113	1.1077
Ration 2	10.6096	0.4473	0.9475
Ration 3	9.2288	0.4953	0.9972
Morning (slope)	−0.2772	0.0084	0.1300

The adjusted treatment means are

Title	Estimate	v_c	Standard Error
Ration 1	9.8234	0.1050	0.4591
Ration 2	8.6637	0.1030	0.4548
Ration 3	7.2829	0.1002	0.4486

20.1 The main caution here is to recognize that mixer, paste method, and tablet press are qualitative variables. However, since mixer has only two levels, it may be regarded as quantitative. In the analysis reported here, paste method is expanded into indicator variables P_1, P_2, and P_3 and tablet press is expanded into P_4, P_5, and P_6. The mediums are labeled M_1 through M_6. The program used would not run if one variable were an exact combination of the other variables and therefore mixer, P_2, P_4, P_5, and M_1 through M_6 are specified as candidate variables. Other sets of candidate variables could have been used. The best models are

$$[X_0 P_2] \quad [X_0 \text{ mixer } P_2] \quad [X_0 P_2 P_5] \quad [X_0 \text{ mixer } M_3 M_5]$$

However, none of the models accounted for as much as 50% of the variability, and a study to identify other factors is in order.

20.2 Variables $x_{11} = x_1 * x_1$, $x_{22} = x_2 * x_2$, $x_{33} = x_3 * x_3$, $x_{12} = x_1 * x_2$, $x_{13} = x_1 * x_3$, and $x_{23} = x_2 * x_3$ were created. The residual sum of squares was calculated for all possible regressions involving the original and created variables. PRESS was calculated for the three models of each size having the smallest residual sum of squares. This technique identified the model

$$[x_0 x_1 x_2 x_{11} x_{22} x_{12}]$$

as having the lowest value of PRESS (30.29). The regression coefficients are

Variable	Estimate	v_c	Standard Error
x_0	19.9005	5939.81	82.1883
x_1	−0.2104	0.0169	0.1385
x_2	9.9625	0.6823	0.8808
$x_1 * x_1$	0.0002	0.0000	0.0001
$x_2 * x_2$	−0.0237	0.0001	0.0102
$x_1 * x_2$	−0.0073	0.0000	0.0008

Another model, with PRESS equal to 30.49, is

$$[x_0 x_1 x_2 x_{11} x_{12} x_{23}]$$

This model may not be appealing because x_3 is present only in a product. Marquart and Snee (see Appendix C) used the model

$$[x_0 x_1 x_2 x_3 x_{12} x_{11}]$$

as a basis for comparison to ridge regression (see Unit 21). This is not a good model with regard to PRESS. However, it does contain the three original variables and can be improved by using the techniques of Unit 21.

21.1 The statement of the question presupposes x_3 is in the model in some form. Since it might be interesting to compare the technique of Unit 21 with that of Marquardt and Snee, the model

$$[x_0 x_1 x_2 x_3 x_{12} x_{11}]$$

is used (see the answer to Exercise 20.2). The objective linear combination is

[1 1100 8 0.012 8800 1210000]

The predicted value for low x_1 and low x_3 would be obtained from the vector

[1 1100 8 0.012 8800 1210000]

The predicted value for high x_1 and high x_3 would be obtained from

[1 1300 8 0.092 10400 1690000]

The prior information that these predicted values are equal is equivalent to their difference being zero. Therefore the supplemental observation is

[0 200 0 0.0800 1600 480000|0]

If we require v_c to be 0.5, then the weight is 0.6687. The prediction is given in the following table:

Title	Estimate	v_c	Standard Error	Residual SS
$W = 0.6687$	33.5260	0.5000	0.8331	15.24
Least squares	27.0979	22.1695	5.4427	13.36

The least squares prediction is given for comparison. Note its very high v_c.

22.1 It is hoped that the blood pressure of patients receiving the drug will decrease during the five-week therapy period. It is expected that the blood pressure of patients receiving the control will have only random fluctuations during this period. The objective of this trial is to see if these hopes and expectations are true. The purpose of having the control is to guard against the effects of an external factor being attributed to the drug. We assume that the subject effects are random and that the only correlation between observations is due to the subject effects. The analysis of variance is

Source	Sum of Squares	d.f.	Mean Square
Mean	818,526.	1	818,526.
Drug	14,058.8	1	14,058.8
Subject	2,198.01	7	314.002
Week	855.553	4	213.888
Drug * week	512.129	4	128.032
Residual	1299.82	22	59.0826

Interaction between drugs and weeks is not significant at the 0.10 level. This result fails to demonstrate the desired action of the drug. Also, the following table of means (not adjusted for missing data) does not indicate the desired result.

	Week 1	Week 2	Week 3	Week 4	Week 5
Drug	127.5	124.0	132.0	127.5	130.0
Control	156.7	173.7	171.2	153.3	173.3

The results might be attributed to an unfortunate allocation of subjects to regimens, resulting in the subjects with higher blood pressure being assigned to the control group.

23.1 The results from fitting the drug model of the text are

Parameter	Estimate	v_c	Standard Error
KA	2.0349	0.1406	0.1405
KE	0.3216	0.0018	0.0160

The standard deviation is 0.3746 and the degrees of freedom are 14. This model seems to fit well. Adding another compartment to the model improves the fit slightly but drastically increases the standard errors of the estimates.

Index

Abbreviated Doolittle (ABDO), 26, 35, 43, 57, 60, 94, 107, 117, 168, 172, 219, 235, 250, 379
ABDO matrix, 117, 125, 127, 129, 169, 284
Analysis of variance, 49, 54, 135, 164

BPMD, 297, 345, 346, 363

Coefficient of multiple determination, 63
Comparison of regression lines, 144, 238
Computing. See Statistical computing.
Confidence interval, 125, 130, 133, 160
Contrast, 123, 133
Contrasts among means, 168, 176, 181, 187, 195, 203, 207, 217, 220, 224, 229, 237, 240, 281, 379, 380, 381
Covariance analysis, 13, 233
Cross classification, 194, 205, 221, 380
Curvilinear relationship, 70, 83, 88, 108, 111, 125, 262

Data analysis strategy, 3, 15, 47, 91, 124, 159, 163, 193, 198, 205, 228, 240, 244, 254, 273
Data augmentation, 263, 266, 273
Data decomposition, 49, 54, 135, 139, 147
Data reduction, 105, 380
Data sets
  arsenic, 32, 36, 40, 44, 50, 66, 313, 346
  barley response, 262, 299
  dairy hypomagnesaemia, 13
  dose dependent blood pressure, 82
  drug and alcohol, 280
  drug assimilation, 299
  electricity load, 141, 320, 363
  fat digestibility, 199, 365

  Finger Lakes, 256
  firefly, 92, 129, 135, 137, 249, 267, 351
  land evaluation, 151
  lymphocyte, 193
  New York rivers, 10, 59
  potato leafhopper survival, 179
  protein nutrition, 215
  soybean physiological, 234, 373
  soymilk, 111
  swamp pH, 221, 368
  thoroughbred auction, 10
  vital statistics, 71

ELIM, 125, 127, 130, 169, 181, 199
Experimental error, 5, 112, 165

Factorial experiment, 194, 196, 205, 210, 221, 228, 240, 245, 278, 280, 288

Inference, 20, 121, 124, 160, 273, 380

Least squares method, 31, 302, 379, 380
Linear combination
  assumptions, 106
  mean and variance, 115, 128
  of means. See Contrasts.
  of observations, 64, 105, 115, 128, 263
  orthogonal, 107, 109, 127, 186
  of regression coefficients, 123, 129, 133, 169, 181, 261, 379
  rules, 106
  weighted, 105
Linear model. See Model.

Index

Matrix, 22
  ABDO, 117, 125, 127, 129, 169, 284
  augmented, 32, 63
  equation, 23, 26, 31
  operations, 22
  transpose, 24
Mean comparison. *See* Contrasts among means.
Minitab, 345, 368, 373
Model. *See also* Inference; Nonlinear model; Polynomial; Regression analysis; Sequence of models.
  additive, 4, 18, 70, 122
  assumptions, 18, 19, 23, 69, 79, 122
  covariance, 233, 238, 245
  dose response, 81
  equal means, 170, 182
  exponential, 70, 84, 296, 300
  fitted, 19, 31, 34, 39, 42, 63, 79, 93
  full, 15, 49, 135
  general, 6, 15, 49, 54, 144, 148, 159
  general means, 164, 179, 204, 215, 221, 240, 281, 380
  intercept and slope, 18, 42, 46, 53, 99, 144
  kinetic, 299
  mean, 32, 34, 37, 52, 54, 94, 170
  mean and slope, 39, 46, 54, 93, 98, 234, 238, 241
  Mitscherlich, 296
  multiplicative, 70, 83
  multivariable, 57
  parameters, 19, 21, 24, 122
  reduced, 15, 49, 54, 135, 144, 148, 160
  regression, 4, 18, 23
  selection, 247, 250, 259
  simpler, 6, 15, 49
  slope, 18
  slope through the origin, 53, 380
  validation, 6, 254, 258
Modeling. *See* Sequence of models.
Multiple regression. *See* Model; Regression analysis.

Nonlinear model, 70, 295, 302, 379

Observational equations, 24, 34, 115, 166
ORTHO, 109, 111, 170, 186, 208, 217
Orthogonality, 106, 109, 115, 127, 170, 186, 218

Partial regression coefficients and sums of squares. *See* Regression analysis.

Polynomial, 108, 111, 114, 125, 262, 379
Predicted value. *See* Regression analysis.
Prediction sum of squares (PRESS), 254, 258
Proportional subclass numbers, 226

Randomization, 5, 164, 210, 277, 289
Randomized block design, 5, 165, 200, 210
Regression analysis. *See also* ABDO; Confidence interval; ELIM; Least squares method; Model; Sequence of models; Standard error; Statistical computing.
  degrees of freedom, 41, 51
  diagnostic plots, 70, 77, 79, 87
  estimation, 21, 32
  extrapolation, 261, 380
  lack of fit, 114, 165, 245
  notation, 22, 62, 63, 167
  partial coefficients, 19, 64, 67, 100, 102, 106, 107, 117, 130, 141, 156, 175, 184, 190, 210, 211, 220, 227, 229, 237
  partial sum of squares, 147, 154, 227
  predicted value, 5, 19, 31, 49, 62, 99, 113, 124, 126, 133, 135, 185, 253, 261, 381
  prediction sum of squares, 254, 258
  residual, 5, 19, 31, 40, 49, 122, 143, 165, 254, 381
  residual mean square, 41, 49, 63, 165, 273, 282
  residual plot, 6, 36, 41, 70, 78, 85, 155, 381
  residual sum of squares, 31, 147, 165, 252, 258
  segmented (spline), 157
  sequential coefficients, 59, 64, 67, 95, 98, 102, 106, 107, 117, 174, 185, 187, 190, 208, 211
  sequential sum of squares, 62, 67, 147, 154, 208, 219, 227, 235, 284
  sum of cross products, 24, 35
  sum of squares, 24, 35, 51, 54, 62, 67, 137, 147, 171, 190
  weighted, 65, 158, 263
Repeated measurements experiment, 277, 289, 379
Residual analysis. *See* Regression analysis.
Ridge regression, 266, 271, 274, 380

SAS, 300, 345, 351, 365
Sequence of models, 6, 10, 13, 15, 49, 54, 61, 94, 101, 113, 135, 138, 142, 144, 148, 160, 169, 176, 182, 200, 206, 224, 228, 242, 248, 259, 279

Sequential regression coefficients and sums of squares. *See* Regression analysis.
Split unit (plot) experiment, 277, 292, 379, 381
STAN (statistical analysis), 28, 133, 311
Standard error, 121, 124, 128, 130, 132, 160, 168, 261, 273
Statistical computing, 26, 28, 107, 109, 114, 117, 125, 132, 133, 147, 170, 174, 186, 188, 189, 198, 203, 209, 220, 227, 229, 237, 248, 250, 251, 254, 263, 273, 292, 297, 300, 302. *See also* ABDO; BPMD; ELIM; Minitab; ORTHO; SAS; STAN; SWEEP.
SWEEP, 248, 250, 251

Transformations, 69, 72, 74, 79, 380, 381

Unequal numbers of observations, 215, 221, 228

Variance. *See also* Linear combination.
   of a linear combination of observations, 106, 115, 117
   of a linear combination of regression coefficients, 124, 127
   of a predicted value, 124, 126, 131, 381
   of a regression coefficient, 117, 124, 130
Variable selection, 247, 250, 251, 259